Social-Ecological Systems (SES)

Mohamed Behnassi · Himangana Gupta ·
Mahjoub El Haiba · Gopichandran Ramachandran
Editors

Social-Ecological Systems (SES)

From Risks and Insecurity to Viability and Resilience

Springer

Editors
Mohamed Behnassi
College of Law, Economics and Social Science of Agadir, Center for Environment, Human Security & Governance (CERES)
Université Ibn Zohr
Agadir, Morocco

Himangana Gupta
JSPS-UNU Postdoctoral Fellow, University of Tokyo & Institute for the Advanced Study of Sustainability
United Nations University
Tokyo, Tokyo, Japan

Mahjoub El Haiba
College of Law, Economics and Social Science of Casablanca
University of Hassan II Casablanca
Rabat, Morocco

Gopichandran Ramachandran
NTPC School of Business
NOIDA, India

ISBN 978-3-030-76246-9 ISBN 978-3-030-76247-6 (eBook)
https://doi.org/10.1007/978-3-030-76247-6

This Springer imprint is published by the registered company Springer Nature Switzerland AG
The registered company address is: Gewerbestrasse 11, 6330 Cham, Switzerland

Preface

One of the most striking dynamics of the current era is the interaction of anthropogenic processes with ecological systems at large scales with increasing intensity, rates, and uncertain consequences. This can probably be correlated with the propensity of human societies to build many forms of disconnected settings to meet a myriad of needs. Most of such needs are not usually fundamental to the survival and positive evolution of human species, yet met by crossing many ecological thresholds, sometimes irreversibly. According to a research published in the journal science in 2015, the pursuit of global economy has already transgressed four planetary boundaries manifest through climate change, deforestation, species extinction, and the runoff of phosphorus and nitrogen into regional watersheds and oceans. These perturbations create the context for an increasingly dangerous, unpredictable, and fragile natural environment that cannot serve the just developmental aspirations of a flourishing human society. Scientists refer to the current geologic era of human-induced environmental change as the *Anthropocene*.

Increasingly sophisticated analyses of socio-economic and Earth system trends confirm that affluent countries worldwide, representing the Global North, are the primary drivers of global change, much as they account for 70% of global economic activity for over seven decades now. Ironically such nations host only 18% of the planet's population. While they reap material benefits of burgeoning economic activities, the environmental consequences are borne disproportionately by the planet's most vulnerable human beings, constituting the Global South. The latter relies on such natural resource-centered sectors as agriculture, fishing, forestry, and tourism for its survival and human security. Importantly, it lacks appropriate capacities to mitigate the damage induced by environmental and other global-scale challenges posed by climate change. These, in turn, exacerbate poverty and inequalities. According to a recent report issued by Oxfam, 20% of the world's

---

The original version of the book was inadvertently published without the postface and the list of contributing authors with the biographical notes. This has been included at the end of the book.

population currently owns approximately 95% of the planet wealth. If current trends continue, the richest 1% of the world's population will control a larger share of the world's wealth in the near future than the remaining 99%. Nearly, 750 million people are unable to access to clean drinking water, and 2.5 billion people lack access to sanitation. Approximately, 820 million people suffer from chronic undernourishment because they lack the resources to grow or purchase sufficient food to satisfy their dietary energy needs. Another 2.6 billion people lack modern energy facilities for cooking, heating, lighting, transportation, or basic mechanical power. The fossil fuel-driven industrial development model, established by the Global North and currently pursued by many emerging economies, has altered the trajectory of the resilience of the Earth system.

The failure to perceive human societies as integral parts of the Earth system (geosphere, atmosphere, biosphere, and hydrosphere) and the tendency to predate resources accordingly lead to a wide variety of mutually reinforcing and unprecedented complex risks. They tend to destabilize the security, viability, and resilience of social and ecological systems at all levels. The current Coronavirus crisis significantly illustrates and exacerbates this failure. Existing regulatory and conceptual frameworks, including principles and practice of international environmental law and sustainability, are not impactful enough to reverse or reduce current trends of deterioration. One of the reasons behind such a failure is the North-South divide that manifests through persistent conflicts between affluent and poor countries. This grid-locks negotiations and accentuates challenges due to non-compliance, thus generating resistance, suspicion, and ill will. The Global North appears to have by and large prioritized the environmental concerns of the affluent, only from a liberal perspective. This attitude to disregard the common but differentiated responsibility tends to ignore environmental and social problems that overly burden the Southern poor further affecting their coping abilities. This is especially the case of indigenous people and small-island states.

Climate change is an excellent example of the North-South conflicts in international environmental law. The North has agreed to assume a leadership role that emphasizes its greater technical and financial resources, yet disavows responsibility for its historic contribution to climate change and other environmental problems.

Failure to tackle impacts of climate change tends to spiral consequences of biodiversity loss, man-made environmental disasters, and epidemics. This compound risks to social systems, including economic development and political integration with many implications for livelihoods, well-being, productivity, and even regional and domestic security. In the past, the human–nature harmony fostered resilience of socio-ecological systems. Such a harmony enabled adaptation through nature-based solutions, conservation and sustainable use of bio-resources through traditional knowledge and practices. Presently, this harmony is being deeply disrupted due to large-scale shifting dynamics.

The thesis held by the Center for Research on Environment, Human Security and Governance (CERES)—which is behind the current contributed volume and the organizer of SES2019 international conference (see the acknowledgment section below)—is that the viability and resilience of social-ecological systems are impossible in a world plagued with significant and growing inequality, insecurity, and risks. This is the reason CERES in its orientation attempts to examine many of the environmental and social challenges of greatest concern as stated to the Global South. The Center questions the veracity of universalization of the Northern consumption-oriented development model. CERES is firm about the need for inclusive conceptual frameworks and response mechanisms to address challenges that riddle social-ecological systems from a resilience and viability perspective. The imperative is to tackle inequities in the international economic order, international law, domestic regulations, and underlying value systems that perpetrate abuse of nature and vulnerable human beings.

This contributed volume has been conceived to stimulate thinking about the nature of interactions between social and ecological systems from alternative perspectives. Social systems currently bear the main responsibility to change their relationship with ecological systems, given the available capacities (i.e., knowledge, technology, power of action, cultural values) that can be mobilized to reverse destructive trends. The shifts to be made should integrate and restructure regulatory frameworks (multilevel norms, actors, values, and deliberations…) from a social-ecological system perspective. In addition, it is imperative to reorganize production and consumption systems away from the consumerist growth model, to focus on mitigation and adaptation, to reinforce resilience, and to mainstream gender approach, human rights, and human security. Enhanced environmental stewardship by private actors can help resolve several of these and related challenges. These relationships are, however, not linear. Uncertainty clouds perceptions. Stakeholders tend to look askance at each other. Scientific evidences of perturbations do not seem to exert themselves adequately in preventive environmental management policies. Importantly, human endeavor to tackle such challenges sustains, albeit at scales that should grow significantly to fulfill sustainable development goals. The CERES accordingly calls attention to the complex contours of decision-making across several levels of organization and undercurrents that determine them in several contexts. The larger purpose is to resolve contradictions and harmonize development agendas.

Twenty-five authors from various disciplines have been provided with the opportunity to share relevant research, insights, and successful practices, and they have explored innovative approaches and provided valuable inputs. Chapters published in this volume deliberate on risk dynamics threatening current social-ecological systems or stimulate thought processes to manage such risks and related negative implications. In addition, the volume's content and approach are multidisciplinary (for a fruitful interaction between numerous scientific fields) and relevant to policy-making processes (enabling interactions among experts and decision makers from different levels and spheres).

In Chapter 1, *Effectiveness of Disaster Risk Governance and Resilience Building: Linkages, Knowledge, Inclusiveness, and Regulation*, Behnassi et al. begin with the evidence that almost all countries regardless of their developmental status are experiencing disasters whose frequency, intensity, and implications have increased over the last decades. These disasters are disruptive of the highest order, costly, uncertain, complex, and occur over a range of temporal and spatial scales. They are outcomes of inevitable hazards that affect highly vulnerable areas and populations, often with low coping capacities and resilience. The authors believe that the capacity to anticipate, mitigate, and adapt to disaster risk determines if a natural hazard turns into a disaster. Therefore, countries imperatively need to invest in disaster-resilience building, especially through appropriate governance. These should be guided by international frameworks, grounded on local context dynamics. They highlight the importance of governance implications as derived from an interplay of approaches to disaster risk reduction, climate change, and sustainable development goals. These are further compounded by consequences of knowledge gap, uncertainty, and science-decision divide; dynamics of stakeholder perceptions, awareness and engagement, laws, policies, and regulations meant to foster human security.

In Chapter 2, *Benefits of Evaluating Ecosystem Services for Implementation of Nature-Based Solutions under the Paris Agreement*, Gupta and Dube highlight knowledge gaps regarding relationships among ecosystems services. Scientists continue to consolidate their understanding of these complex relationships from a social-ecological perspective given its vital importance for landscape management, decision-making, and policy development. Currently, ecosystem-based adaptation (EbA) and nature-based solutions (NbS) are increasingly gaining importance in climate debates. They are yet to be juxtaposed with Paris Agreement's market and non-market mechanisms in addition to sustainable development goals (SDGs), despite their cross-cutting positive implications for the green climate funds portfolios post-2020. They accordingly examine the emergence of NbS and assess the need to develop appropriate indicators and evaluation of ecosystem services to better monitor NbS projects.

In Chapter 3, *The Common Heritage of Humankind (CHH): Visions of Developed Countries vs. Developing Countries*, Chougrani and El Amrani focus on the concept of 'common heritage of humankind (CHH)' perceived as shared resources and systems that should aim at benefitting the human society across the globe, hence the need to foster their sustainability and balance. The authors uncover the problems and difficulties that prevent the development of a common vision about the concept and practice of CHH, especially the globalizing dynamics and the conflict of interests among countries. The Global South advocates a strict international regulation of the commons while the Global North prefers the free-market approach. The latter enables over-exploitation of developing countries' natural resources and the weakening of their sovereignty over their resources. The authors believe that a framework for sharing responsibilities and burdens is needed. Such a framework should enable the needed shifts from the ultra-liberal '*laisser faire, laisser passer*' model and the free trade ideology to a model where environmentally destructive activities are reduced or eliminated.

In Chapter 4, *Climate Change Disaster, Gender Conflicts, Risks, and Insecurity in Lower-Gweru Community of Zimbabwe: Toward Capacity Building and Resilience in the Face of Social-Ecological Threats*, Winniefridah claims that climate change is severely hitting the poorest and the most vulnerable resulting in changing patterns of interactions between societal and ecological entities. Through the social-ecological theoretical framework, and based on an empirical work, the author attempts to determine the unequal impact of climate change-induced disasters on men and women in Lower-Gweru community of Zimbabwe. The chapter delves on the nexus between climate change, gender conflicts, risks, and insecurities, and to the scope to assess existing response mechanisms to foster resilience and coping capacities of women in particular. The results presented by the author reveal that climate change has gender-differentiated impacts and triggers gender-sensitive conflicts, risks, and insecurities in a context marked by the lack of viable response mechanisms to tackle them. Alternatively, the author recommends the design of a joint, coordinated, and viable model for resilience that integrates gender analysis and climate change disaster resilience frameworks.

In the same vein, Manuku considers in Chapter 5, *Resilience of Social-Ecological Systems: At the Limits of Hegemonic Masculinity,* advancing human development without crossing ecological thresholds and undermining critical ecological services as a main challenge to be managed in the Anthropocene era. However, the author remains doubtful about the capacity of current societies to enable ecological resilience and viability if the underlying value systems, especially hegemonic masculinity patterns, are not altered. By reference to the Raewyn Connell's (1987) theory of hegemonic masculinity, the author shows hegemonic masculinity as a structural and systemic driver of social-ecological systems' insecurity, vulnerability, and risks. In other terms, the author proposes a shift to a gender transformative paradigm in social-ecological systems resilience that targets hegemonic masculinity.

Similarly, Linus in Chapter 6, *Gendered Perspective: Climate Change Adaptation Strategy in Malawi,* highlights the importance of understanding adaptation to climate change from a gender perspective. The author highlights the need to integrate gender dimension into existing relevant frameworks, address research gaps and linkages between climatic shock, attributes of affected systems, and response characteristics. Accordingly, the author attempts to contextualize gender as a conceptual framework by testing the hypothesis— through Chi-squared test of independence and regression analysis—that gender is independent from adaptation strategy and investigating the factors influencing such a strategy. The author recommends that the adaptive capacity of women should be focused on improving adaptation rather than infuse external support. This is because the regression analysis revealed that: women with basic and advanced education are more likely to adopt autonomous adaptation strategy compared to women without education and access to information resulted in adoption of autonomous adaptation; and women with access to credit are less likely to adapt autonomously.

In Chapter 7, *The Effects of Land Cover Change on Sustainability: Human Security and Environmental Change in Semi-arid Ecosystems*, Deafalla et al. empirically define the links between environmental change and human security—especially the outbreak of violent conflicts—based on land use/land cover (LU/LC) changes. They elaborate that pressures imposed resulted in a complex of spatial and temporal interactions within topographical systems in the study area, where it has led both to a new rapidity and depth in rural transformation and a significant impact on urban areas as well. The authors suggest that stakeholders should be empowered —especially through more information and knowledge exchange—to overcome the current coping capacity gaps which constitute impediments to the efficient management of environmental change and its various implications for human security.

In Chapter 8, *Resilience Building Initiatives to Counter Shocks and Stressors Affecting Rural Communities in Chiredzi District, Zimbabwe*, Defe and Matsa assess the effectiveness of various resilience-building initiatives pertaining to rural communities in the study area while protecting their livelihoods and enhancing the recovery gains. The findings reveal the existence of three main types of resilience-building initiatives in the study area related to crop production, livestock, and social and well-being, respectively. They helped deal with recurrent shocks and stressors affecting concerned rural communities over a ten-year horizon. Training of farmers on resilience building and the need for the initiatives' implementers to cover the whole concerned territory are important given the significant difference in the style and quality of life inwards under the projects.

In Chapter 9, *Chances of Conflict on Account of Large-Scale Influx of Climate Refugees in India*, Gupta and Sharma demonstrate focus on the rise of global mean sea level and increased human displacements in many vulnerable regions. By investigating the case of India, the authors claim that increased climate-induced human displacements in neighboring countries—especially Nepal, Bangladesh, Sri Lanka, and Maldives—have the potential to increase the stress and strife the country is already undergoing. (The authors try to deeply understand such dynamics and suggest some relevant response mechanisms.

In Chapter 10, *Key Knowledge Gaps at the Interface of Chemical Ecology and Ecosystem-Based Adaptation Planning: A Public Policy Perspective*, Gopichandran draws a picture of thrust areas at the interface of chemical ecology and ecosystem-based planning that have not received adequate attention, despite significant progress in the respective fields. Importantly, several facets of bio-resources management across the soil and plant continuum feature in this gaps framework. These include, according to the author, the need to focus on mineralization mediators in soils, age correlated changes in biochemical profiles with implications for tri-trophic interactions, and therefore productivity and related interpretation as signals of resilience. As part of the public policy framework upon which this eclectic research is centered, the author refers to recent advances in science communication that will enable target-oriented engagement with stakeholders to fulfill the agenda of science-led decision-making. Additional insights about the relevance of chemical ecological aspects stated and the felt need to embed them in policies that guide action had been discussed by the author.

In Chapter 11, *When Environmental Inequalities Lead to Social Inequalities*, El Amrani investigates the links between environmental change and social inequalities. This is with reference to the context wherein those who consume and pollute less are often those who suffer more. This growing paradox deepens gaps between the rich minority and the poor majority. The author claims that decision-making processes, in a desperate attempt to combat such inequalities, are often unable to find efficient and long-lasting solutions, despite political will. It is, therefore, important to address the underlying causes, moving away from superficial solutions. As a consequence, the fight against inequalities is still at an early stage and achieving equality often requires the adoption of unconventional response mechanisms. For the author, the viability and resilience of social-ecological system depend on the possibility to simultaneously advance both ecological and social equality, hence the need to make transformative changes.

In Chapter 12, *Approaching Sonic Devices from a Multicriteria Sustainable Approach to Enhance the Resilience of Social-Ecological Systems*, Gueiros Teixeira et al. analyze aspects of noise and vibration impacts on living beings and their scale of damage, depending on physical and psychophysical conditions of beings and environment. The authors also question, based on a multicriteria sustainable approach, the concepts of noise repellants, sonic non-lethal weapons, and other devices generating the soundscape scenarios over human health and other non-human species. This research is the result of many insights developed from different approaches brought together for the creation of a new research group addressing vibration, sustainable management, and strategic planning, with special emphasis on sonic devices under a defense approach. The authors propose in this research a multicriteria approach investigation, to provide the product engineering area—and the many other stakeholders which partner with the subject, including strategic planning actors, associated with defense market—an appropriate scientific review regarding sonic products innovation, use-phase impacts, and consequences over public health, defense (health, safety, and security) and legislation, environmental values, bioethics and bio-policy, psychoacoustics and physio-acoustics, metrology and standards, and biomimetics.

Gopichandran et al. in Chapter 13, *The Anatomy of Uncertainty and Implications for Preventive Environmental Management*, focus on the dynamics of uncertainty. This is seen as a modulator of decisions on the arena of environmental management and often an excuse to evade preventive management. The authors refer to scientific evidences of global-scale perturbations to reinforce the need for concerted action to internalize externalities and the hope of some collective action by industry in particular. These telescope into reference to the latest environmental, social, and governance (ESG) framework and the emerging imperatives of environmental protection through a call by financial institutions and the growing influence of the log arm of law around the world. These are seen as potential signals of positive change that should be fostered to restore some semblance of much-needed balance with social-ecological systems.

Finally, through all these chapters, this volume intends to contribute to the debate about the various risk dynamics and forms of insecurity currently threatening social-ecological systems and the potential pathways to viability and resilience of such systems, especially in the Anthropocene era. The analysis focuses on some countries from the Global South—especially Africa and Asia—given their specific situations in terms of environmental and climate vulnerability, resource decline, and security implications. Through the insights shared by different authors, the book is an added value to the growing academic literature on social-ecological systems with the potential to shape relevant research and political agendas. Interested scholars, students, practitioners, and decision makers from concerned regions and across the world will find the publication a useful and instructive reference.

Mohamed Behnassi
Himangana Gupta
Mahjoub El Haiba
Gopichandran Ramachandran

Acknowledgements

This contributed volume is partly based on the outcome of the International Conference on *Social-Ecological Systems – From Risks and Insecurity to Viability and Resilience (SES2019)*, organized in October 24–25, 2019, in Marrakech, Morocco, by the Research Center for Environment, Human Security, and Governance (CERES). More than 50 researchers, practitioners, decision makers, and experts from different parts of the world, representing a myriad of disciplines, institutions, and spheres of influence participated in this conference. The event fostered a multidisciplinary debate about social-ecological systems (SES) in a changing and uncertain risk-infested context. The approach assessed structural drivers of vulnerability, crises, and insecurity of SES; identified areas of integration and synergy given the interdependence of SES; focused on the ability and limitations of existing response mechanisms (governance frameworks, conceptual referential, cultural patterns and values) to foster resilience and viability of SES and reverse undesirable trends.

It was my deep honor to chair the SES2019 and share the editorship of this volume with my colleagues Dr. Himangana Gupta, JSPS-UNU Postdoctoral Fellow at the United Nations University Institute for the Advanced Study of Sustainability (UNU-IAS) and the University of Tokyo, Japan; Dr. Mahjoub El Haiba, Full Professor at the Faculty of Law, Economics, and Social Sciences, Hassan II University of Casablanca, Morocco; and Dr. Gopichandran R., Professor at the NTPC School of Business in India. I warmly thank all of them for their collaboration and support during the publishing process. Their professionalism, expertise, and intellectual capacity made the editing process an exciting and instructive experience, and definitely enriched the quality of this publication.

I seize this opportunity as well to pay tribute to all chapters' authors without whom this valuable and original publication could not have been produced. Their collaboration, responses, and openness during the process were remarkable and impressive.

The chapters in this volume are also the result of the invaluable contribution made by peer-reviewers, who generously gave their time and energy to provide insight and expertise to the selection and editing process. On behalf of my co-editors, who actively participated in the peer-reviewed process, I would specifically like to acknowledge, with sincere and deepest thanks, the following peer-reviewers from India: Dr. Hardev Singh, Department of School Education, Jammu; Dr. Patterson Edward J. K., Suganthi Devadason Marine Research Institute (SDMRI); Dr. Lakhvinder Kaur, Panjab University Chandigarh; Dr. Hareesh Chandra P., National Academy of Agricultural Research Management (NAARM); Lokesh Chandra Dube, Ministry of Environment, Forest, and Climate Change, Government of India India; Moharana Choudhary, Voice of Environment (NGO); Raj Kumar Gupta, Senior journalist; Rebekah Singh, Ernst & Young; Biba Jasmine, Ministry of Environment, Forest, and Climate Change, Government of India; Barath Mahadevan, Independent Climate Change Consultant; Bhupesh Verma, Center for Study of Science, Technology and Policy; Arjun Rawat, Amazon; Mr. Raghuvansh Saxena, Director, Earthwatch Institute India, Haryana; Mr Mahesh Pandya, Director Paryavaran Mitra, Ahmedabad, Gujarat; and Dr. Thupalli Ravishankar, International Community Forest Biodiversity Specialist, Kakinada, Andhra Pradesh. In addition, my thanks go to Alis Daniela Torres, ICLEI—Local Governments for Sustainability, European Secretariat, Ecuador/ Germany, and Dr. Fiaz Ahmad, Senior Scientific Officer, Head Physiology/Chemistry Section, Central Cotton Research Institute Multan, Pakistan.

Mohamed Behnassi

About the Publishing Institution

The Center for Research on Environment, Human Security, and Governance (CERES)

The CERES, previously the North-South Center for Social Sciences (NRCS), 2008–2015, is an independent and not-for-profit research institute founded by a group of researchers and experts from Morocco and other countries. The CERES aims to develop research and expertise relevant to environment and human security and their governance from a multidimensional and interdisciplinary perspective. As a think tank, CERES aspires to serve as a reference point, both locally and globally through rigorous research and active engagement with policy-making processes. Currently, CERES is a member of MedThink 5+5 which aims at shaping relevant research and decision agendas. Through its research and expertise program, the CERES aims to investigate the links between environmental/climate change, their implications for human security and the needed shifts to be undertaken in both research and policy. The CERES, led by Dr. Mohamed Behnassi and mobilizing a large international network of researchers and experts, aims to undertake original research, provide expertise, and contribute to effective science and policy interactions through its publications, events, capacity building, and expertise.

Contents

About the Editors

Dr. Mohamed Behnassi is a professor at the Faculty of Law, Economics, and Social Sciences, Ibn Zohr University of Agadir, Morocco. He is as well as a senior researcher of International Law and Politics of Environment and Human Security focusing on some specific regions such as the MENA and the Mediterranean. He has a PhD in International Environmental Law and Governance (Hassan II University of Casablanca, 2003) and a Diploma in International Environmental Law and Diplomacy (University of Eastern Finland and UNEP, 2015). He is currently the founding director of the Center for Environment, Human Security, and Governance (CERES)—Former North-South Center for Social Sciences (NRCS)—which is a member of MedThink 5+5 aiming at shaping relevant research and decision agendas. From 2015 to 2018, he was the director of the Research Laboratory for Territorial Governance, Human Security, and Sustainability (LAGOS) in the same university. Recently, he was appointed as an expert for the Intergovernmental Science-Policy Platform on Biodiversity and Ecosystem Services (IPBES), the National Center for Scientific and Technical Research (CNRST/Morocco), and

Mediterranean Experts on Climate and Environmental Change (MEDECC). He is among the lead authors who elaborate the 1st Assessment Report (MAR1): *Climate and Environmental Change in the Mediterranean Basin - Current Situation and Risks for the Future* (MEDECC, 2021). He has published 15 books, including *Human and Environmental Security in the Era of Global Risks* (Springer, 2019); *Climate Change, Food Security and Natural Resource Management: Perspectives from Africa, Asia and the Pacific Islands* (Springer, 2019); *Environmental Change and Human Security in Africa and the Middle East* (Springer, 2017); *Sustainable Food Security in the Era of Local and Global Environmental Change* (Springer, 2013). In addition, He has organized many international conferences covering the above research areas and managed many research and expertise projects on behalf of various national and international organizations. He is regularly requested to provide scientific expertise nationally and internationally. Other professional activities include social compliance auditing and consultancy by monitoring human rights at work and the sustainability of the global supply chain.

Dr. Himangana Gupta is a JSPS-UNU postdoctoral fellow at the United Nations University Institute for the Advanced Study of Sustainability (UNU-IAS) and the University of Tokyo, Japan. She received her doctorate in Environment Science from Panjab University, India, in 2015. She has worked on climate change and biodiversity policy and diplomacy and is currently working on linkages between biodiversity, climate, and communities in socio-ecological production landscapes. Before this, she was a part of the National Communication Cell (NATCOM) of the Indian Ministry of Environment, Forest, and Climate Change. She contributed to India's Second Biennial Update Report and several other publications of the ministry. She is a certified expert in Climate Adaptation Finance. She is a University Gold Medallist and a recipient of Academic Excellence Award. She has published three edited books with Springer and has also written research papers in reputed international and national journals on climate policy, forestry, biodiversity, and women in climate change mitigation and adaptation.

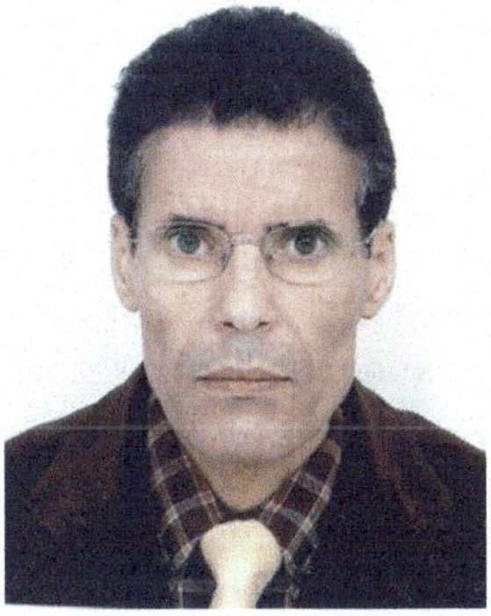

Dr. Mahjoub El Haiba is a full professor at the Faculty of Law, Economics, and Social Sciences, Hassan II University of Casablanca, Morocco. He graduated from the Hague Academy of International Law, the Center for Studies and Research in International Law and International Relations in 1984, and in 1994, he had State Doctorate in Public Law and Political Science from the same university. He held many academic and official positions, including being: Former Member of the National Consultative Human Rights Council (2002–2010), and his Secretary General (2005–2010); first Inter-ministerial Delegate for Human Rights (March 2011–December 2018); former Member ex officio of the Higher Council for Education, Training, and Scientific Research, and of its Executive Office; nominated Member of the Equity and Reconciliation Instance (IER) (2004–2005); former and current Member of the United Nations' Human Rights Committee (2008–2011/2020-at present); former Member and President of the Jury to the League of Arab States' Environment Award, Cairo; former Member and President of the Jury to the Hassan II Environment Award; former Member of the Jury to the Kingdom of Saudi Arabia's Award on Environmental Administration; Founding Member of the Moroccan Human Rights Organization (OMDH); Founding Member of the Arab Network for Environment and Development (RAED), Cairo; Founding Member of the Center for Studies on Migration, Faculty of Law, Economics, and Social Sciences, Hassan II University of Casablanca; and former Deputy-Dean of the same faculty. His areas of teaching, research, and expertise include human rights, transitional justice, and environmental law. He has a publication record in these areas, supervised numerous doctorate and master theses, and made many presentations in national and international scientific meetings.

Dr. Gopichandran Ramachandran is a professor teaching various aspects of mitigation, adaptation, preventive environmental management and business communication at the NTPC School of Business in India. These are important elements of public policy, pertaining to which, he has an overall work profile that spans thirty-two years. He has recently contributed to the review of the first and second-order drafts of the upcoming sixth assessment report of the IPCC, with a special emphasis on chemical ecology for improved adaptation strategies. His contributions through the Compliance Assistance Program, OzonAction of the UNEP, have been quite significant and consistently so for more than two decades at the regional and global levels. This created the opportunity for him to serve as a member of the Inter-Ministerial Empowered Steering Committee constituted by the Ministry of Environment, Forest, and Climate Change, Government of India, on aspects of the Montreal Protocol. He is a well-known specialist in the areas of science and, technology management communication, with a large number of theme-specific editorials and other publications to his credit. He holds two doctoral degrees in the fields of microbial and chemical ecology respectively, with a degree in law and is an alumnus of the International Visitors Leadership Program of the Department of State, USA.

List of Abbreviations and Acronyms

ACLCA	American Committee of Life Cycle Assessment
ADB	Asian Development Banks
AFOLU	Agriculture, Forestry, and Other Land Use
ALPA	Adaptation Learning Programme for Africa
ASTER	Advanced Spaceborne Thermal Emission and Reflection Radiometer
CARLA	Climate Adaptation for Rural Livelihood and Agriculture
CBD	Convention on Biological Diversity
CBDRR	Community-Based Disaster Risk Reduction
CBNRM	Community-Based Natural Resources Management
CCCM	Camp Coordination and Camp Management Cluster
CCS	Carbon Capture and Storage
CDD	Community Development Department
CDM	Clean Development Mechanisms
CDP	Carbon Disclosure Project
CERES	Center for Research on Environment, Human Security and Governance
CHH	Common Heritages of Humankind
COP	Conference of Parties
CRI	Climate Risk Index
CSR	Corporate Social Responsibility
DEM	Digital Elevation Model
DRM	Disaster Risk Management
DRR	Disaster Risk Reduction
EbA	Ecosystem-Based Adaptation
EbA	Ecosystem-Based Adaptation
EBDRR	Ecosystem-Based Disaster Risk Reduction
EbM	Ecosystem-based Mitigation
EC	Environmental Change
ENTRUST	Energy System Transition Through Stakeholder Activation, Education and Skills Development

EPA	Environmental Protection Agency
EPD	Environmental Product Declaration
ERDAS	Earth Resources Data Analysis System
ES	Ecosystem Services
ESG	Environmental, Social and Governance
ESMAP	Energy Sector Management Assistance Program
FAO	Food and Agriculture Organization
GEO	Global Environment Outlook
GEOBIA	Geographic Object-Based Image Analysis
GFDRR	Global Facility for Disaster Risk Reduction
GHGs	Greenhouse Gases
GIZ	Deutsche Gesellschaft für Internationale Zusammenarbeit
GRI	Global Reporting Initiative
HFA	Hyogo Framework for Action
IASC	Inter-Agency Standing Committee's
ICRC	International Committee of the Red Cross
IDMC	Internal Displacement Monitoring Centre (India)
IDP	Internally displaced persons
IEA	International Energy Agency
IEP	Instituto de Ecologica Poluticia
IFRC	International Federation of Red Cross and Red Crescent Societies
ILO	International Labor Organization
INDC	Intended Nationally Determined Contribution
IOM	International Organization for Migration
IPCC	Intergovernmental Panel on Climate Change
IPIECA	International Petroleum Industry Environmental Conservation Association
IRC	International Rescue Committee
ISA	International Seabed Authority
ISDR	International Strategy for Disaster Reduction
ISO	International Organization for Standardization
IUCN	International Union for Conservation of Nature
JMA	Joint Mitigation and Adaptation
LCA	Life Cycle Assessment
LEG	Least Developed Countries Expert Group
LU/LC	Use/Land Cover
MRV	Measurement, Reporting and Verification
NAPA	National Adaptation and Programmes of Action
NASA	National Aeronautics and Space Administration (USA)
NbS	Nature-Based Solutions
NDCs	Nationally Determined Contributions
NGOs	Non-Governmental Organizations
NMAs	Non-Market Approaches
NOAA	National Oceanic and Atmospheric Administration
OHCHR	Office of the United Nations High Commissioner for Human Rights

OSCE	Organization for Security and Cooperation in Europe
PCR	Product Category Rules
PCRs	Product Category Rules
PPS	Population Proportional to Size
SAARC	South Asian Association for Regional Cooperation
SDGs	Sustainable Development Goals
SIDA	Swedish International Development Cooperation Agency
SMEs	Small and Mid-Size Enterprises
SPCGI	Shurugwi Partners Community Garden Initiative
TEEB	The Economics of Ecosystems and Biodiversity
TNCs	Transnational Corporations
UNCLOS	United Nations Convention on the Law of the Sea
UNDESA	United Nations Department of Economic and Social Affairs
UNDP	United Nations Development Programme
UNDRR	United Nations Office for Disaster Risk Reduction
UNEP	United Nations Environment Program
UNESCAP	United Nations Economic and Social Commission for Asia and the Pacific
UNESCO	United Nations Educational, Scientific and Cultural Organization
UNFCCC	United Nations Framework Convention on Climate Change
UNGA	United Nations General Assembly
UNHCR	United Nations High Commissioner for Refugees
UNHCR	UN High Commissioner on Refugees
UNIDO	United Nations Industrial Development Organization
UNIFEM	United Nations Development Fund for Women
UNISDR	United Nations International Strategy for Disaster Reduction
USAID	United States Agency for International Development
VSL	Village Savings and Lendings
WBSCD	World Business Council for Sustainable Development recently
WEF	World Economic Forum
WFP	World Food Programme
WHO	World Health Organization
WHO	World Organization of Health
ZRBF	Zimbabwe Resilience building Fund

List of Figures

Gendered Perspective: Climate Change Adaptation Strategy in Malawi

The Effects of Land Cover Change on Sustainability: Human Security and Environmental Change in Semi-arid Ecosystems

Resilience building Initiatives to Counter Shocks and Stressors Affecting Rural Communities in Chiredzi District, Zimbabwe

When Environmental Inequalities Lead to Social Inequalities!

Approaching Sonic Devices from a Multicriteria Sustainable Approach to Enhance the Resilience of Social-Ecological Systems

The Anatomy of Uncertainty and Implications for Preventive Environmental Management

List of Tables

When Environmental Inequalities Lead to Social Inequalities!

Approaching Sonic Devices from a Multicriteria Sustainable Approach to Enhance the Resilience of Social-Ecological Systems

Effectiveness of Disaster Risk Governance and Resilience Building: Linkages, Knowledge, Inclusiveness, and Regulation

Mohamed Behnassi, Himangana Gupta, Nira Ramachandran, Matsa Winniefridah, Gopichandran Ramachandran, Silvana Lakeman, and Mohammad Ashfaq

Abstract Almost all countries are experiencing disasters whose frequency and intensity have increased over the last decades due to many natural and anthropogenic factors, including climate change. These disasters are increasingly deadly, costly, uncertain, complex, and occurring over a range of temporal and spatial scales. They are the outcomes of inevitable hazards that affect highly vulnerable areas and populations with low coping capacities and resilience. The capacity to anticipate, mitigate

N. Ramachandran—Currently Independent Researcher and Consultant.

M. Behnassi (✉)
Faculty of Law, Economics and Social Sciences, Ibn Zohr University of Agadir, Center for Research on Environment, Human Security and Governance (CERES), Agadir, Morocco
e-mail: m.behnassi@uiz.ac.ma; behnassi@gmail.com

H. Gupta
The United Nations University Institute for the Advanced Study of Sustainability (UNU-IAS), The University of Tokyo, Tokyo, Japan
e-mail: gupta@unu.edu

N. Ramachandran
Institute of Economic Growth, Delhi University, New Delhi, India
e-mail: nira_ramachandran@hotmail.com

M. Winniefridah
Gender Institute, Midlands State University, Gweru, Zimbabwe
e-mail: matsaw@staff.msu.ac.zw

G. Ramachandran
NTPC School of Business, Noida, U.P., India
e-mail: gopichandran@nsb.ac.in

S. Lakeman
Political Science, University of Bremen and Climate-Security Researcher, Marie Curie Fellow at the Bremen International Graduate School of Social Sciences, Bremen, Germany
e-mail: silvanalakeman@gmail.com

M. Ashfaq
AGM (Corporate Planning) NTPC Ltd., New Delhi, India
World Energy Council – India Secretariat, New Delhi, India
e-mail: mdashfaq@ntpc.co.in; mdashfaq@wecindia.in

M. Behnassi et al. (eds.), *Social-Ecological Systems (SES)*,
https://doi.org/10.1007/978-3-030-76247-6_1

and adapt to disaster risk is currently making a significant difference as to whether or not a natural hazard turns into a disaster. Presently, there is no region immune to the impacts of hazards and a country's vulnerability to disaster risk is the outcome of several factors, mainly the failure of its related governance system. Therefore, countries imperatively need to invest in disaster-resilience building, especially through the development of appropriate governance arrangements according to international frameworks while considering local context dynamics. Disaster resilience is understood here as the ability to adapt to and recover from hazards, shocks or stresses without compromising long-term prospects for development. This process, in order to be effective, requires the consideration of many issues, which will be investigated throughout this chapter, such as: the governance implications of the linkages between disaster risk reduction, climate change and sustainable development goals; the impacts of knowledge gap, uncertainty and science-decision divide; the dynamics through which stakeholder perceptions, awareness and involvement are shaped; the different relevant approaches to be mainstreamed; and the role of laws, policies, and regulations as critical tools in reducing and preventing disaster risk, thus fostering human security.

Keywords Disaster risk · Governance · Resilience · Climate change · SDGs · Knowledge · Gender · Risk communication and perception · Laws and regulations

1 Introduction

Since the dawn of time, human beings have had to face various natural disasters that threatened their very existence on this planet. Ancient and historical stories have detailed recordings about the great natural disasters, their causes, and their impacts on human societies. Many of the great civilizations in history – such as the Mayan, the Minoan and the old Egyptian Empire – were ultimately brought to their knees not by their enemies, but by the effects of floods, drought, famine, earthquake, volcanic eruption, tsunamis and other widespread disasters.[1] Today, human societies are experiencing more frequent, deadly and costly disasters which are increasingly uncertain and complex and arising from a variety of sources and occurring over a range of temporal and spatial scales. In addition to natural disasters,[2] societies are increasingly experiencing human-induced ones. Anthropogenic disasters include armed conflicts and wars, famine, ethnic conflicts and displaced populations, industrial and transport accidents, environmental degradation such as pollution, as well as intentional hazards

[1]For example, the extinction of the Tang Dynasty in China is believed to be a result of the yearly shift of the monsoon, which resulted in mass crop failure during the eighth and ninth centuries (Coppola, 2011).

[2]Naturally occurring disasters can be classified into five groups: *geophysical* (earthquakes, landslides, tsunamis, and volcanic activity); *hydrological* (avalanches and floods); *climatological* (extreme temperatures, droughts, and wildfires); *meteorological* (cyclones and storms/wave surges); and *biological* (disease epidemics) (IFRC 2015a, 2015b).

such as terrorism or weapons of mass destruction (IFRC 2015a; 2015b; Chen et al. 2018).

The frequency and intensity of these disasters have increased over the last decades due to many factors including global change (such as climate-induced disasters). In addition to their severe impacts on ecosystems, disasters affect the human security of billions of people and cause substantial economic losses worldwide. These disasters not only brought death and destruction, they did so disproportionately to the poor and marginalized (UNDP and IFRC 2014), and it might take years for a community or a country hit by a disaster to recover environmentally, socially and economically. For instance, rural and peri-urban communities in Japan face risks of discrete event natural phenomena – including earthquakes, floods, and landslides – in addition to the persistent disruptive stress due to risks that remain active over long durations, such as the loss of community capacities because of an aging population (Chabay 2018). Disasters have even become one of the main risks with the potential to significantly undo previous achievements in the areas of human security and development.

According to the International Federation of Red Cross and Red Crescent Societies (IFRC), "today, it is well accepted that our own actions – as individuals, communities and nations – make all the difference between a natural event and a natural disaster. We may not be able to stop the earth from shaking, or storms from striking, but our choices can determine the extent of death and damage they cause" (IFRC 2013). Therefore, countries cannot effectively stop disasters, but they can mitigate disaster risks and reduce their negative impacts (UN 2015). To do so, countries need to reduce their vulnerabilities to disasters, especially through the development of appropriate policy and institutional frameworks and efficient governance systems. Currently, these capacities make a significant difference as to whether or not a natural hazard turns into a disaster.

Against this background, this chapter is built on the assumption that fostering disaster risk governance is a prerequisite for building disaster resilience, which is the ability to adapt to and recover from hazards, shocks or stresses without compromising long-term prospects for development. The analysis starts with the identification and definition of key concepts and dynamics which will be further analyzed throughout the chapter. In the second part, the linkages between disaster risk, climate change and SDGs are unpacked while highlighting their governance implications. The third part is devoted to the analysis of the impacts of the knowledge gap, uncertainty and science-decision divide on disaster risk governance for resilience. The fourth part deals with elements of stakeholder perception with reference to the dynamics of risk communication. It cites some outreach initiatives and adds a special focus on energy sector related risk communication as a reference for targeted engagement. It also refers to the emerging call for the meta precautionary principle as part of citizen groundswell. The fifth part emphasizes the importance of mainstreaming many relevant approaches in disaster risk governance such as the community- and ecosystem-based approaches, the multi-stakeholder and participatory approaches, and the gender approach. The last part highlights the vital importance of laws, regulations, and policies in fostering disaster riskgovernance and building resilience.

2 Disaster Risk Governance and Resilience: Key Concepts and Dynamics

While the risk is the probability of damaging events that is derived from the confrontation of risks, social vulnerability and nature (Smith 2013), a disaster can be defined as a catastrophe resulting in injury, loss of life, and social and economic disruption that exceed the coping capacity of the affected people and the ecology (UN 2015). The United Nations International Strategy for Disaster Reduction (UNISDR) defines disaster as a serious disruption of the functioning of a community or a society at any scale due to hazardous events interacting with conditions of exposure, vulnerability and capacity, leading to one or more of the following: human, material, economic and environmental losses and impacts (UNISDR 2018). According to Zhoux et al. (2017), a hazard is a phenomenon that may cause loss of life, injury or other health impacts, property damage, social and economic disruption or environmental degradation. Hazards cause disaster risk which is the potential loss of life, injury, damage of assets which could occur to a system, society, or community in a specific period of time, determined as a function of hazard exposure, vulnerability and capacity.

In the same perspective, the United Nations Economic and Social Commission for Asia and the Pacific (UNESCAP) notes that disasters – like hazards – place people, especially the poor, at risks. Hazards are a hindrance to the process of economic development and, because they are inevitable, they cannot be prevented but only anticipated, mitigated and adapted to (UNESCAP 2017). Indeed, people cannot effectively stop natural disasters but they can mitigate disaster risks and reduce their negative impacts through information sharing, behaviour change, effective governance and technological innovations (UN 2015). In fact, disasters are the outcomes of natural hazards that affect highly vulnerable areas and populations with low capacities and resilience for disaster risk. Moreover, lack of information and data on risks, lack of resources, weak or non-existent early warning systems and fragile infrastructure all contribute to the magnitude of disaster losses in lives, livelihoods, assets, economy, environment, etc. (UNDP et al. 2013). In other terms, several factors contribute to a country's vulnerability to disasters, including inappropriate governance systems and failure of policy and institutional frameworks. Today, it is well accepted that the actions and decisions of individuals, communities and nations make a significant difference as to whether or not a natural hazard turns into a disaster. For Chmutina and von Meding (2019), regulations and building codes, urban planning, risk management and awareness raising, politics, governance, media, development, growth, and culture, among others, play a central role in reducing vulnerability to disaster risk and building resilience.

Appropriate and effective governance systems should aim at fostering resilience. There is no universal definition of resilience and attempts to define it, according to Walsh-Dilley and Wolford (2015), have confined it to the parameters for specific study areas that seek to investigate resilience to what, how, and by whom. For Zhoux et al. (2017), resilience should be understood as the ability of a system, community or society exposed to hazards to resist, absorb, accommodate, adapt to, transform, and recover from the effects of a hazard and restoration of its essential basic structures and risks. For Acosta et al. (2017), resilience can be defined as the capabilities – in isolation or in tandem at the individual, household, and collective level – to leverage the environment to expect, withstand, recover, and adjust to disaster-induced distress.

Response mechanisms adopted and implemented with the aim of reducing impacts of natural or anthropogenic hazards are commonly described as Disaster Risk Reduction (DRR). The goal of DRR is to design and create a context for reducing human losses and protecting assets against hazards (Blaikie et al. 2014). For Peters et al. (2019), DRR is a systematic approach to identifying, assessing and reducing the risk of disaster and socio-economic vulnerabilities while dealing at the same time with the environmental factors and other hazards that trigger risks. In addition, response mechanisms adopted and implemented from a DRR perspective are multiscale and diverse, especially that these response mechanisms are currently originating from a growing number of stakeholders and relevant areas. Moreover, even if human societies are experiencing common disasters arising from a variety of sources and occurring over a range of temporal and spatial scales, how the risks of disastrous events are perceived and acted upon by different sectors of these societies in different contexts and cultures varies greatly. In addition, some risks are related to changes on a global scale and manifest on that scale, yet even these global scale risks have significant local and regional impacts. In order to mitigate or adapt to risks, and to recover from and restore functionality after disasters, whether actual or potential events are global or local in origin, people and institutions at risk must engage in risk governance. This process – which includes assessment, preparation, communication, response, recovery, and restoration appropriate to and effective in local conditions and cultures – and the decisions and actions taken by different institutions and actors in communities at risk before, during, and after an event show considerable variation (Chabay 2018). This variation may result from divergent perceptions of the same risks by different stakeholders. Klein et al. (2019) show (Fig. 1) that cross-level partnerships among multiple stakeholders can facilitate the integration of community and ecosystem objectives in disaster resilience and address paradoxes that pose challenges for DRR, especially in the mountains.

To harmonize the stakeholders' perception of risks, effective communication of disaster risks should be enhanced since it has the potential to mediate between the affected people and public or individual action while triggering actions that juxtapose findings from risk research with the social and political will required to achieve safety of lives and property. In addition, risk-communication-mediated pathways have the potential to enhance risk assessment and development of multi-hazard early-warning systems. For Volenzo and Odiyo (2019), risk communication offers a seamless support system for the integration of the precautionary and DRR principles in the pursuit of adaptation and sustainable development agenda.

In addition, it is increasingly demonstrated that risks often interact with each other and cannot be dealt with in isolation. In this regard, 'systemic risks' gained prominence and refer to unconventional risks of increasing importance, which cannot be tackled by conventional techniques of risk management and governance. According

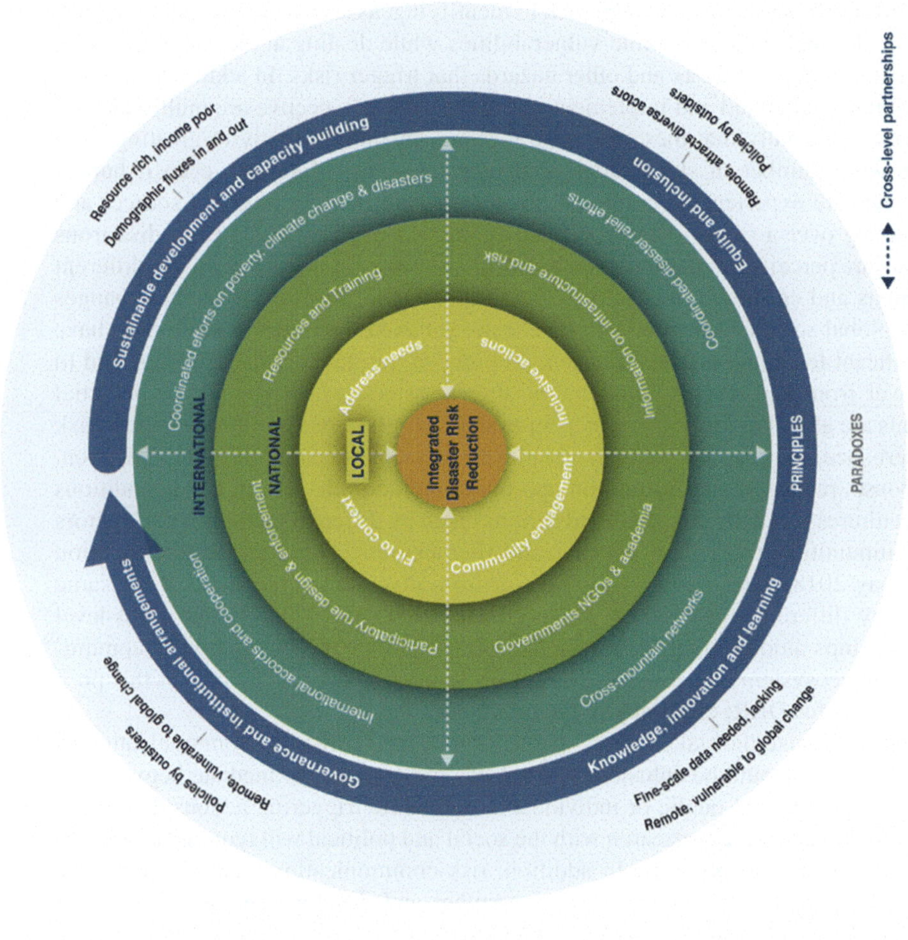

Fig. 1 Cross-level partnerships for integrated disaster risk reduction Source: Klein et al. (2019)

to Lucas et al. (2018), a critical mechanism related to systemic risks is the breakdown of macroscopic patterns of whole systems due to feedback reinforcing actions of agents on the micro-level, where the reinforcement is triggered by boundary conditions moving beyond critical tipping points. As a means to identify, assess, and manage this kind of risks, the concept of 'risk governance' has been developed within the perspective to provide integrative and comprehensive tools to deal with the complex risk situations. Encompassing both formal and informal institutional arrangements, the risk governance comprises generally five phases: pre-estimation, interdisciplinary assessment, risk evaluation, risk management, and risk communication (Renn et al. 2018).

In addition, the state of knowledge in the area of DRR is still problematic given the existing levels of uncertainty, the knowledge gap pertaining to the linkages between DRR, climate change, and Sustainable Development Goals (SDGs), and the existence of different sources of knowledge, often perceived as exclusive. This state of arts has many negative impacts on disaster risk governance and resilience building, and should be addressed to ensure the efficiency of the process. One way forward may indeed be to promote a holistic approach for comprehensive DRR measures, where both scientific and traditional knowledge systems can work together.

Another issue to be addressed within disaster risk governance is the academic-practitioner divides in disaster management and research which can be persistent and pernicious, bearing consequences for disaster survivors and future affected populations. Indeed, the gap between academic researchers and disaster professionals is often treated as an unavoidable structural problem or a neutral accident of professional silos and circumstance. Yet, these gaps are not neutral, and that they can and must be overcome in order to avoid expensive inappropriate actions and, most importantly, to decrease human suffering.

In addition, the evolution in the paradigms of disaster management concept and framework from 'response and relief' to 'mitigation and preparedness' has brought to light many approaches, which should be mainstreamed in disaster risk governance. For the purpose of this research, focus will be made below on some relevant approaches such as the community- and ecosystem-based approaches, the multi-stakeholder and participatory approaches, and the gender approach.

3 DRR, Climate Change and SDGs: Linkages and Governance Implications

This section will unpack the linkages between DRR, climate change and SDGs while highlighting their governance implications. The correlation between the SDGs and target actions with the priorities for action and targets of the Sendai Framework for Disaster Risk Reduction (UNDRR) will be demonstrated. The analysis will examine as well the SDGs through the lens of reducing disaster risk and building resilience and draw attention to measures that can contribute to the implementation and achievement of many relevant goals. Combined with climate policy, namely adaptation, this means the current context offers a chance to integrate three key international frameworks to guide policy and action more effectively and coherently, thus fostering the governance of these three interrelated areas.

Debates on climate change can be traced back to the Rio Summit in 1992. DRR and resilience date back to the Hyogo Framework for Action (HFA) 2005 to 2015: Building Resilience of Nations and Communities to Disasters. After realizing that disasters have grave consequences for survival, dignity, and livelihoods and threatens the world economy, the HFA provides full action-oriented policy guidance based on a comprehensive understanding of disaster risks that arise from human vulnerability to natural hazards (Valibeigi et al. 2019). The goal was to integrate DRR in development policy, planning and programmes at all levels while focusing on prevention, mitigation, preparedness, vulnerability reduction, and the strengthening of institutional mechanisms and resilience against hazards, especially at the community level (UNISDR 2017).

In 2015, the international community adopted almost simultaneously the Paris Agreement and the Sendai framework for Disaster Risk Reduction (2015–2030). The Paris Agreement sets out a global framework to avoid dangerous climate change by limiting global warming to well below 2 °C and pursuing efforts to limit it to 1.5 °C. It also aims to strengthen countries' ability to deal with the impacts of climate change and shift to low carbon economy and support them in their efforts through financial and technical assistance (UNFCCC 2015). Like HFA, the Paris Agreement focuses on prevention, mitigation, adaptation, preparedness and reduction of vulnerability to hazards. Mitigation by cutting GHG emissions is for instance a form of DRR and prevention of new risks. The Sendai Framework comprises four priorities and seven targets. The engagement in disaster risk gained momentum after the adoption of this framework which aims to increase understanding of disaster risk, to strengthen disaster risk governance systems, to enhance disaster preparedness for effective response, and to incorporate DRR in resilience building. Undeniably, the Sendai Framework focuses on multiscale anticipatory DRR actions and is guided by the principle that no country is immune to the impacts of hazards which are increasing in frequency and intensity as a result of environmental and climate changes. It focuses, among others, on DRR and resilience in livelihoods, health systems, infrastructure, educational facilities, technology, and ecosystems and natural resources. The Sendai Framework runs concurrently with SDGs and puts DRR at the core of the implementation process through resilience building and breaking the cycle of exposure, thereby reducing the risk of disasters and helps eliminate hunger and protect livelihoods (UNISDR 2017). Although DRR started with HFA, its strength as a systematic approach had been revealed during the formulation of both the Sendai Framework and the 2030 Agenda (ILO 2017).

The Intergovernmental Panel on Climate Change (IPCC), and before the adoption of both Paris Agreement and the Sendai Framework, has also called for greater alignment of national actions on climate change and national strategies for DRR (IPCC 2015). There was also an attempt not to isolate weather elements from other risks like poverty, rapid and unplanned urbanization, loss of protective ecosystems and population growth in hazard exposed areas (UNISDR 2018); thus showing the linkage between weather elements, environmental issues, and their risks to human security.

Therefore, there are strong linkages between climate change, DRR and SDGs and related frameworks as far as goals, targets and indicators converge. Climate change has the potential to undermine the efforts undertaken to achieve many SDGs, hence the vital importance of adaptation. In addition, it is increasing the risks of disasters and reversing decades of social and economic gains in many regions. More particularly, climate-induced disasters result in loss of food supplies and livelihoods (SDGs 2, 3), damage to infrastructure (SDGs 9, 11), increase competition over diminishing resources (SDG 15), worsen inequalities (SDGs 5, 10), and generate or amplify conflicts, violence and poverty (SDGs 1, 6, 16), among others (see Fig. 2). Furthermore, DRR and resilience are also crucial as enabling factors to achieving many SDGs, namely poverty eradication, economic growth, the reduction of inequality and the development of sustainable cities and settlements. Indeed, progress in DRR can contribute to the achievement of many SDGs and, similarly, efforts to achieve many SDGs can potentially enable, constrain or even cancel progress in DRR.

In addition, SDGs are inherently linked to each other, hence the need to work across and with all of them. Changes in environmental and climate patterns – for

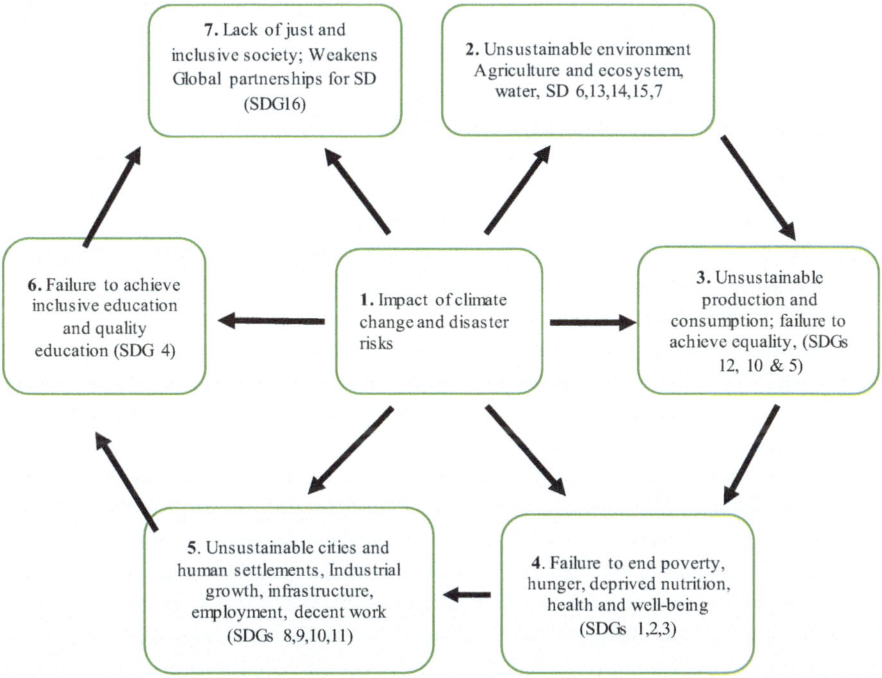

Fig. 2 Inter-linkages between climate change, DRR and SDGs *Source: Developed by authors.*

instance greater variability of rainfall, loss of farmlands, spread of food and waterborne diseases, and extreme events such as hurricanes and droughts – often result in less food supplies and decreased livelihoods (SDGs 1, 2, 3), and lead to malnutrition and spread of diseases. Damage to infrastructure and stress on the poor societies (SDG 3) often result in increased competition over diminishing resources (SDG 15). The ripple effects in turn cause increased ethnic, class tension, violence and conflicts (SDGs 6, 16). The poorest are often hard hit, thus worsening economic and gender inequalities (SDGs 5, 10). Similarly, unsustained attempts to achieve SDGs can increase disasters: for example, increased agricultural land-use and production to promote food access (SDG 2), using unsustainable methods, may increase losses from disasters (SDG 1). SDGs can be mutually reinforcing, therefore, improvement in one SDG can positively affect other SDGs (UNESCAP 2017): for instance, food increase in a sustainable way leads to improvements in health, nutrition, and well-being (SDG 3) while reducing inequalities and building resilience (SDGs 5, 10). On the contrary, SDGs can be in conflict when not implemented properly: for instance, interventions in infrastructure, land-use planning, ecosystems rehabilitation, and resource management system can potentially transfer risks from one area to another or can create behaviors that might be risky in the long term.

Furthermore, DRR cuts across different aspects and sectors of sustainable development. There are 25 targets related to DRR in 10 of the 17 SDGs, firmly establishing the role of DRR as a core development strategy. Therefore, the SDGs should reflect this by incorporating DRR targets into at least four goals, namely no poverty (1), industry, innovation and infrastructure (9), sustainable cities (11), and climate action (13). A coherent approach to policymaking helps avoid the risk of achieving progress in one goal at the expense of another. It will ensure that disaster risks are addressed to contribute positively to achieving these and other related goals. A systems approach to understanding the integrated nature of SDGs can generate high value addition for policy coherence and can help us look across the goals at the possible synergies and trade-offs to prioritize policies and investments that are mutually beneficial to build disaster resilience. It can help us see how to change policy systems more effectively to further sustainable development.

At another level, efforts to reduce disaster risks and climate risks have co-existed for a long time, and in the last two decades, they have increasingly been linked. In particular, the DRR and climate adaptation nexus has received significant attention, leading to calls for greater policy convergence between the two areas. It is true that efforts to reduce disaster risks and adaptation have co-existed for several years, but recently more sincere efforts at integrating the two in policy and practice have become common. Historically, there has been some frustration within the DRR community about the growing attention and funding being given to adaptation, without the recognition that DRR performs several of the same functions. On the climate side, as multilateral negotiations have stalled on many key issues, the growing impact of climate-related disasters has become a major motivator for trying to break the inaction. The relationship between disaster risk and climate change – and between response mechanisms to address them – is thus a very timely and policy-relevant issue. However, despite many overlaps, DRR and adaptation have evolved separately, with distinct differences, and these differences may affect the governance of these areas. Enabling DRR and adaptation to operate more closely requires a better understanding of the relationship between disaster and climate risks, their underlying causes, and the links between DRR and adaptation. Not only are there considerable similarities and overlaps in the types of actions needed to reduce both kinds of risks, but there is also great scope for mutual learning between DRR and adaptation researchers, practitioners and policy-makers.

Based on the above, laws, policies, and plans relating to disaster risk, climate change, and SDGs should go hand in hand since they have many similar goals and targets. This should apply as well to the governance processes and mechanisms in these areas (Rao 2013). The integration of climate change adaptation and SDGs in core DRR policies and operations is thus crucial to provide simultaneous benefits for both social and ecological systems. Therefore, integration and coherence at all levels should be promoted (UNESCAP 2018) in order to avoid contradicting laws, policies and plans during the formulation and implementation which may make the process inefficient or shift risks from one area to another.

4 Impacts of Knowledge Gap, Uncertainty and Science-Decision Divide

The state of knowledge in the area of disaster risk is still problematic given the existing levels of uncertainty, the knowledge gap pertaining to the linkages between DRR, climate change, and SDGs, and the existence of different sources of knowledge, often perceived as exclusive. This section emphasizes the need to make an end to the conflict between traditional and scientific knowledge and to reduce the divide between scientists and decision-makers/practitioners given the negative impacts of these problems on disaster risk governance.

Ancient and historical stories have detailed recordings about the great natural disasters, their causes and their impacts on human civilizations. The same stories have also highlighted human societies' response mechanisms during the said disastrous events. For example, the story of Noah's ark (cited in the Kur'an and the Bible) has depicted the very image of a huge flood on the earth's surface and has given the account of early prediction, preparedness and mitigation strategies. For Rai and Khawas (2019), traditional knowledge holds the element of those ancient stories, myths and folklores which, according to the modern scientific worldview, do not have any rational explanation because the scientific knowledge system does not believe in subjective reality. Measurement, experimentation and verification are one of the prime steps of modern scientific knowledge. However, modern scientific society forgets that those ancient stories and folklores have been developed through centuries-old experiences with the natural environment. Indeed, Iloka (2016) noted that traditional or indigenous knowledge is valuable knowledge that has helped local communities all over the world survive for generations. This knowledge originates from the interaction between members of the community and the environment in which they live.

In recent times, many studies identified rich indigenous and local knowledge and practices as a key determinant of resilience (i.e., Renzaho and Kamara 2016; Ngwenya et al. 2017). In a case study about the Mogalakwena community in Limpopo, South Africa, Rankoana (2016a) noted that the indigenous knowledge of seasons and early warnings, as well as traditional practices such as mixed cropping, the use of livestock manure, and the use of early maturing seeds are used to adapt to drought. Similarly, another study on drought by the same author pointed out that women use indigenous knowledge and culture to promote coping and adaptation practices, such as seed dressing, traditional crop maintenance, and rain-making rituals that promote food security and a constant seed supply (Rankoana 2016b). Renzaho and Kamara (2016), in a case study about resilience in Swaziland and Lesotho, observed that the application of traditional knowledge complemented with contemporary knowledge and skills improved health-related behavior and practices. The authors found that knowledgeable communities were those informed about their inherent capacities and services; they knew when and where to seek assistance. In the same perspective, a case study about agro-ecological knowledge among Ovambo farmers in north-central Namibia suggested that mixing traditional agro-ecological knowledge with scientific agricultural knowledge co-produced a hybrid agro-ecological knowledge without the limitations of either knowledge base. This hybrid knowledge imbued farming with drought adaptation practices such as early maturing crops, the use of donkey traction to plough large expanses of land, destocking, hunting, gathering, and sharing food among households (Newsham and Thomas 2011).

Kolawole et al. (2016), in a case study about disaster-prone communities in Ngamiland District, Botswana, suggested that adverse weather effects such as droughts led to a reduction in farming output and food availability, ultimately decreasing human welfare. Indigenous knowledge of weather forecasting (ethnometeorology) was a critical factor in adaptation and resilience. Households relied on the old tradition of observing the natural phenomena within their environment to inform their agricultural decisions. Some of the natural phenomena highlighted were the behavior of particular plants, the presence of particular insects, wild animal migrations, and the position and brightness of particular stars. Changes in the patterns of natural phenomena symbolized good rains, poor rains, or droughts, and forewarned communities to prepare for adversity and adjust their livelihoods accordingly.

In addition to traditional knowledge, some studies identified traditional community structures as strong mechanisms for resilience building. According to Ngwenya et al. (2017), traditional institutions mediate access to resources and promote adaptation based on local ecological knowledge. Such institutions may arrange and facilitate local learning processes that empower people to adapt or adjust their livelihoods based on the nature of adversity. By reference to a case study in the Okavango Delta in north-western Botswana, the authors claimed that households with diversified livelihoods were protected from the extreme impacts of floods, human and livestock disease outbreaks, and frequent droughts. Traditional institutions – such as Kgotla which is a traditional meeting place and chieftaincy – incentivized collective action and community participation in decision-making to mitigate periodic droughts and floods that negatively affected livelihoods and well-being. The institutions guided the development committees and volunteer associations for socio-health activities such as malaria and tsetse fly prevention. They mobilized farmer groups into traditional malapo (flood recession) farming which was an important livelihood in the harsh environment. Farmers adapted to planting quick maturing crops in the flood recession plains or dry river beds where soils retained moisture from seasonal floods. Traditional institutions and knowledge were credited with maintaining community cohesion and promoting adaptation to climate risks and were the central government's heartbeat of consultation on public policy.

Although much has been written about indigenous knowledge, its documentation in the area of DRR and climate change, especially in Africa, has been very limited. The wealth of this knowledge has not been well-recognized in the DRR area, as policy-makers still rely on mitigation strategies based on scientific knowledge. According to Iloka (2016), colonialism and lack of proper documentation of indigenous knowledge are some of the contributing factors to this state of affairs. Ignoring or underestimating the importance of understanding local communities' adaptive strategies has led to failed projects. For the same author, understanding how local communities in Africa have managed to survive and adapt for generations before the arrival of Western education, may be the key to developing effective policies to mitigate future challenges. Other researchers claimed that, in many contexts, this adaptation capacity has been slowly eroded by western-modelled education and external assistance (Newsham and Thomas 2011; Renzaho and Kamara 2016; Ngwenya et al. 2017). People with modern education often undervalued traditional early warnings and some simply dismissed them in favor of early scientific warnings. This is compounded by many NGO humanitarian and development interventions that inadvertently undermined intrinsic community capacities. The interventions often discouraged community efforts in self-sustainment in anticipation of handouts. For example, Renzaho and Kamara (2016) observed that disregard for intrinsic capacities and poor community engagement by humanitarian actors discouraged communities from anticipating and preparing for disasters.

According to Rai and Khawas (2019), despite the development of advanced technological innovation for DRR – with the invention of high-resolution satellite imagery, digital cartography and modern engineering building techniques to high-yielding agricultural production – their efficiency in reducing the impact of disasters remains questionable. The significant factor for the limited success of modern scientific society, according to the authors, is that it perceives the world from a temporal perspective where humans are believed to be an active agent in modifying every natural possibility into opportunity. The very composite environmental system is simplified whilst extracting resources, resulting in resource depletion and environmental degradation, consequently opening the door for disasters. Gad-el-Hak (2008) postulates that nature is supreme in governing any natural law and natural laws govern the evolution of any disaster. Although humans advanced themselves from primitive civilizations to modern scientific civilizations, they still lack the understanding of nature's activities or ability to predict its outcome. Modern science, for example, does not have precise laws in predicting earthquake hazards, thus making earthquake prediction more or less a black art. The unpredictability of nature's work still shows mankind that nature is the actual boss or dominant player.

Therefore, technocratic science must recognize the need for a relational or holistic approach rather than believing in reductionist approaches alone whilst dealing with disasters. In this context, the traditional knowledge has the potential to fill up the existing gaps generated by the modern science. This form of knowledge has different sets of ingredients to foster the development of the relational or holistic approach as it involves, interacts and interconnects humans, non-humans and nature together, setting a perfect balance for sustainability and DRR. It has vast undocumented observational data of changing natural phenomena, and in today's scenario of climate change and uncertainty, it can create a path for reliable adaptation measures related to climate-induced disasters. Therefore, a holistic approach is needed for comprehensive DRR measures where both scientific and traditional knowledge systems can work together (Rai and Khawas 2019).

On another level, Browne et al. (2018) claim that academic-practitioner divide in disaster management and research can be persistent and pernicious, bearing consequences for disaster survivors and future affected populations. The impetus to bridge such divide has drawn scholarly attention for decades (Eriksson and Sundelius 2005; Trainor et al. 2018). Focusing on the American context, Browne et al. (2018) note that the gap between academic researchers and disaster professionals is often treated as an unavoidable structural problem or a neutral accident of professional silos and circumstance. The authors reject the neutrality of this gap which is often the outcome of common misconceptions linked to different cultures of work, practice, goals, language, some professional animosity, and the non-alignment of the scale and temporality for practitioner and academic work. Therefore, there is urgency in connecting researcher and practitioner knowledges to avoid expensive inappropriate actions. This helps create a much-needed paradigm shift in how the work of DRR and recovery is achieved. With insight and clarity from bridging the work of experts, there is immense suffering that could be reduced or eliminated. This will surely foster the disaster risk governance for resilience on the long run.

5 The Role of Disaster Risk Communication in Shaping Stakeholders' Perceptions, Awareness and Participation

The link between disaster risk governance and communication outcomes is quite complex. Researches around the world identified a wide variety of factors that determine stakeholders' perceptions and often highlighted the need to holistically consider the individual, synergistic or antagonistic influences on outcomes. The role of risk communication is to address such pre-set conditions and create the context for collective action. Such important aspects as beliefs, cultural background, academic, technical and institutional preparedness, have been elaborated in this context in other sections of the present Chapter. The most recent Assessment Report on DRR (UN Office of DRR Assessment Report on Disaster Risk reduction 2019) and the review by Cazeau (2019) present a detailed overview of related dynamics. It also draws attention to the increasing complexity of disasters, related perceptions and a continually expanding agenda of communication and governance. Changing mindsets is one of the most important targets. Kondo et al. (2019) also deliberate on the science of societal safety with reference to a wide variety of risks and disasters and related risk communication.

Perceptions determine the importance conferred to the disaster, its impacts, confidence on mitigation and adaptation options, and most importantly, the credibility of the communicator. Access to information in a timely manner and the clarity of messages are equally important determinants of perceptions. This also pertains to the consequences of action or inaction stakeholders foresee across immediate, medium- and long-term horizons and conflicts that compound ambiguities. The robustness of predictive analyses and background information stakeholders use determines their responses to calls for concerted action. Communication strategies should, therefore, address strategically important thrusts and knowledge gaps to empower stakeholders to exercise their own discretion. These are essential risk communication elements pertaining to risk governance with cross-cutting implications for human, material, economic, environmental and related safeguards.

The following narrative focuses on some important insights that were consolidated especially over the present decade by several investigators and institutional processes. These are about the dynamics of perceptions and, therefore, the opportunity for risk communication strategies to address them systematically as part of emerging risk communication and governance processes.

A classic review of risk perception and communication dynamics with a special emphasis on artificially-induced perturbations was resented by Davis et al. (2019). Yet another on successes and failures involving government agencies with respect to risk communication was presented recently by Boholm (2019).

With special reference to climate change, the IPCC (2018) highlights risk communication as a low-regret cyclical and iterative intervention. It should, however, explicitly characterize uncertainty across four mutually-reinforcing phases of engagement, namely pre-assessment, appraisal, evaluation, and management. Its role as an evidence-based approach consists in promptly delivering appropriate information to tackle false risk perceptions and eliminate maladaptation. Communication strategies could motivate private and collective adaptation based on public interest, understanding and acceptance. Stakeholders could use climate risk information in decision making to minimize exposure, thereby reducing vulnerability. This is true also for policy makers, implementers, and risk specific outreach initiatives. National level agro-meteorology and coastal zone management programmes are typical cases in point.

Complexities inherent in risk communication have been elaborated by Adekola et al. (2020) with special reference to community resilience and increased ownership of risk by all concerned stakeholders. They refer to several technology-based approaches and elements of the International Disaster Database that present deep insights on resilience and related policy approach.

The UNISDR (2017) comprehensive framework for disaster risk assessment contextualized communication for engagement as discussed above. More recently the National Oceanic and Atmospheric Administration (ERG and NOAA 2019) presented a basket of eight best practices for optimal outcomes through risk communication. These are embedded also in varying measures in the framework of risk-informed development presented by Stapleton et al. (2019). The report refers to the agenda of resilience, especially under the influence of continually evolving forms and functions of threats, shocks and hazards. Importantly, it elaborates on the links of resilience with risk-based decision-making processes and the absolute need for transparent communication, building on the convergence provided by the Sendai Framework for resilience across all sectors of development. Similarly, the World Health Organizations has recently launched its risk communication strategy focusing on emergencies, especially in SE Asia (WHO 2019). They are comparable with FAO's elements and guiding principles of risk communication. The Capacity Building Initiative for Transparency is another case in point.

Ilona et al. (2011) deliberated on the mental model of stakeholder engagement. They indicated that differences in perception and governance of adaptation to climate change and extreme weather events are related to sets of beliefs and related concepts. Learning and exchange might be facilitated by participatory approaches and early involvement of stakeholder groups in the policy process. Accordingly, Bubek et al. (2012) reviewed risk perceptions with reference to flood mitigation behavior. Such aspects as coping appraisal and resultant preparedness to engage in mitigation are reportedly interrelated. They refer to the protection motivational hypothesis as the context to determine the ease of adopting precautionary measures and the need to optimize on such interventions through risk governance. Surprisingly, they indicated that high risk perception does not necessarily result in improved mitigation behavior. An excellent analysis of the dynamics of threat appraisal and coping appraisal follows in relation to perceived vulnerability, severity and related thresholds to define a coping response. These have implications for risk awareness raising.

The May 2012 Sackler Colloquium on The Science of Science Communication, presented a deep understanding of the dynamics of perceptions. It highlighted the convergence aspect of communication for people to share their understanding of facts, deliberate on values, and therefore the opportunity to prioritize thrust areas for action. Fischoff (2013) considers four tasks that should be performed by well-targeted science communication to deliver best. It is also important to take note of behavioral principles of judgment and choice within the larger framework of risk communication. Risk analysis, assessment and management are embedded in this framework.

von Winterfeldt (2013) argued that aspects of science embedded in risk communication should be used to shape values. These leads are interpreted with respect to health issues and environmental hazards wherein controversial beliefs and uncertainties could distort perceptions and engagement as part of the way ahead. Nkiaka and Lovett (2019) raise an interesting dilemma that societally-relevant scientific research results do not always contribute to policy development to solve societal problems. Timeliness and segmentation of information relevant to the immediate spatio-temporal contexts seem to determine the success of communication. The authors present an inclusive model of communication with special reference to public health-related adaptive capacity. The 2019 Global Risks Report of the World Economic Forum (WEF) presents the likelihood-impact-categories matrix and changing perceptions of risks over time. It appears to indicate there is no linear relationship between perception of risk and collective will to tackle them. While environmental risks and extreme weather continue to dominate thinking, environmental policy failure was implicated along with results of inaction. The WEF duly acknowledged the interplay of economic, geopolitical, societal and technological transitions amongst others in determining perceptions.

van Valkengoed and Steg (2019) presented a psychological perspective to adaptation by individuals and households to categorize behavior. Cuthbertson et al. (2019) highlighted the need to identify and profile current and emerging disaster risks as part of robust risk management practice. Several of the above stated elements of perception and management are embedded even in the most recent 'Submission to the Environment Agency's consultation on the Draft National Flood and Coastal Erosion Risk Management Strategy for England' (Surminski et al. 2019).

Franks et al. (2014) focused on private sector actors within the framework of socio-ecological change. Market instruments and regulations appear to shape their perceptions. They tend to convert environmental and social risks into business costs. This tendency, however, provides a useful window of opportunity to enlist their support for preventive management. Some interesting elements of outreach within the energy sector, in particular, are presented in the following. This is due to the centrality of the energy sector for global climate related externalities and the vulnerability of its infrastructure to challenges posed by climate change. A specific focus on local and transboundary nature of impacts and economic and trade benefits of reducing vulnerabilities appears to motivate collective action. The need to address elements of the Industrial Accidents Convention building on the increasing number of signatories is addressed in the above referred Global Assessment Report. Some initiatives by energy sector-related platforms including governments are accordingly considered in this context. These appear to have the potential to evolve as important elements of sector-specific disaster risk governance (UN Office of DRR 2019).

The World Bank Group's Energy Sector Management Assistance Program (ESMAP) collaborated with the Global Facility for DRR (GFDRR), to raise awareness within the sector (World Bank Group 2016). The specific focus was on the challenges posed by extreme weather, several other natural perturbations, and geological risks to the power sector. This interfaced with various forms, functions, age, and spatial distribution of power networks, energy services, and emergent economic health. It engaged with such stakeholder groups as utilities, policymakers, regulators and financial institutions worldwide. The purpose was to enhance capacities to mitigate risks and cope with impacts following disasters. Stakeholder perceptions predominantly highlighted the need for proactive maintenance to reduce risk and enhance natural disaster management interventions. Developing countries called for a special focus on five pillars of standard risk-management practices. The fifth of these, namely resilient recovery, pertained to the quality and timeliness of recovery and reconstruction efforts. Many power sector organizations are reportedly unaware of integrated risk management practices and, therefore, the opportunity to develop and implement locally-adapted system specific risk treatment options.

The 2019 Lifelines Report of the World Bank (Claire et al. 2019) further substantiated its earlier findings more emphatically. This time round, it included the renewables sector too. It indicated poor levels of awareness amongst several stakeholder groups about exposure of power infrastructure to hazards due to climate change and other natural disasters. Disaster risk management practices are reportedly often quite inadequate due to this aspect. Importantly, utilities, in order to become more resilient, should focus on a multi-step, multi-target approach covering technical, regulatory and operational elements in addition to institutional, and financial aspects. These target groups have their own unique communication needs and approaches of engagement. Utilities connect strongly with down-stream logistics and many value chains, with massive implications for the economy. It further reinforced the fact that, in most countries, the power system is designed to cope with high frequency, but relatively low impact events. Natural disasters as low-frequency and high-impact events are reportedly rarely considered. This is true as well for several industrial sectors (UNIDO 2019).

The Organization for Security and Cooperation in Europe (OSCE) in 2016 highlighted the need to protect electricity networks from natural hazards. It raised awareness through a handbook considering a wide variety of disasters and associated impacts. The Oil and Gas sector's International Petroleum Industry Environmental Conservation Association (IPIECA) defined opportunities to fulfill SDGs (IFC, IPIECA and UNDP, undated) duly recognizing the importance of workforce perceptions and multi-stakeholder dialogues in this process. It took note of the cross-cutting nature of the goals and approaches. The Global Reporting Initiative (GRI) Global Sustainability and Standards Board (2020) is presently developing the GRI Sector Standard for the oil, coal and gas sectors. It lays significant emphasis on stakeholder expectations with special reference to climate change impacts management and relevant reporting of such concerns. This is true also for the entire chemical industries' sector. Energy-intensive processes and operations lend themselves to best practices that can deliver integrated emission reduction and energy efficiency enhancement gains (OECD et al. and UN Environment 2015).

The website of the US Climate resilience tool kit[3] provides access to energy sector-case examples about climate-resilience enhancement by communities and businesses. The European Climate Adaptation Platform (Climate ADAPT), a partnership of the European Commission and the European Environment Agency,[4] also highlights infrastructure-related adaptation dynamics. Energy industry is an integral part of the Establishing Synergies for Disaster Prevention in the EU (ESPRESSO) initiative (Dilanti et al. 2018). This initiative aims at harmonizing DRR of the Sendai

[3]https://toolkit.climate.gov/#case-studies.

[4]https://climate-adapt.eea.europa.eu/eu-adaptation-policy/sector-policies/energy.

Framework and Paris Agreement's CCA thrusts. This harmonization is also reflected in the Energy system transition Through Stakeholder Activation, Education and Skills Development (ENTRUST) initiative of the EU (Dallamaggiore et al. 2019).

At the country-level, the case of India is inspiring; the Ministry of Power (2017) is currently finalizing its crisis and disaster management plan for the power sector. This is in direct response to an increasing need to tackle the challenges posed by natural disasters, climate change and human-induced crises that inflict significant losses on infrastructure, including human resources. This is aligned with the 2005 Disaster Management Act of the country. Interestingly, it has created a formal mechanism to address stakeholders' perceptions as agents of change for immediate local relevance. The IEA India energy policy 2020 (IEA 2020) reinforces this integrated approach, aligned with SDG 7 'Ensure access to affordable, reliable, sustainable and modern energy for all'. Detailed assessments of the sustainability of outcomes from communication as a forerunner for action are needed to theorize about the relative influence that governments and bilateral and multilateral institutions exert on engagement.

One of the most recent trends in risk communication is the groundswell of public reactions to perceived lackadaisical approach by political leaders to climate change impacts management. Setzel and Byrnes (2019) highlight growing climate change litigation from a human rights and science-centered perspective. Citizens look for real life mitigation and adaptation outcomes beyond the court room proceedings. This tends to emphasize the meta-precautionary principle whereby uncertainty is not taken as a cause for inaction (Patterson and McLean 2019). Knowledge condition, harm condition, recommended precaution and proportionality determine perceptions in this matrix. Knowledge on factors that determine perception and responses to risks related to risk framing is central to the design of a robust communication strategy as a social process.

6 Key Approaches to Disaster Risk Governance for Resilience

Disaster risk governance requires the mainstreaming of many relevant approaches at all levels. For the purpose of this research, the analysis focuses in the below section on the community- and ecosystem-based approaches, the multi-stakeholder and participatory approaches, and the gender approach. This focus is justified by the fact that these approaches are highly relevant and may reinforce the effectiveness of the disaster risk governance at all levels.

6.1 Ecosystem- and Community-Based Approaches

Increased disaster risk undermines both ecosystem and community resilience (Munang et al. 2013). Over the past few years, two approaches to tackle disaster risk have gained importance: Community-Based Disaster Risk Reduction (CBDRR); and Ecosystem-Based Disaster Risk Reduction (EBDRR). The first is meant to empower local communities to reduce their vulnerabilities while the second focuses on harnessing the management of ecosystems as a means to provide goods and services (Dhyani and Thummarukudy 2016), which is seen as a cost-effective, no regret and a win–win solution (Renaud et al. 2013; McVittie et al. 2018; CBD 2018).

The Sendai Framework, under Priority 2, promotes disaster risk governance at the national, regional and global levels with transboundary cooperation for policy and planning to implement ecosystem-based approaches within river basins and along coastlines. It also focuses on broader and a more people-centric preventive approach to disaster risk, enhancement of collaboration among people at the local level to disseminate disaster risk information, and empowerment of local authorities through regulatory and financial means (United Nations 2015). It is well known that ecosystem-based adaptation (EbA) strategies offer significant opportunities to strengthen the links between conservation, agriculture, water and land management, and drought prevention (UNCCD, undated). In fact, about 12% of the EbA projects listed on the UNFCCC database primarily aim to mitigate the effects of natural hazards (Milman and Jagannathan 2017).

The dynamic landscape of ecosystem-based pathways is a conceptual framework that uses an integrated ecosystem management approach within interlinked socio-ecological systems to enhance resilience and maintain ecological functions and services at the landscape scale (Devisscher 2010). These pathways may include protected area networks management, vulnerability assessment, restoration, rehabilitation, and water resource management. As risk is not confined to any territorial or political division, it is also critical that DRR strategies or plans consider transboundary and regional solutions, such as basin- or ecosystems-based management (UNDRR 2019). Examples of EBDRR include mangrove afforestation in coastal areas to tackle storm surge, saline intrusion and coastal erosion (Ahammad et al. 2013; Dhyani and Thummarukudy 2016), river basin restoration (CBD 2018), buffer strips (GIZ et al. 2018), wetland restoration for coastal protection, and protection of forest for slope stabilization (McVittie et al. 2018).

Although the current developments in EBDRR tend to focus on scientific projections, engineering techniques, and their respective roles in shaping economic benefits, effective implementation of such solutions depends on the governance practices and interactions between relevant actors, interests, and institutional structures (Triyanti and Chu 2018). However, local decisions that support appropriate actions are scarce (Young et al. 2019).

In effect, CBDRR is a shift from reactive emergency management to DRR as it focuses more on pre-disaster interventions such as prevention, mitigation, and preparedness related activities (Shaw 2016), hence enhancing resilience (Lassa et al.

2018). EBDRR too requires active and inclusive involvement of a range of stakeholders in order to enhance the knowledge base (Murti and Mathez-Stiefel 2019). CBDRR and EBDRR are often connected to each other. CBDRR is aimed at recognizing the roles of local communities and local capacities (Lassa et al. 2018), providing an opportunity to utilize indigenous knowledge for DRR and settle strategy for risk reduction at the community level (Abedin 2013). In comparison, EBDRR requires engagement of local communities as co-adaptive managers (Murti and Mathez-Stiefel 2019) and enables people to adapt to climate change impacts and disasters through sustainable management, conservation and restoration of ecosystems (McVittie et al., 2018). However, an understanding of linkage or environment-disasters interface is required to incorporate ecosystem-based approaches into development planning (Dhyani et al. 2018). One way is to integrate indigenous and scientific knowledge and apply the principles of inclusion and collaboration in adaptation intervention for a resilient and empowered community (Bacud 2018).

Natural resource managers, disaster risk managers, humanitarian aid relief actors, sustainable development practitioners, investors and issue-specific experts are some examples of practitioners who need to come together to operationalize EBDRR (Murti and Mathez-Stiefel 2019). This participation can subsequently help local communities create and maintain a sustainable interaction with the environment while enabling them to use the natural resources to sustain their livelihoods (Lin 2019). Coordinated community- and ecosystem-based actions offer a potential path to achieve DRR, climate adaptation, sustainable development, and biodiversity conservation (Klein et al. 2019).

The positive causal link from resilience to the community's ability to recover from a conflict or disaster creates an additional causal pathway to promote individual and familial well-being, creating the conditions for further participation in community greening initiatives (Tidball et al. 2018). Failure to acknowledge community resilience undermines effective, socially sustainable disaster management and rural development planning, and thus, there is a need to understand, recognize and strengthen the capacities of local communities (Imperiale and Vanclay 2016). Research on drought resilience highlights the importance of community capacities. The main elements identified were sharing of information, costs, and risks, social networks and interaction, solidarity, participation, cooperation, joint efforts, and connectedness (Kamara et al. 2018). For example, Renzaho and Kamara (2016) underscored community cohesiveness and support, social networks, empowerment and participation, psychological well-being and community ownership of disaster preparedness as collective elements that enhanced resilience. In all cases, community- and ecosystem-based approaches foster the communities' environmental and social resilience and prepare them to respond to future disasters (Lin 2019).

However, there have been so far no corresponding methodologies for assessing the political, social, and institutional dimensions of Eco-DRR or ecological engineering (Triyanti and Chu 2018). According to Young et al. (2019), DRR does not include some important aspects such as: the use of ecosystem services; the impact of local climate on ecosystem services; the impact of human behavior on ecosystem services; and the benefits of nature-based solutions. It is important to understand as well that successful adaptation depends not only on governments, but also on the active and sustained engagement of stakeholders, including national, regional, and local organizations, the public and private sectors, civil society, and other relevant stakeholders (Ahmed and Suphachalasai 2014) (Table 1).

6.2 Making Disaster Risk Governance Gender Sensitive

Disasters, natural or man-made, are by nature, non-discriminatory, then why the need for gender sensitivity? As McEntire (2001) puts it, any disaster is a combination of a triggering agent and a set of vulnerabilities; and it is these vulnerabilities, the conditions which affect the capacity of a community or a society to respond to the triggering agent, which represent the controllable component of a disaster. Looked at from this viewpoint, disasters may be unbiased, but their impacts are most certainly not.

It has long been acknowledged that women are unequally affected by disasters, because they are largely confined to their domestic space, lack access to information or even disaster warnings on radios, mobiles or in public places, have often been conditioned not to leave their homes without male support, and ultimately shoulder the responsibility for children, the aged and the infirm, and even for domestic animals. Following the 2004 Asian tsunami, Oxfam International (2005) found that in many villages in Aceh, Indonesia, and in parts of India, females accounted for over 70% of the dead. In the 1991 cyclone disaster that killed 140,000 in Bangladesh, 90% of the victims were women and girls. A study of 141 countries over a twenty-year period (1982–2001) found that more women than men are killed during disasters (Neumayer and Plümper, 2007). While two decades have passed since, there has been little change, for instance 70% of the lives lost during the 7.5 magnitude Hindu Kush earthquake in Afghanistan (2015) were those of women and young girls.

While 'gender' is not synonymous with women, and in rare instances, more male lives have been lost during disasters – as in the case of Hurricane Mitch in El Salvador and Gautemala, in 1998, where social pressures required men to take more risks – this is not the case in general, and women continue to be unequal victims of both natural and manmade disasters. However, the understanding that women are more vulnerable to loss of life during disasters and face greater hardships in the aftermath, has led to the narrow perspective of women as 'hapless victims' needing special focus and care packages. While this viewpoint is a welcome first step in disaster response and rehabilitation, it disregards the positive role that women can play in disaster risk governance given the specialized knowledge and skills which they possess as caretakers of homes, fields and the environment in which they live. In fact, the International Strategy for Disaster Reduction (ISDR) non-equivocally states that "gender equality in disaster reduction requires empowering women to have an increasing role in leadership, management and decision-making positions because

Table 1 Ecosystem-based adaptation and Eco-DRR measures by land use and land cover category

Category	Broad measure	Example measures	Impact addressed
Agriculture	Agricultural habitats	– Agro-forestry and crop diversification – Buffer strips and hedgerows – Improved water retention in agricultural areas – Meadows and pastures	– Floods – Flash floods – Drought
	Agricultural management	– Crop rotation – Low till agriculture – No till agriculture – Green cover	
Forestry	Forest planting	– Reforestation – Afforestation – Forest riparian buffers – Land use conversion – Maintenance of forest cover in headwater areas	– Climate change mitigation – Landslides – Floods – Heat waves
	Forest management	– Water sensitive forest management – Coarse woody debris – Continuous cover forestry	
Coastal		– Beach nourishment – Coastal managed realignment – Dune reinforcement and strengthening – Cliff stabilization	– Sea level rise – Storm surges – Landslides
Urban	Green infrastructure	– Green roofs – Rain gardens – Soakaways – Swales – Urban greenspace – Urban forest parks – Urban trees and forests	– Heat waves – Flash floods – Floods
	Blue infrastructure	– Basins and ponds – Channels and rills – Detention Basins – Filter Strips – Infiltration basins – Permeable surfaces – Retention Ponds Sediment capture ponds – SUDS – Temporary flood water storage	
Water management	River restoration	– Elimination of riverbank protection – Natural bank stabilization – Re-meandering – Reconnection of oxbow lakes and similar features – River restoration and rehabilitation – Riverbed material re-naturalization – Stream bed re-naturalization	– Floods – Flash floods
	Floodplain restoration		
	Groundwater restoration		
	Lake restoration		
	Wetland restoration		

Source: Adapted fromMcVittie et al. (2018)

women are not only victims of disasters but they can act as agents of change in disaster reduction planning" (ISDR 2002).

Enarson and Meyreles (2004) found that the exclusion of women from the disaster management and decision-making processes, especially in developing countries, has contributed to the higher rates of female victimization in disasters. A case study from Afghanistan finds that quite often, women tend to lose their lives as they make several trips back home to secure essential provisions, after carrying their children to safety, and are usually unaware of the duration or severity of the expected disaster. It is, however, heartening to note that even in rural Afghanistan, where women rarely take decisions without the permission of a male relative, the advent of technology has brought in positive change. Village women have formed groups to send warnings, via mobile phones, of impending natural disasters like floods. While women, in their individual capacities, may not have the confidence or the authority to flaunt conventions and take active measures for DRR, as members of an empowered group, they can effectively contain the loss of life and property and reduce the time and cost of rehabilitation (Hamidazada et al. 2019).

The participatory exercises with women and men in Afghanistan, revealed that the major issues that women face, among others, are: where to shelter during a disaster; how to prepare for different types of disasters; and how to install early warning systems in the event that mobile phones do not operate in some mountainous areas (Hamidazada et al. 2019). These concerns are likely to echo across all disaster-prone remote areas. Training of women in DRR has been seen to have a significant impact. Oxfam reports that in a recent landslide in Tajikistan, where 35 households were at risk of being buried, a female community mobilizer had prepared the community so well that the risk of imminent landslide was noticed, a warning given out, the area evacuated, and lives spared. Forty years earlier, 134 people had died from a similar landslide in the village (Oxfam 2009).

To facilitate grassroots DRR, collaboration with NGOs, local women leaders, religious and other mentors is necessary to ensure that: designated shelters/safe areas are identified; warning systems, both traditional and modern are in place; basic provisions are stored and maintained in good condition at the shelter; key women are selected and trained to evacuate the habitation quickly and efficiently; reliable means of communication are installed at the shelter; and training and regular drills in disaster response are conducted.

At the global level, the awareness of gender issues in DRR has been slowly increasing with 51 of 62 national reports to UNISDR in 2007 acknowledging gender as important to DRR, but there is still very little concrete gender approach mainstreaming in policies and programmes (ISDR, UNDP and IUCN 2009). While women's representation at all levels of disaster risk governance is essential to ensure that women-specific concerns are addressed during the policy-making process, it is the active collaboration of women at the grassroots which will, in reality, significantly cut down on loss of life, livelihoods and belongings, and effectively reduce the impacts of disasters and reinforce the resilience to disaster risks.

6.3 Participatory and Multi-stakeholder Approaches: The IASC Cluster Approach

The Inter-Agency Standing Committee's (IASC) Cluster Approach is an existing and well-established multi-stakeholder approach to emergency preparedness, response and recovery that has been employed routinely in both natural disasters and conflict settings for over fifteen years, as part of a UN-wide Humanitarian Reform Agenda 'to enhance predictability, accountability and partnership' (UN OCHA 2018). Activated in response to a natural disaster for the first time in 2005 following flooding in Pakistan, the approach represents a definitive attempt at both horizontal and vertical coordination between inter-governmental organizations, NGOs and governments (UN OCHA 2018). The Cluster Approach may be understood as a coordination mechanism through which agencies and organizations act as pre-determined cluster/sector leads or co-leads for eleven clusters (along with partner agencies from national and local governments in the affected State) on a variety of issue areas such as health, logistics, food security and protection. These clusters are, in theory, ready to jump into action (often beginning work with local authorities prior to disasters) once a disaster strikes and States invite or accept international humanitarian support.

Whilst the Cluster Approach has faced criticism, particularly regarding its bureaucratic structure which has been seen as detracting both attention and resources from affected communities, coordination during disasters and crises for many researchers is an essential component of any successful intervention[5] (Ong et al. 2015; Sanderson 2019:1). Its capacity for information sharing, reducing resource waste and coordinated preparedness across a wide range of actors means the Cluster Approach merits at least some consideration in the discussion on DRR (Abbaya et al. 2019). This is especially true when considering that either as a response to, or as a direct consequence of, having previously activated the Cluster Approach, some countries have installed 'cluster-like' or 'sector'-based approaches into their own national frameworks for disaster mitigation, preparedness and response, a famous example of which may be seen in the Philippines where a comprehensive national DRR plan mirroring a similar sector-based system has been in place for a decade (Dy and Stephens 2016:6; Abbaya et al. 2019). As such, in addition to representing a disaster response mechanism, the Approach can also be understood as a contribution to DRR practices.

To disregard the complexities and relationships surrounding and within the Cluster Approach is to both ignore its strengths and consider how the Approach could be made more participatory and inclusive, as well as consider what the field of DRR may learn from it regarding coordination (Guadagno 2016; Jahre and Jensen 2010). Perhaps the largest critique of the Approach, as raised by many researchers and

[5]The COVID-19 crisis has demonstrated the importance of coordination at all governance levels as an efficient mechanism to ensure that response mechanisms adopted by different actors to face the pandemic and its related consequences result in generating the expected outcomes. For instance, in the case of the European Union and the USA, the weak levels of coordination result in shifting risks between states, divergent decisions and actions more or less science-based, weak levels of information sharing, the competition over resources, insufficient solidarity, etc.

assessments, has been its lack of inclusiveness – participatory approaches – and failure to strengthen national and local capacities (Hanley et al. 2014; Steets et al. 2010). Yet, perhaps as a response to such criticisms, years later, agencies have been seen working to empower locals and trigger better preparedness measures in the context of the Cluster Approach and beyond. Such work often takes place in the recovery phase following a disaster, however given the cyclical nature of many types of disasters (such as yearly flooding or tropical cyclones), such efforts might be seen as DRR and preparedness efforts for the future (UN Environment 2005). This is especially true when considering that a central component of the Cluster Approach is for the global cluster lead to ensure capacity building on multiple levels, which can include 'training personnel or participating in mitigation efforts for future disasters in exposed areas' (Jahre and Jensen 2010: 660–661).

For example, within and around it's typical role as lead of the Camp Coordination and Camp Management Cluster (CCCM), the International Organization for Migration (IOM) has invested heavily in recent decades in capacity building and training activities at the local level in a number of locations (IMO 2008; 2011; 2013a; 2013b; 2014; 2017). This goes against the critique that the Approach has bypassed local capacity building, which further disregards 'behind-the-scenes' work agencies may be doing because of, or in addition to, their Cluster-related roles (Noureddine Tag-Eldeen 2017). Although the IOM's engagement may be best understood contextually and via the agency's role as a development and project-based actor reliant on the financial support of Member States, preparation and risk reduction practices are a logical way in which IOM and other agencies can contribute to better practices at the local level.

Given that disasters often require a coordinated response, it seems logical that conceptualizations such as DRR do so as well. Continuing with the IOM as an example case, whilst the agency may be active at the field level and highly familiar with community engagement in disaster contexts, it lacks the expertise, authority and accountability mechanisms in numerous areas of concern for those affected by disasters, particularly when it comes to issues of human rights and protection – areas to which other agencies such as the OHCHR and UNHCR are able to contribute (Human Rights Watch 2004; OHCHR 2019; UNHCR 2019). As such, coordination across agencies on multiple levels is required for both effective DRR and response, and for creating resilient communities for the future. Although there is certainly a strong argument – and need – for bottom-up approaches propelled by civil society and technological advances regarding adaptation, to disregard existing structures within which DRR is taking place is to disregard what we may learn from the past, as well as to close the door to considering what these approaches may contribute towards burgeoning work in the field of DRR. The Cluster Approach is evidence of both horizontal and vertical collaboration, which should not be forgotten as a means through which to provide checks and balances across a variety of actors who are all reliant on one another to carry out collaborative work in this area.

7 The Role of Laws, Regulations, and Policies

As mentioned above, the capacity of countries to reduce disaster risk and building resilience is increasingly dependent on effective governance arrangements. In this regard, there is a widespread agreement that laws and regulations are critical tools for governments to shape their options in this regard. Within this perspective, Elhadj As Sy (IFRC) and Helen Clark (UNDP) believed that:

> "strong legal frameworks to manage and reduce risks will underpin efforts to tackle the challenges that lie ahead. A risk-informed legal framework can make the difference between a strong building left standing and one reduced to rubble when a natural hazard strikes. It can mean the difference between the safety of one neighbourhood and the devastating flooding of another, or a child able to protect herself when a disaster strikes, and one who does not know how to get to safety" (IFRC and UNDP 2015a:4).

In 2005, the HFA highlighted the important role laws and regulations play in reducing existing disaster risks, preventing new ones from arising and making people safer. In light of this first international guidance, several countries have sought to strengthen their laws and regulations for DRR. However, some DRR experts and activists have expressed doubts and disappointment with the legislative route, arguing that the many new laws and policies that have been developed to address DRR seem not to have made the difference they promised, citing in particular gaps in implementation at the community level. It was found for instance that communities were not well enough informed, engaged or resourced to take an active part in DRR, and that rules to prevent risky behavior, particularly in construction and land use, often go unenforced (UNDP and IFRC, 2014). Numerous reports relating to HFA implementation have also indicated slow progress in reducing disaster risk at the community level, and a lack of clear information and analysis on the role of laws and regulations.

In 2011, the Resolution 7 of the 31st International Conference of the Red Cross and Red Crescent encouraged states – with support from their National Red Cross and Red Crescent Societies (National Societies), the IFRC, UNDP and other relevant partners – to review existing legislative frameworks in light of the key gap areas identified in the IFRC report to the Conference (IFRC 2011). Together with the HFA, Resolution 7 set the framework for the country studies to be undertaken for the multi-country report, in particular by stressing the value of civil society and community participation, emphasizing the importance of building codes and land-use planning to reduce underlying risks, and considering accountability and legal liability as potential legal incentives for DRR.

In a multi-country assessment of legal frameworks for DRR in 31 countries, undertaken by UNDP and IFRC in 2014, it was found that in order to support a whole-of-society approach, legal frameworks for DRR should include institutional mandates, allocate dedicated resources, facilitate the participation of communities, civil society and vulnerable groups, and establish the responsibility and accountability of relevant actors. Effective frameworks facilitate the mainstreaming of DRR into relevant sectors, are sustainable within the available resources and capacity of government at national and local levels, and fit within the overall legal and institutional structure of the country. In addition, the findings indicate that DRR is a more distinct priority in policies, plans and strategies than in legal frameworks. However, the interaction between law and policy, whilst complex, is often essential for successful implementation. Countries rarely tackle the fundamental reform towards DRR without a specific legal framework, since disaster risk management (DRM) laws are essential for setting the DRR priorities and mandates of implementing institutions. Even countries with successful DRR regimes that are based on policy only, eventually look to codify key elements through legislative provisions. However, they also use policy processes to advance new law reforms. Hence, policy can both set the agenda for the law reform process and be a key tool to guide the implementation of laws. Furthermore, most disaster risk management laws established a single agency (such as a national disaster management agency or a civil defense office) as the central national focal point for cultivating a whole-of-society approach to DRR, and providing national leadership and policy direction. However, it was found that these institutions often need to strengthen their coordination with other sectors and stakeholders, especially those related to development planning and climate change adaptation (UNDP and IFRC 2013).

Based on this outcome, the Sendai Framework in 2015 reiterates the call for a renewed focus on reviewing and strengthening legal frameworks to ensure that DRR is integrated across all sectors and at all levels with clear mandates and accountability frameworks (IFRC and UNDP 2015a). More precisely, the Framework invites governments to:

- review and promote the coherence and further development, as appropriate, of national and local frameworks of laws, regulations and public policies;
- encourage the establishment of necessary mechanisms and incentives to ensure high levels of compliance with existing safety-enhancing provisions of sectoral laws and regulations, including those addressing land use and urban planning, building codes, environmental and resource management, and health and safety standards, and update them, where needed, to ensure an adequate focus on disaster risk management;
- assign, as appropriate, clear roles and tasks to community representatives within disaster risk management institutions, processes and decision making through relevant legal frameworks. Undertake comprehensive public and community consultations during the development of such laws and regulations to support their implementation;
- Encourage parliamentarians to support the implementation of DRR through developing new or amending relevant legislation and setting budget allocations;
- allocate the necessary resources, including finance and logistics, as appropriate, at all levels of administration for the development and the implementation of DRR strategies policies, plans, laws and regulations in all relevant sectors; and
- review and strengthen national laws and procedures on international cooperation, based on the Guidelines for the domestic facilitation and regulation of international disaster relief and initial recovery assistance (UN 2015).

Table 2 Checklist on law, regulations, and DRR

1	Do you have a dedicated law for disaster risk management that prioritizes risk reduction and is tailored to your country context?
2	Do your laws establish clear roles and responsibilities related to risk reduction for all relevant institutions from national to local level?
3	Do your key sectoral laws incorporate provisions to increase safety and reduce vulnerability?
4	Do your laws ensure that sufficient resources are budgeted for disaster risk reduction?
5	Do your laws establish clear procedures and responsibilities for risk assessments and ensure risk information is considered in development processes?
6	Do your laws establish clear procedures and responsibilities for early warning?
7	Do your laws require education, training and awareness-raising to promote a whole-of-society approach to disaster risk reduction?
8	Do your laws ensure the engagement of civil society, the private sector, scientific institutions and communities in risk reduction decisions and activities?
9	Do your laws adequately address gender considerations and the special needs of particularly vulnerable categories of persons?

Source: IFRC and UNDP (2015b)

To guide countries throughout the process, the IFRC and the UNDP developed a Checklist (see Table 2) which provides ten prioritized questions that lawmakers, officials, practitioners and those supporting them need to consider in order to ensure that their laws and regulations provide the best support for DRR. The Checklist is intended to ensure that risk reduction and resilience are well integrated into and supported by legal systems. However, it is not designed to comprehensively address all issues related to law and disaster management. It does not as well specifically address disaster preparedness, response or recovery, though plans were underway to develop similar tools for these areas. Similarly, the Checklist is not a model DRM law. The guidance provided in this Checklist is designed to help prioritize DRR in dedicated DRM laws on one hand, and sectoral laws on the other. Both areas of legislation are equally as important, and as highlighted in the Checklist questions, no single law is able to completely address all aspects of DRR (IFRC and UNDP 2015b). In addition, the Checklist covers not only dedicated DRM laws, but also other sectoral laws and regulations – covering issues such as the environment, land and natural resource management – that are critical for building safety and resilience. Therefore, the Checklist aims to foster a more integrated approach to DRR by incorporating climate change and sustainable development considerations into its review of legislation. The Checklist mainly focuses on disasters caused by natural hazards, and does not include specific considerations related to industrial accidents, public health emergencies and situations of generalized violence (IFRC and UNDP 2015a).

The process by which the Checklist questions are discussed and agreed upon may be as important as the content of the answers themselves. Given the multi-sectoral and inter-disciplinary nature of DRR, responding to the Checklist questions will require an efficient governance system enabling the engagement and contribution of a range of stakeholders from national to local levels, including government, civil society and community representatives. It is expected that the process of bringing these stakeholders together in a common dialogue to respond to the Checklist questions will be an additional benefit of the process. Overall, the methodology for using the Checklist should be tailored to each country context and their respective needs. However, the expected outputs of the process should provide: an analysis which has the potential to provide a clear overview of the strengths and gaps in the legal framework, both in terms of the content of the legislation and its implementation; the ability to determine and identify priority areas to address in order to bring the legal framework in line with international standards, and more specifically the Sendai Framework for DRR; and the opportunity to strengthen dialogue and understanding between different actors involved in the regulation of DRR (IFRC and UNDP 2015b).

8 Conclusion

The adoption of the Sendai Framework, the SDGs, and Paris Agreement in 2015 was supposed to have a profound and far-reaching impact on our ability to reduce disaster risk. However, the shared challenge remains the translation of these frameworks over which there is a global consensus into meaningful actions, as the only true measure of success is the impact they should generate at the local level, in the affected communities around the world (IFRC and UNDP 2015a). Within such a perspective, it was shown in the above analysis that the capacity of countries to reduce disaster risk is increasingly dependent on effective governance arrangements. Based on this assumption, the analysis focused on many topics considered highly important for the effectiveness of the disaster risk governance and resilience building process.

Regarding the linkages' governance implications, it was demonstrated that the integration of climate change adaptation and SDGs in core DRR policies and operations is crucial to provide several simultaneous co-benefits. These areas have many similar goals and targets whose achievement may be enabled through the promotion of a certain coherence in their relevant governance processes and response mechanisms at all levels. Failure to do so may result in making these processes and mechanisms inefficient or shifting risks from an area to another. Therefore, the nexus between these areas is a very timely and policy-relevant issue.

Regarding the knowledge gap and uncertainty and their impacts on disaster risk governance, the analysis demonstrated the vital importance of traditional knowledge, which has the potential to fill up existing gaps and uncertainties generated by modern science since it involves, interacts and interconnects humans, non-humans and nature together, setting a perfect balance for sustainability and DRR. Indeed, in today's scenario of climate change and uncertainty, traditional knowledge can create a path for reliable adaptation measures related to climate-induced disasters. Therefore, a relational or holistic approach with the potential to connect modern science with traditional knowledge and institutions may make disaster risk governance more effective, especially on the local level. In addition to this, it was demonstrated that the integration of critical research and the realities of practice should be stimulated and reflective practitioners and engaged academics should be encouraged to invest in bridging the divide between them. This helps create a much-needed paradigm shift in how the work of DRR and recovery is achieved, therefore fostering the disaster risk governance for resilience in the long run.

Risk communication as a forerunner for collective action has been used extensively in several spheres with implications for risk management. These target the whole range of natural and artificially induced risks, disasters and related externalities. Despite significant outreach, challenges continue to prevail about the spread and depth of engagement with stakeholders. Perceptions of risk and response that follow, appear to modulate outcomes of engagement. Several characteristics of perception could be similar across sectors, in addition to thrust area specific drivers. Detailed investigations are needed to theorize links between perceptions and preparedness to engage, in order to tackle barriers in engagement.

Additionally, it was demonstrated that disaster risk governance requires the mainstreaming of many relevant approaches at all levels. These approaches have the potential to make disaster risk governance more inclusive, gender-sensitive and receptive to both community- and ecosystem-based approaches; thus ensuring a sustainable disaster resilience.

Finally, it was shown that the capacity of countries to reduce disaster risk and building resilience, especially at the community level, is increasingly dependent on their ability to develop appropriate laws, regulations, and policies. International framework and guidance exist but the progress in this area is still insufficient to manage emerging challenges efficiently in many countries. Therefore, there is an urgent need for countries to review and strengthen their legal and policy frameworks to ensure that DRR is integrated across all sectors and at all levels with clear mandates and accountability frameworks.

References

Abbaya TMR, Le Dé L, Lopez Y (2019) Localizing the UN cluster approach: the Philippines as a case study. Environ Hazards. https://doi.org/10.1080/17477891.2019.1677209

Abedin M (2013) Community-based disaster risk reduction approaches in Bangladesh. In: Shaw R, Fuad M, Aminul I (eds) Disaster Risk Reduction Approaches in Bangladesh, pp 259–279

Acosta JD, Chandra A, Madrigano J (2017) An agenda to advance integrative resilience research and practice: Key themes from a resilience roundtable. RAND Health Q 7:5

Josephine A, Fischbacher-Smith D, Fischbacher-Smith M (2020) Inherent complexities of a multi-stakeholder approach to building community resilience. Int J Disaster Risk Sci 11(1):32–45. https://doi.org/10.1007/s13753-020-00246-1

Ahammad R, Nandy P, Husnain P (2013) Unlocking ecosystem-based adaptation opportunities in coastal Bangladesh. J Coast Conserv 17:833–840. https://doi.org/10.1007/s11852-013-0284-x

Ahmed M, Suphachalasai S (2014) Assessing the costs of climate change and adaptation in South Asia. Asian Development Bank, Mandaluyong City, Philippines

Bacud ST (2018) Integration of indigenous and scientific knowledge in disaster risk reduction: resilience building of a marginalized sampaguita growing community in the Philippines. Procedia Eng 212:511–518. https://doi.org/10.1016/j.proeng.2018.01.066

Blaikie P, Cannon T, Davis I, Wisner B (2014) At Risk: Natural Hazards, People's Vulnerability and Disasters. Routledge, London

Boholm A (2019) Lessons of success and failure: practicing risk communication at government agencies. Saf Sci 118:158–167

Browne KE, Caela O, Laura MY (2018) Journey through the groan zone with academics and practitioners: bridging conflict and difference to strengthen disaster risk reduction and recovery work. Int J Disaster Risk Sci 9:421–428

Bubeck P, Botzen WJW, Aerts JCJH (2012) A review of risk perceptions and other factors that influence flood mitigation behavior. Risk Anal 32(9):1481–1495

Cazeau JW (2019) Review of the integration of disaster risk reduction in the work of the United Nations system in the context of the 2030 Agenda for Sustainable Development, p 75. Joint Inspection Unit. UN, Geneva. https://digitallibrary.un.org/record/3840225?ln=en

Chabay I (2018) Taking time, sharing spaces: adaptive risk governance processes in rural Japan. Int J Disaster Risk Sci 9:464–471

Convention for Biodiversity (CBD) (2018) Guidelines for Ecosystem-based Approaches to Climate Change Adaptation and Disaster Risk Reduction

Chen R, Zhang Y, Xu D, Liu M (2018) Climate change and coastal megacities: disaster risk assessment and responses in shanghai city. In: Climate Change, Extreme Events and Disaster Risk Reduction. Springer, Germany, pp 203–216. ISBN 978-203-319-56468-56465

Chmutina Ksenia, von Meding Jason (2019) A Dilemma of language: "natural disasters" in academic literature. Int J Disaster Risk Sci 10(3):283–292

Claire N, Jun R, Albertine P, Van L, Sam O, Amy S, Mark D, Elco K, Christopher A, Diana C, Li j (2109) Stronger power - improving power sector resilience to natural hazards. In: LIFELINES: The Resilient Infrastructure Opportunity 2019. The World Bank, p 46. https://openknowledge.worldbank.org/bitstream/handle/10986/31910/Stronger-Power-Improving-Power-Sector-Resilience-to-Natural-Hazards.pdf?sequence=1&isAllowed=y

Coppola DP (2011) Introduction to International Disaster Management. Elsevier, Burlington

Cuthbertson J, Rodriguez-Llanes JM, Robertson A, Archer F (2019) Current and emerging disaster risks perceptions in oceania: key stakeholders recommendations for disaster management and resilience building. Int J Envt Res Pub Health 16(460), 13. https://www.researchgate.net/publication/330882352_Current_and_Emerging_Disaster_Risks_Perceptions_in_Oceania_Key_Stakeholders_Recommendations_for_Disaster_Management_and_Resilience_Building/link/5c5a249892851c48a9bd68cc/download

Dallamaggiore E, Eva B, François A, Breffní L, Rosemarie MS, Christine G, Niall D, Alberto L, Josep O (2019) ENTRUST Energy System Stakeholder Characterisation Deliverable D2.1 154p. EU Horizon 2020. http://www.entrust-h2020.eu/wp-content/uploads/2017/01/D2.1_Energy-System-Charact_-release-v2.pdf

Davis LE, Tom LT, David EM, Lois MD, David RH (2019) Appendix D. A review of the risk – perception and risk communication literature. In: Individual Preparedness and Response to Chemical, Radiological, Nuclear, and Biological Terrorist Attacks, RAND Corporation, pp 147–152. https://www.jstor.org/stable/pdf/10.7249/j.ctt1q60dh.16.pdf?refreqid=excelsior%3Ad36ee60fa5e36e1df0ac8224e40769c5

Devisscher T (2010) Ecosystem-based Adaptation in Africa: Rationale, Pathways, and Cost Estimates. Stockholm Environment Institute

Dhyani S, Lahoti S, Khare S et al (2018) Ecosystem based disaster risk reduction approaches (EbDRR) as a prerequisite for inclusive urban transformation of Nagpur City, India. Int J Disaster Risk Reduct 32:95–105. https://doi.org/10.1016/j.ijdrr.2018.01.018

Dhyani S, Thummarukudy M (2016) Ecosystem based approaches to disaster risk reduction (EbDRR) and climate change adaptation (EbA). InBackground Paper. CSIR-NEERI and UNEP collaborative National consultative workshop Ecological Engineering for Disaster Risk Reduction

Dilanthi A, Richard H, Nuwan D, Chamindi M (2018) ESPRESSO Synthesis report of existing legal, policy and science approaches in relation to DRR and CCA (Deliverable 2.1) Compiled. http://www.espressoproject.eu/images/deliverables/ESPREssO_D2.1.pdf

Dy, P, Stephens T (2016) The Typhoon Haiyan Response: Strengthening Coordination among Philippine Government, Civil Society, and International Actors. https://www.hks.harvard.edu/sites/default/files/centers/research-initiatives/crisisleadership/files/Dy_and_Stephens.pdf
Eastern Research Group, Inc. (ERG) and the National Oceanic and Atmospheric Administration (NOAA) Social Science Committee (2019). A Practical Guide for Natural Hazard Risk Communication. http://performance.noaa.gov/economics. https://www.performance.noaa.gov/wp-content/uploads/practical-guide-07-24-19-fillable508.pdf
Eriksson J, Sundelius B (2005) Molding minds that form policy: How to make research useful. Int Stud Perspect 6(1):51–71
Fischoff B (2013) The sciences of science communication. PNAS 110(3): 14033–14039. https://www.pnas.org/content/pnas/110/Supplement_3/14033.full.pdf
Franks MD, Davi R, Bebbington AJ, Ali SH, Kemp D, Scurrah M (2014) Conflict translates environmental and social risk into business costs. PNAS 111(21):7576–7581. www.pnas.org/cgi/, https://doi.org/10.1073/pnas.1405135111
Gad-el-Hak M (2008) Large-scale disasters: Prediction, controls and mitigation. Cambridge University Press, New York
GIZ, EURAC, UNU-EHS (2018) Climate Risk Assessment for Ecosystem-based Adaptation: A guidebook for planners and practitioners
GRI Global Sustainability Standards Board (2020) Draft GRI Sector Standard: Oil, Gas and Coal, p 96. https://www.globalreporting.org/standards/media/2545/item-05-draft-gri-sector-standard-oil-gas-coal.pdf
Guadagno L (2016) Human mobility in the sendai framework for disaster risk reduction. Int J Disaster Risk Reduct 7:30–40. https://doi.org/10.1007/s13753-016-0077-6
Marina H, Cruz AM, Yokomatsu M (2019) Vulnerability factors of Afghan rural women to disasters. Int J Disaster Risk Sci 10:573–590. https://link.springer.com/content/pdf/, https://doi.org/10.1007/s13753-019-00227-z.pdf
Hanley T, Binas R, Murray J, Tribunalo B (2014) IASC Inter-agency Humanitarian Evaluation of the Typhoon Haiyan Response. Accessed Oct 2014, www.unhcr.org/research/evalreports/583305867/iasc-inter-agency-humanitarian-evaluation-typhoon-haiyan-response.html?query=typhoon%20haiyan
Human Rights Watch (2004) Human Rights Watch's Statement to the IOM Governing Council 30 Nov—3 Dec 2004 (88th Session). https://governingbodies.iom.int/system/files/jahia/webdav/shared/shared/mainsite/about_iom/en/council/88/humanrights.pdf
IEA (2020) India 2020 Energy policy review, p 305, https://niti.gov.in/sites/default/files/2020-01/IEA-India%202020-In-depth-EnergyPolicy_0.pdf
IFC, IPIECA & UNDP (undated) Mapping the Oil and Gas industry to the Sustainable Development Goals: An Atlas, p 112. IPIECA Oil and Gas, https://www.ipieca.org/media/4404/online-mappingoilandgastosdgatlas.pdf
ILO (2017) Building on the Sendai Framework in support of the Paris Agreement: Implementation and monitoring. https://www.ilo.org
Iloka N (2016) Indigenous knowledge for disaster risk reduction: an African perspective. J. Disaster Risk Studi. 8(1). https://doi.org/10.4102/jamba.v8i1.272
Ilona OB, Matczak P, Wesseler J, Wechsung F (2011) Different perceptions of adaptation to climate change: a mental model approach applied to the evidence from expert interviews. Reg Environ Change 11:217–228. https://link.springer.com/content/pdf/, https://doi.org/10.1007/s10113-010-0144-2.pdf
Imperiale AJ, Vanclay F (2016) Experiencing local community resilience in action: learning from post-disaster communities. J Rural Stud 47:204–219. https://doi.org/10.1016/j.jrurstud.2016.08.002
International Federation of Red Cross and Red Crescent Societies (IFRC) and United Nations Development Programme (UNDP) (2015a) The Handbook on Law and Disaster Risk Reduction. https://www.undp.org/content/dam/undp/library/Climate%20and%20Disaster%20Resilience/Handbook%20on%20law%20and%20DRR_dec2015.pdf

International Federation of Red Cross and Red Crescent Societies (IFRC) and United Nations Development Programme (UNDP) (2015b) The checklist on law and disaster risk reduction - Pilot Version. https://www.ifrc.org/PageFiles/115542/The-checklist-on-law-and-drr.pdf
International Federation of Red Cross and Red Crescent Societies (IFRC) (2013). Better laws, safer communities? Emerging themes on how legislation can support disaster risk reduction. https://www.ifrc.org/PageFiles/118981/IFRC_better-laws-safer-communities_2013.pdf
International Federation of the Red Cross and Red Crescent Societies (IFRC) (2011) Law and disaster risk reduction: background report for the 31st International Conference of the Red Cross and Red Crescent.
International Organization for Migration (2008) Building capabilities and capacities at all levels for timely humanitarian assistance, including disaster risk reduction. https://www.iom.int/speeches-and-talks/building-capabilities-and-capacities-all-levels-timely-humanitarian-assistance
International Organization for Migration (IOM) (2011) IOM Co-Hosts Camp Management Training in Timor-Leste. https://www.iom.int/news/iom-co-hosts-camp-management-training-timor-leste
International Organization for Migration (IOM). (2013a). Building community resilience in Papua New Guinea. https://www.iom.int/news/building-community-resilience-papua-new-guinea
International Organization for Migration (IOM). (2013b) Training workshop on managing population in natural disasters, Dakar. https://www.iom.int/news/dakar-training-workshop-managing-population-natural-disasters
International Organization for Migration (IOM) (2014) Angola Builds Displacement Camp Management Capacity. https://www.iom.int/news/angola-builds-displacement-camp-management-capacity
International Organization for Migration (IOM) (2017) IOM Builds Protection Capacity of Korean Humanitarian Workers. https://www.iom.int/news/iom-builds-protection-capacity-korean-humanitarian-workers
International Strategy for Disaster Reduction (2002)
IPCC (2018) Managing the risks of extreme events and disasters to advance climate change adaptation, p 594. https://www.ipcc.ch/site/assets/uploads/2018/03/SREX_Full_Report-1.pdf
IPCC (2015) Summary for Policy makers in climate change. Policy Brief. Part A. pp 63. International Strategy for disaster reduction (ISDR) (2006). Hyogo Framework for Action 2005–2015: Building Resilience of Nations and communities to disasters. UN World conference on disaster reduction
ISDR, UNDP, IUCN (2009) Making Disaster Risk Reduction Gender Sensitive: Policy and Practical Guidelines, United Nations
Jahre M, Jensen L-M (2010) Coordination in humanitarian logistics through clusters. Int J Phys Distrib Logist Manag 40(8/9):657–674. https://doi.org/10.1108/09600031011079319
Kamara JK, Akombi BJ, Agho K, Renzaho AMN (2018) Resilience to climate-induced disasters and its overall relationship to well-being in southern Africa: a mixed-methods systematic review. Int J Environ Res Public Health 15:2375
Klein JA, Tucker CM, Steger CE et al (2019) An integrated community and ecosystem-based approach to disaster risk reduction in mountain systems. Environ Sci Policy 94:143–152. https://doi.org/10.1016/j.envsci.2018.12.034
Kolawole OD, Moseki RM, Ngwenya BN, Thakadu O, Mmopelwa G, Letsholo KD (2016) Climate variability and rural livelihoods: How households perceive and adapt to climatic shocks in the Okavango Delta. Botswana WCAS 8:131–145
Kondo S, Yukio H, Hideyuki S (2019) Risk communication and disaster information. In: Abe S, Ozawa M, Kawata Y (eds) Science of Societal Safety. Living at the times of risks and disasters, Trust: Interdisciplinary Perspectives 2. Springer Nature, Singapore Pte Ltd., pp 129–140. https://doi.org/10.1007/978-981-13-2775-9_12
Lassa JA, Boli Y, Nakmofa Y, et al (2018) Twenty years of community-based disaster risk reduction experience from a dryland village in Indonesia. Jàmbá J Disaster Risk Stud 10. https://doi.org/10.4102/jamba.v10i1.502

Lin P-SS (2019) Building resilience through ecosystem restoration and community participation: post-disaster recovery in coastal island communities. Int J Disaster Risk Reduct 39:101249. https://doi.org/10.1016/j.ijdrr.2019.101249

Klaus L, Ortwin R, Carlo J, Saini Y (2018) Systemic risks: a homomorphic approach on the basis of complexity science. Int J Disaster Risk Sci 9:292–305

McEntire DA (2001) Triggering agents, vulnerabilities and disaster reduction: towards a holistic paradigm. Disaster Prev Manag 10(3):189–196

McVittie A, Cole L, Wreford A et al (2018) Ecosystem-based solutions for disaster risk reduction: lessons from European applications of ecosystem-based adaptation measures. Int J Disaster Risk Reduct 32:42–54. https://doi.org/10.1016/j.ijdrr.2017.12.014

Milman A, Jagannathan K (2017) Conceptualization and implementation of ecosystems-based adaptation. Clim Change 142:113–127. https://doi.org/10.1007/s10584-017-1933-0

Morrison J (2004) Gender and Development. Routledge, New York

Munang R, Thiaw I, Alverson K et al (2013) The role of ecosystem services in climate change adaptation and disaster risk reduction. Curr Opin Environ Sustain 5:47–52. https://doi.org/10.1016/j.cosust.2013.02.002

Murti R, Mathez-Stiefel S (2019) Social learning approaches for ecosystem-based disaster risk reduction. Int J Disaster Risk Reduct 33:433–440. https://doi.org/10.1016/j.ijdrr.2018.09.018

Neumayer Eric, Plümper Thomas (2007) The gendered nature of natural disasters: the impact of catastrophic events on the gender gap in life expectancy, 1981–2002. Ann Assoc Am Geogr 97(3):551–566

Newsham Andrew J, Thomas David S.G. (2011) Knowing, farming and CLIMATE change adaptation in North-Central Namibia. Glob Environ Change 21(2):761–770

Ngwenya BN, Thakadu OT, Magole L, Chimbari MJ (2017) Memories of environmental change and local adaptations among molapo farming communities in the Okavango Delta, Botswana—A gender perspective. Acta Tropica 175:31–41

Nkiaka Elias, Lovett Jon C (2019) Strengthening the science-policy interface for climate adaptation: stakeholder perceptions in Cameroon. Reg Environ Change 19(4):1047–1057

Noureddine Tag-Eldeen Z (2017) Bridging urban planning knowledge into post-disaster response: early recovery road map within the international humanitarian cluster system. Int. J. Disaster Risk Reduct. 24:399–410

OECD, UNECE, UNEP, WHO, INISDR, UN Environment/OCHA Joint Unit OCHA, and EPSC (undated) International efforts for industrial and chemical accidents prevention, preparedness and response, p 34. https://www.oecd.org/chemicalsafety/chemical-accidents/Brochure%20chemical%20accidents%20prevention,%20preparedness%20and%20response.pdf

OHCHR (2019) The UN Human Rights Office in Humanitarian Action. https://www2.ohchr.org/english/OHCHRreport2015/allegati/9_OHCHR_s_approach_to_field_work_2015.pdf

Ong JC, Flores JM, Combinido P (2015) Obliged to Be Grateful: How Local Communities Experienced Humanitarian Actors in the Haiyan Response. https://plan-international.org/obliged-be-grateful http://hdl.handle.net/2381/33421

Organization for Security and Cooperation in Europe (OSCE) (2016) Protecting electricity networks form natural hazards, p 124. https://www.osce.org/secretariat/242651?download=true

Oxfam International (2005) The Tsunami's impact on women. Oxfam Briefing Note, March 2005. Oxfam International, Oxford. http://www.oxfam.org.uk/resources

Oxfam Tajikistan case studies (2009) 'I feel so free', 'Riding High', and 'Inclusive', internal communication briefings

Peters K, Peters LER, Twigg J, Walch C (2019) Disaster Risk Reduction Strategies. Working Paper 555

Peters K, Peters LR, Tanner (2016) Resilience across the post-2015 Frameworks: How to create greater coherence. Overseas development Institute (ODI), London

Rai P, Khawas V (2019) Traditional knowledge system in disaster risk reduction: exploration, acknowledgement and proposition. Jàmbá J Disaster Risk Stud 11(1):a484. https://doi.org/10.4102/jamba.v11i1.484

Rankoana SA (2016a) Perceptions of climate change and the potential for adaptation in a rural community in Limpopo province, South Africa. Sustainability 8:672
Rankoana SA (2016b) Rainfall scarcity and its impacts on subsistence farming: the role of gender and religious rituals in adaptation to change. Agenda 30:124131
Sumedh R (2013). Disaster risk governance at national and sub-national levels, Applied Knowledge Services (GSDRC)
Renaud FG, Sudmeier-Rieux K, Estrella M (2013) The Role of Ecosystems in Disaster Risk Reduction. United Nations University Press
Ortwin R, Andreas K, Pia-Johanna S (2018) Risk governance: application to urban challenges. Int J Disaster Risk Sci 9:434–444
Renzaho AM, Kamara JK (2016). Building Resilience in Southern Africa: A Case Study of World Vision in Swaziland and Lesotho. World Vision International: Johannesburg, South Africa
Sanderson D (2019) Coordination in urban humanitarian response. Progr Disaster Sci 1:1–4. https://doi.org/10.1016/j.pdisas.2019.100004
Setzer J, Byrnes R (2019) Global trends in climate change litigation: 2019 Snapshot. Policy report 12p. LSE. Granthan Research Institute on Climate Change and the environment, Columbia Law School, Sabin Centre for Climate Change Law, University of Leeds, Centre for Climate Change Economics & Policy. http://www.lse.ac.uk/GranthamInstitute/wp-content/uploads/2019/07/GRI_Global-trends-in-climate-change-litigation-2019-snapshot-2.pdf
Shaw R (2016) Community-based disaster risk reduction. Oxf Res Encycl Nat Hazard Sci. https://doi.org/10.1093/acrefore/9780199389407.013.47
Smith K (2013) Environmental hazards: Assessing risk and reducing disaster. Routledge, London
Stapleton SO, Rebecca N, Jan KT, Margherita C, Adriana Q, Katie P, Leigh M (2019) Risk informed development – From crisis to resilience, p 52. ODI, UNDP & SDC, UNDP, USA. https://www.undp.org/content/dam/undp/library/Climate%20and%20Disaster%20Resilience/Risk-informed%20development_WEB_final.pdf
Steets J, Grünewald F, Binder A, de Geoffroy V, Kauffmann D, Krüger S, Sokpoh B (2010). Cluster Approach Evaluation 2 Synthesis Report. https://www.humanitarianresponse.info/sites/www.humanitarianresponse.info/files/documents/files/Cluster%20Approach%20Evaluation%202.pdf
Surminski S, Viktor R, Sara M, Rebecca B, Bob W, Josh B (2019) Submission to the Environment Agency's consultation on the Draft National Flood and Coastal Erosion Risk Management Strategy for England, p 12. http://www.lse.ac.uk/GranthamInstitute/wp-content/uploads/2019/10/GRI-Submission-to-EA-consultation-on-draft-national-flood-and-coastal-erosion-risk-management-strategy-for-England.pdf
Tidball KG, Metcalf S, Bain M, Elmqvist T (2018) Community-led reforestation: cultivating the potential of virtuous cycles to confer resilience in disaster disrupted social–ecological systems. Sustain Sci 13:797–813. https://doi.org/10.1007/s11625-017-0506-5
Trainor JE, Stern EK, Subbio T (2018) On bridging research and practice in disaster science and management: Unified system or impossible mission? In: Rodr´ıguez H, Donner W, Trainor JE (eds.) Handbook of disaster research, pp 161–178. Springer, Berlin
Triyanti A, Chu E (2018) A survey of governance approaches to ecosystem-based disaster risk reduction: Current gaps and future directions. Int J Disaster Risk Reduct 32:11–21. https://doi.org/10.1016/j.ijdrr.2017.11.005
UN Environment (2015) APELL A process for improving community awareness and preparedness for technological hazards and environmental emergencies, p 108. https://www.preventionweb.net/files/45469_unepawarenesspreparednessemergencie.pdf
UN Environment. (2005) Environmental management and disaster preparedness: Lessons learnt from the Tokage Typhoon in Japan. https://wedocs.unep.org/bitstream/handle/20.500.11822/8838/-Environmental%20Management%20and%20Disaster%20Preparedness_%20Lessons%20Learnt%20from%20the%20Tokage%20Typhoon%20(Typhoon%2023%20of%202004)%20in%20Japan-2005468.pdf?sequence=3&amp%3BisAllowed=
UN OCHA (2018) What is the Cluster Approach? https://www.humanitarianresponse.info/en/about-clusters/what-is-the-cluster-approach

UNCCD (undated), Ecosystem-based-adaptation. http://www.unccd.int/en/programmes/Event-and-campaigns/WDCD/Pages/What-is-Ecosystem-Based-Adaptation.aspx, Accessed 1 Nov 2014
UNDP, UNISDR et al (2013) Concept Note - The First Arab Regional Conference for Disaster Risk Reduction, Aqaba, Jordan, 19–21 March 2013
UNFCCC (2015) The Paris Agreement. Summary climate focus. Washington: Climate focus. https://climatefocus.com
UNIDO (2019) International Conference on ensuring industrial safety. The role of government, regulations, standards and new technologies, p 92. https://www.unido.org/sites/default/files/files/2020-01/International%20Conference%20on%20Ensuring%20Industrial%20Safety.pdf
United Nations (2015) Sendai Framework for Disaster Risk Reduction 2015–2030
United Nations Development Programme (UNDP) and International Federation of Red Cross and Red Crescent Societies (IFRC) (2014) Effective law and regulation for disaster risk reduction: a multi-country report – Summary. https://www.ifrc.org/Global/Publications/IDRL/country%20studies/summary_report_final_single_page.pdf
United Nations Economic and Social Commission for Asia and the Pacific (UNESCAP) (2017) Building resilience to disasters in Asia and the Pacific: Resilience in global Development Frameworks. https://www.unescap.org/resources.
United Nations Economic and Social Commission for Asia and the Pacific (UNESCAP) (2018) Framing Policy Coherence. Module 1. https://www.UNESCAP.org
United Nations High Commissioner for Refugees (UNHCR) (2019) Protection Cluster (UNHCR Emergency Handbook). http://www.ifrc.org/PageFiles/95884/D.01.03.%20Handbook%20for%20Emergencies_UNHCR.pdf
United Nations International Strategy for Disaster Reduction (UNISDR) (2018) Paris Agreement. https://www.unidsr.org/archive
United Nations International Strategy for Disaster Reduction (UNISDR) (2017) Think Piece on national disaster risk reduction strategy requirements. UNIDSR, Geneva
United Nations International Strategy for Disaster Reduction (UNISDR) (2018) Terminology. https://www.unisdr.org/we/inform/terminology#letter-d.
United Nations Office for Disaster Risk Reduction (UNDRR) (2019) Global Assessment Report on Disaster Risk Reduction. United Nations Office for Disaster Risk Reduction, p 472. https://gar.unisdr.org
Valibeigi M, Feshari M, Zivari F, Motamedi A (2019) How to improve public participation in disaster risk management: A case study of Buein Zahra, a small city in Iran. Jàmbá J Disaster Risk Stud 11(1):a741. https://doi.org/10.4102/jamba.v11i1.741
Volenzo TE, Odiyo JO (2019) Linking risk communication and sustainable climate change action: a conceptual framework. Jàmbá J Disaster Risk Stud **11**(1):a703. https://doi.org/10.4102/jamba.v11i1.703.
Von Winterfeldt D (2013) Bridging the gap between science and decision making. Proc Nat Acad Sci 110(Supplement_3). https://doi.org/10.1073/pnas.1213532110
Walsh-Dilley M, Wolford W, (2015) (Un) defining resilience: Subjective understandings of 'resilience' from the field. Resilience 3:173–182
World Bank Group (2016) Enhancing power sector resilience: Emerging practices to manage weather and geological risks. ESMAP, p 125. http://documents.worldbank.org/curated/en/469681490855955624/pdf/113894-ESMAP-PUBLIC-FINALEnhancingPowerSectorResilienceMar.pdf
Young AF, Marengo JA, Martins Coelho JO et al (2019) The role of nature-based solutions in disaster risk reduction: the decision maker's perspectives on urban resilience in São Paulo state. Int J Disaster Risk Reduct 39:101219. https://doi.org/10.1016/j.ijdrr.2019.101219
Zhoux X, Moinuddin M, Xu M (2017) SDGS, Inter-linkages and network analysis: A practical tool for SDG integration and policy coherence. Institute for Global environment strategy (IGES)

Benefits of Evaluating Ecosystem Services for Implementation of Nature-based Solutions Under the Paris Agreement

Himangana Gupta and Lokesh Chandra Dube

Abstract Ecosystems provide a variety of provisioning, supporting, regulating and cultural services which are key to climate change adaptation and environmental sustainability. Current trends of biodiversity loss and overburdened natural ecosystems further challenge climate adaptation in the developing world. Scientists are trying to develop their understanding of the relationships among different ecosystem services provided by diverse ecosystems given its vital importance for landscape management, decision-making and policy development. The mechanisms in complex socio-ecological systems and their mutual impact have still not been studied, especially from the climate perspective. However, Ecosystem-based Adaptation (EbA) and Nature-based Solutions (NbS) are increasingly gaining importance in climate debates, but methods to bring them into the Paris Agreement's market and non-market mechanisms are still unclear. Evaluating the ecosystem services in nature-based solutions would instill confidence into the climate strategists, investors, financers and buyers for market and non-market approaches. Such evaluation could include socio-economic parameters so as to also benefit the livelihoods of the local communities. Monitoring of carbon regulating services and provisioning services will help in establishing robust market mechanisms. On the other hand, evaluating social and cultural benefits of the ecosystems are more beneficial for SDGs and non-market mechanism of the Paris Agreement. UNFCCC COP-26 will be instrumental in deciding the guidelines for the much-awaited Paris Mechanisms. In light of the rising concerns regarding the NbS, this chapter discusses the emergence of such a mechanism and assesses the need for the development of appropriate indicators and evaluation of ecosystem services to better monitor NbS projects. Such monitoring and evaluation could be helpful in drawing the benefits of the Paris Mechanisms and climate finance through avenues such as the green climate fund in the post-2020 period.

H. Gupta (✉)
United Nations University Institute for the Advanced Study of Sustainability (UNU-IAS) and The University of Tokyo, Tokyo, Japan

L. C. Dube
National Communication Cell (NATCOM), Ministry of Environment, Forest and Climate Change, Government of India, New Delhi, India

M. Behnassi et al. (eds.), *Social-Ecological Systems (SES)*,
https://doi.org/10.1007/978-3-030-76247-6_2

Keywords Nature-based solutions · Ecosystem services · Monitoring and evaluation · Climate change · Multiple benefits

1 Introduction

After trying and testing numerous climate mitigation and adaptation methods in the 28 years of the history of the United Nations Framework Convention on Climate Change (UNFCCC), the world now sees Nature-based Solutions (NbS) as a potential mechanism to tackle climate change along with several other co-benefits. The need for such a mechanism arose much earlier in the form of Ecosystem-based Adaptation (EbA), which is defined as the use of biodiversity and ecosystem services to adapt to the adverse effects of climate change whilst strengthening the resilience of communities and ecosystems (Secretariat of CBD 2009a; UNFCCC 2017).

Ecosystems provide a variety of provisioning, supporting, regulating and cultural services which are crucial for climate change adaptation and environment sustainability. If conserved and managed sustainably, ecosystems and their ability to provide services can play a vital role in helping people adapt to climate change (IISD 2018). EbA approaches rely on this concept and thus include sustainable management, conservation and restoration of ecosystems, as part of an overall adaptation strategy that takes into account the multiple social, economic and cultural co-benefits for local communities (Secretariat of CBD 2009b; Doswald and Osti 2011). Accordingly, it is embedded in the concepts of ecosystem services and climate change adaptation (Chong 2014; Wamsler et al. 2014). Such approaches integrate biodiversity and the provision of ecosystem services into the overall climate change adaptation strategies, are cost-effective and help maintain resilient ecosystems, besides potential benefits for indigenous peoples and local communities (Secretariat of CBD 2009b). IUCN's Commission on Ecosystem Management also strengthened the case for ecosystem-based options (Barrow et al. 2012), as it offers significant opportunities to strengthen the links between conservation, agriculture, water and land management and drought prevention (UNCCD). The conservation and sustainable use of ecosystem services has the potential to contribute to mitigating climate change and to help human societies adapt to its impacts.

The projected climatic changes will influence population dynamics, species distribution patterns, species interactions and ecosystem services, as a result of spatial or temporal reorganisation (Bellard et al. 2012). Climate change will also influence several factors of importance to habitat quality and biodiversity. It is one of the most potential drivers of the landscape change (Schirpke et al. 2017; Seidl et al. 2019), which can significantly affect the nature and value of ecosystem services. While EbA initiatives contribute to climate change mitigation by reducing net emissions from ecosystem degradation and by enhancing carbon sequestration (UNFCCC 2017), continuous loss of ecosystem services can be a threat to future climate adaptation goals. As of now, about 131 nations (i.e., 66% of all signatories to the Paris Agreement) have chosen EbA, in one form or the other, as an implementation strategy in their Nationally Determined Contributions (NDCs) (Seddon et al. 2020b). The world is contemplating options that allow and assist nature to heal itself for sustenance of nature's benefits to the society. Such options included EbA which is now being categorised into NbS that might include a broader set of activities including ecosystem-based mitigation.

The emergence of the concept of NbS was a step-by-step process of test and trial with ecosystem-based methods. However, putting an economic value on ecosystems is a difficult task for trading under market mechanisms of the UNFCCC. Market mechanisms like Clean Development Mechanism (CDM) of the Kyoto Protocol focused solely on climate mitigation while the ecosystem conservation aspects were sidelined (Gupta and Dube 2018). This approach of valuing only carbon sequestration among a myriad of ecosystem services, in fact, discouraged NbS projects by various countries as it was almost impossible to earn credits with non-quantitative benefits. CDM included land-based activities such as forestry as eligible options, that did not consider benefits other than GHG reduction/removals. This also meant that monocultures could easily win carbon credits, even if they failed to address the overall ecological well-being (Gupta 2015). The rising demand for ecosystem protection and addition of " + " in the REDD + made it clear that mindlessly targeting to remove emissions by creating carbon sinks will not solve the problem of climate change, and thus, more robust approaches combining mitigation and adaptation would be needed (Gupta and Dube 2018). This effort has already kick-started with the Paris Agreement which incentivizes joint mitigation and adaptation approaches for the integral and sustainable management of forests, while reaffirming the importance of incentivising, as appropriate, non-carbon benefits associated with such approaches (United Nations United Nations 2015). This step opened several doors for mechanisms like NbS and EbA. The momentum for NbS continued to grow in the 25th Conference of the Parties (COP) to the UNFCCC (Nature4Climate 2019). The Convention on Biological Diversity (CBD) had realised this fact earlier and the decision taken in its 10th session of the Conference of Parties (COP 10) on 'biodiversity and climate change' invited countries to consider the ways to conserve, sustainably use and restore biodiversity and ecosystem services while contributing to climate change mitigation and adaptation (COP-10 2010). It also encouraged to implement ecosystem-based approaches for adaptation, including sustainable management, conservation and restoration of ecosystems, as part of an overall adaptation strategy that takes into account the multiple social, economic and cultural co-benefits for local communities.

It seems that the world is now willing to implement NbS but the question is how? The time has come to deliberate on this question as the modalities of the new market mechanisms of the Paris Agreement are still to be decided and making rules that allow for successful implementation of NbS under the Paris Agreement is now necessary. This also means that we need robust methods to quantify the benefits of ecosystem services while not forgetting the non-quantitative benefits, such as the intrinsic value of biodiversity. The development of ecosystem services indicators is progressing rapidly (Schirpke et al. 2016), but their role in new market mechanisms of the Paris Agreement are not well considered. In fact, the requirements for EbA implementation and benefits under future climate change regime are also unclear (Milman and Jagannathan 2017). For successful implementation of NbS, ecosystem services must be valued in terms of climate mitigation and adaptation, along with multiple benefits such as food security, water resource management, and conservation.

This chapter discusses the emergence and need of NbS, while assessing the possibilities of including NbS measures in the new market mechanisms of the Paris Agreement. It assesses the type of projects and sectors that could potentially be a part of the crediting mechanism in addition to non-market benefits and how ecosystem evaluation could help them become more marketable.

2 Emergence and Need of Nature-Based Solutions

In the past, an array of societal challenges were solved in a manner that led to the worldwide biodiversity and climate crises, increased inequalities, and global human vulnerability (IUCN 2019). This calls for new and sustainable methods of problem-solving and NbS is being considered as an effective framework for reversing the above trend, by increasing the alignment between conservation and sustainable development objectives (Cohen-Shacham et al. 2019). Maes and Jacobs (2017) defined NbS as "any transition to a use of ecosystem services with decreased input of non-renewable natural capital and increased investment in renewable natural processes." The concept is distinguished from more traditional and top-down conservation, *e.g.* via protected areas, towards finding solutions that aim to meet the needs of a diverse range of stakeholders (Kabisch et al. 2016). NbS focuses on the benefits to people and the environment itself, to allow for sustainable solutions that are able to respond to environmental change and hazards in the long-term (Eggermont et al. 2015). As per the European Commission (EU 2020), NbS are inspired and supported by nature and are cost-effective, simultaneously providing environmental, social and economic benefits and building resilience.

NbS have two aspects in the context of climate change: On the one hand, these solutions help adapt to the changing climate by reducing exposure of communities, ecosystems and sectors or by improving adaptive capacity; on the other hand, these solutions tend to either reduce greenhouse gas emissions or remove them by sink action, thereby mitigating climate change. Griscom et al. (2020) found that in half of the tropical countries, cost-effective Natural Climate Solutions (NCS) could mitigate over half of national emissions. Their estimates do not fully account for the potential economic benefits of ecosystem services which can be very high. NbS are estimated to provide 37% of climate change mitigation until 2030 and could help meet the goal of keeping climate warming below 2 °C with co-benefits for biodiversity (IPBES 2019). NbS also address major societal challenges such as food security, climate change, water security, human health, disaster risk, and social and economic development (IUCN 2016). Since NbS needs to serve multiple purposes, it would engage multiple stakeholders (Fig. 1).

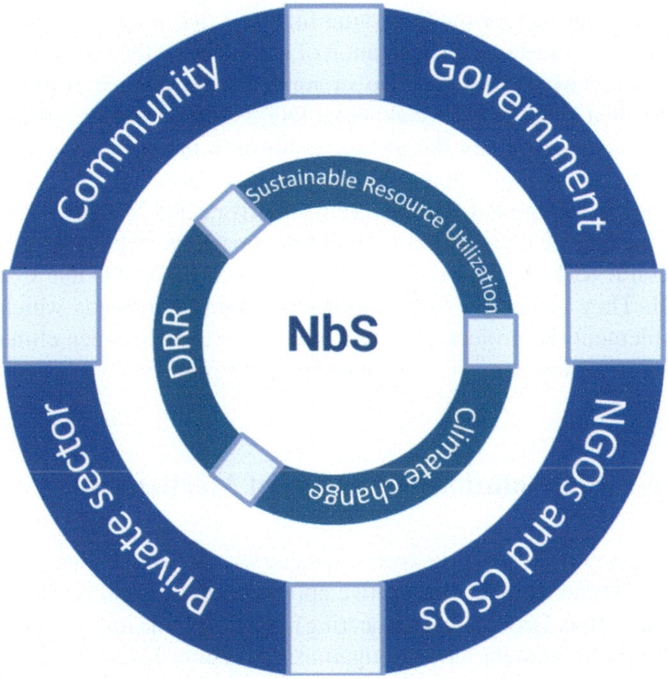

Fig. 1 Stakeholder engagement in Nature-based Solutions (NbS) Source: Developed by the authors

In addition to climate change (mitigation, adaptation or both), NbS could serve multiple purposes such as sustainable utilization of resources, biodiversity conservation and disaster risk reduction (DRR). So far, NbS concept does not explicitly address whether the conservation and protection of biodiversity is a goal or simply a prerequisite for NbS (Nesshöver et al. 2017). For DRR, Young et al. (2019) mentions that current methods of DRR do not include the use of ecosystem services, the impact of local climate on ecosystem services, and the benefits of NbS. However, in future, NbS could act as a linkage that can fill gaps in parallel streams of environmental action. It is emerging as an integrated approach that can solve the climate and biodiversity crises, whilst closing the gap between SDGs (Seddon et al. 2020a). As per IPBES (2019), NbS could turn out to be a cost-effective tool to meet the Sustainable Development Goals (SDGs). This was an important subject of discussion during a UNFCCC COP 25 side event which showcased collaborative actions to address climate issues through the lens of the SDGs, particularly SDG 15 (life on land). It stressed on the need to address land issues in the next round of Nationally Determined Contributions (NDCs) as land conservation can provide 30% of mitigation potential required to meet the 2 °C target (Mead 2019).

However, community participation will be the key to the success of NbS. It can help increase livelihood opportunities for the community, besides environmental protection. This has been proved true through some initiatives like, for example, the

International Partnership for the Satoyama Initiative that focuses on nature-human harmony and aims at sustainable utilization of resources by the community as a way of life (IPSI Secretariat). NbS can also strengthen the resilience of the community by improving their productivity (Nakanuku-Diggs and Hesselink 2014). The Paris Agreement aims to strengthen the global response to the threat of climate change by making finance flow consistent with a pathway towards low GHG emission and climate-resilient development. The government, private and NGO sector partnership can enhance the finance flow and help in building capacities of the local communities. Many successful REDD+ projects were implemented by NGOs like Conservation International. They are closer to the community and their needs which makes it easier to implement the project. NbS can reduce the gaps between climate change adaptation, biodiversity conservation, and rural development.

3 Nature-Based Solutions and Market Mechanisms

Cohen-Shacham et al. (2019) suggest five categories of approaches which can be a part of NbS. These include: Restorative approach (Ecological restoration, Forest landscape restoration, Ecological engineering); Issue-specific approach (Ecosystem-based adaptation, Ecosystem-based mitigation, Ecosystem-based disaster risk reduction, Climate adaptation services); Infrastructure approach (Natural infrastructure, Green infrastructure); Management approach (Integrated coastal zone management, Integrated water resources management); and Protection (Area-based conservation approaches, including protected area management and other effective area-based conservation measures). In the context of climate change, these approaches could offer several activities which primarily target mitigation but have adaptation benefits as spin-offs and vice versa. If we particularly focus on climate change, we can categorize NbS into three distinct categories (Fig. 2): (i) Ecosystem-based Adaptation (EbA); (ii) Ecosystem-based Mitigation (EbM); and (iii) Joint Mitigation and Adaptation (JMA).

Figure 2 shows the sectors in which NbS projects can be implemented. EbM can make a powerful contribution by preventing the degradation and loss of natural ecosystems, in addition to acting as natural carbon sinks (Cohen-Shacham et al. 2016). For mitigation, NbS can include other sectors like energy, waste and agriculture in addition to forestry. For adaptation, there are several options in different sectors that include water, agriculture, forestry, and coastal sector. Some of such measures are already being implemented as part of national measures in many countries but are not likely recognized as climate action for adaptation. EbA can help vulnerable communities, especially those who depend on natural resources, to better adapt and become more resilient to the adverse effects of climate change, including extreme weather events and climate-related disasters (Cohen-Shacham et al. 2016).

An integrated ecosystem-based approach has emerged under the Paris Agreement that presents an opportunity to explore and operationalize Joint Mitigation and Adaptation which has the ability to serve multiple purposes (Gupta and Dube 2018). JMA

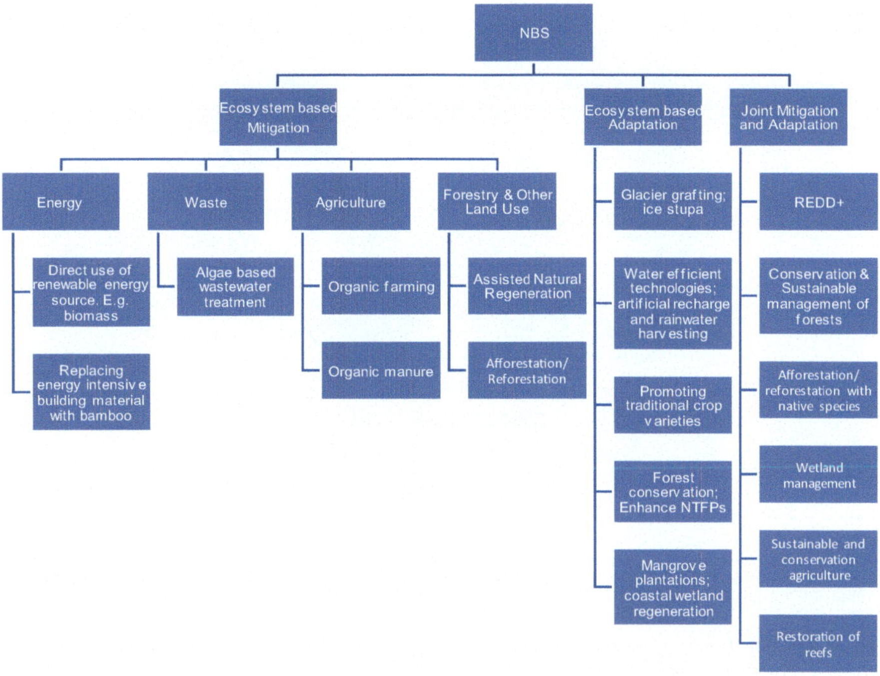

Fig. 2 Examples of Nature-based Solutions (NbS) under various sectors Source: Developed by the authors

projects must show explicit adaptation benefits, and can include forestry and agriculture sectors to a large extent. However, the demand for such measures is currently poor due to the lack of monitoring and verification. In the scenario of adaptation funding, monitoring and evaluation will be key in reducing uncertainty by progress measurements through appropriate indicators and proxies that can measure effectiveness (Ford et al. 2013). Milman and Jagannathan (2017) analysed projects identified by the UNFCCC as examples of EbA and found that monitoring and evaluation procedures undertaken by the EbA projects are limited. 39% projects reported metrics that tracked both progress and effectiveness of the adaptation measures undertaken, 19% measured only progress, and 19% of projects measured only effectiveness, and 23% did not report either progress or effectiveness metrics. This makes such projects difficult to gain currency in the market environment.

There are learnings from EbA which can be used for the implementation of NbS as a market mechanism. EbA is perceived as a policy mix that promotes adaptive transition, which is a step towards sustainability transitions (Scarano 2017). Reid et al. (2018) also proposed a framework for assessing the effectiveness of ecosystem-based approaches to adaptation. However, this framework cannot be applied on an ex-ante basis. Although such studies can be helpful in planning for NbS and assessing its effectiveness if implemented. The framework of Friends of Ecosystem-based

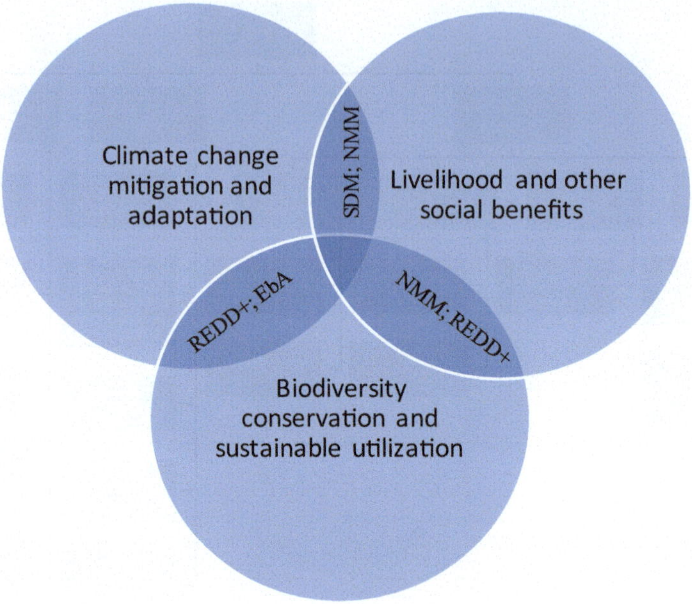

Fig. 3 Role of different market mechanisms as a part of nature-based solutions (NbS) Source: Developed by the authors

Adaptation (FEBA) for defining qualification criteria and quality standards for EbA can be a starting point for setting up NbS qualification criteria. According to the framework, EbA should: (i) reduce social and environmental vulnerabilities; (ii) generate societal benefits in the context of climate change adaptation; (iii) restore, maintain or improve ecosystem health; (iv) be supported by policies at multiple levels; and (v) support equitable governance and enhances capacities (FEBA 2017). All these criteria could be a starting point for NbS, of which EbA will be an important part.

The negotiations to include NbS in the market mechanisms have already started. The draft negotiating texts of the Article 6 of the Paris Agreement which deals with the new market mechanisms of the Agreement contains the reference to NbS. Some Parties proposed keeping sink activities like NbS only in the non-market approaches. However, the operational text as of now retains NbS in Article 6.4 which deals with market mechanisms. The NbS coalition presented a manifesto at the Climate Action Summit in September 2019. One of the priority areas of the manifesto is increasing and mainstreaming NbS within national governance, climate action, and climate policy-related instruments, including NDCs, Adaptation Communications, long-term low greenhouse gas emission development strategies, spatial planning, national development plans, and business plans (NBS Coalition 2019). This shows that it is possible to get multiple benefits from NbS by making it a part of different mechanisms such as non-market mechanisms (NMM) and sustainable development mechanism (SDM) proposed by the Paris Agreement (Fig. 3). In all the market approaches, eligibility must be defined by the level of transparency of accounting, monitoring and baseline of the project. Importance must be given to incentives to local community, benefit sharing, and sustainable use of ecosystem services. The information should be available on a web-based platform maintained by the UNFCCC.

In addition to market mechanisms, scaling-up NbS within the AFOLU (Agriculture, Forestry, and Other Land Use) sector—the protection, restoration and sustainable use of forests, grasslands and wetlands—can represent a cost-effective opportunity for countries to enhance their nationally determined contributions (NDCs) (UNDP 2019). So far, out of the total NDCs submitted, 104 countries include NbS as an adaptation component, 77 have it as both adaptation and mitigation components, while only 27 Parties include such actions as part of mitigation actions only (Seddon et al. 2020b). In the run-up to UNFCCC COP25 in Madrid, Spain, 32 countries agreed on a set of principles, known as the San Jose Principles for High Ambition and Integrity in International Carbon Markets, that constitute the basis upon which a fair and robust carbon market should be built (Ortiz 2019). Meant for the implementation of a framework for Article 6 mechanism, it incorporates 11 principles, one of which is to "Maintain High Ambition", ensuring environmental integrity and enabling the highest possible mitigation ambition. Environmental integrity also means maintaining system resilience in addition to getting mitigation benefits. A literal high ambition fulfilling all major goals of the Paris Agreement requires focusing on both adaptation and mitigation. However, there is also a need to evaluate the ecosystem services that are a part of the NbS to win it an appropriate place in the market mechanisms. It would also be necessary to focus on ecosystem services to safeguard biodiversity and dependent community from any negative impacts of the solutions if they are just focused on climate mitigation. IUCN's Global Standard for NbS somehow offers to solve this problem through its self-assessment tool to consistently design effective NbS projects that are ambitious in scale and sustainability (IUCN 2019). Its eight major criteria are shown in Fig. 4.

Any evaluation may thus consider biodiversity net-gain, sustainability and adaptive management, for all of which ecosystem services evaluation play an important part. Adaptive management has been identified as one way to reduce uncertainties and increase the likelihood of achievement of adaptation outcomes (IISD 2018).

4 Evaluation of Ecosystem Services for Implementation of NbS

NbS must be designed to benefit biodiversity and support the delivery of a range of ecosystem services (EU 2020). Healthy ecosystems provide a diverse range of services (Seddon et al. 2020a). These ecosystem services support adaptation in a variety of ways, ultimately reducing human vulnerability to climate change enabled by resilient ecosystems and sustainable delivery of ecosystem services (IISD 2018). While the idea of environmental or ecosystem services (ES) began to establish itself in the 1970's, it was realised that a more systematic approach was required to document and understand the relationship between people and nature in 1990's

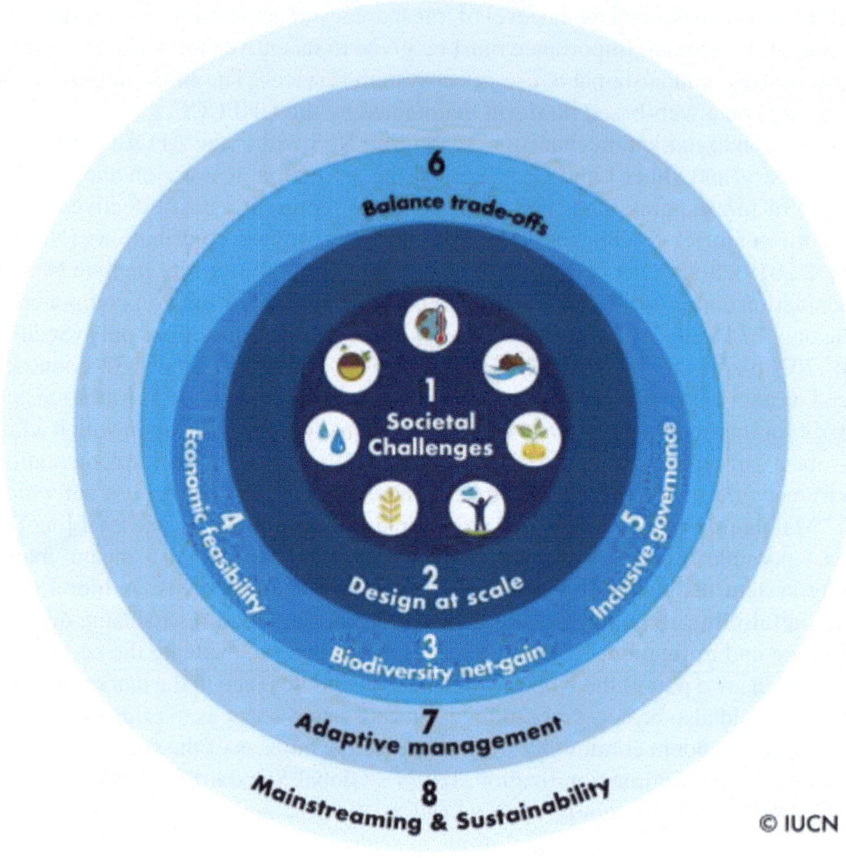

Fig. 4 Criteria for self-assessment of nature-based solutions (NbS) project design. Source: IUCN 2019

(Cohen-Shacham et al. 2016). After the Millennium Ecosystem Assessment, which explained this relationship and provided a conceptual framework for ES (Millennium Ecosystem Assessment 2005), The Economics of Ecosystems and Biodiversity (TEEB) focused on making nature's values visible. It aimed at mainstreaming the values of biodiversity and ecosystem services into decision-making at all levels (TEEB 2013). As per the TEEB guidelines for country studies, it is important to identify the most relevant ecosystem services, define information needs, assess the value of ecosystem services, and then assess the pros and cons of policy options (TEEB 2013). This is similar to what needs to be done to monitor the impact of NbS in the context of ecosystem services. Predicting and measuring the effectiveness of NbS is a challenge that can be overcome by the use of context-specific metrics with reduced chances of posing unwarranted impacts (Seddon et al. 2020a).

4.1 *Effectiveness and Quantification of Adaptation and Mitigation Actions*

For effective measurement of the success of the project, a baseline of current action needs to be created from which the progress can be evaluated (Ford et al. 2013). The potential of NbS to provide intended benefits is not yet assessed (Seddon et al. 2020a). A compilation of 170 case studies of EbA shows that they had been widely applied to improve ecosystem resilience and to maintain and enhance the quality and quantity of ecosystem services (UNFCCC 2015), however in most cases there was no quantification of adaptation benefits. To quantify the benefits of a project and also for mutual learning, the monitoring and evaluation (M&E) mechanism is necessary (Nesshöver et al. 2017). Ecosystem accounting, which has been developed to integrate ecosystems and ecosystem services into national accounts, includes the compilation of an ecosystem services supply and use account, which reflects actual flows of ecosystem services, and the ecosystem capacity account, which reflects the capacity of ecosystems to sustainably supply ecosystem services. A capacity assessment requires detailed data on ecosystem processes, which are often not available over large scales (Vargas et al. 2019). It is therefore important to gather and review data, calculate indicators, whilst developing monitoring and reporting systems (Brown et al. 2014). National statistics, remote sensing data, numerical simulation models and field estimations can be used for data gathering (Brown et al. 2014).

The three key concepts of ecosystem assets include ecosystem condition, ecosystem extent, and ecosystem monetary asset (World Bank 2017). This baseline information helps in evaluating the ecosystem services and how an intervention could change the extent to ecosystem services. UNEP's Adaptation, Livelihoods and Ecosystems Planning Tool lists the steps needed to monitor and evaluate EbA options which include: identification of long-term indicators to measure adaptation outcomes, and short-term indicators to measure EbA options; description of baseline situation for each adaptation outcome; and data collection and methods for both monitoring and evaluation (IISD 2018). Further, the GIZ's Guidebook for Monitoring and Evaluating EbA Interventions mentions ecosystem services as one of the indicators for EbA interventions and stresses on a good evaluation design in the beginning itself (GIZ et al. 2020). M&E also depends on the level of intervention with the ecosystem. Eggermont et al. (2015) describe two typologies: (i) how much engineering of biodiversity and ecosystems is involved in NbS?; and (ii) how many ecosystem services and stakeholder groups are targeted by a given NbS? This will define the impact of the measure and, thus, its monitoring. The monitoring of mitigation is yet possible, but adaptation monitoring suffers from three main challenges: (i) development of indicators to track adaptation; (ii) measurement of effectiveness of adaptations to reduce vulnerability; and (iii) identification of data sources for monitoring and evaluation (Ford et al. 2013). To overcome these challenges, it would be necessary to assess the linkages between the system (understanding of the socio-ecological system), ecosystem functions and its services, and the actual value and benefits to the society (Nature Based Solutions undated). In terms of climate change, accounting for carbon regulating services would be enough but that would be true only for mitigation. It is clear that for adaptation, a more robust and tougher assessment of all indicators, including impact on ecosystem services, would be needed to take full advantage of NbS. It is also highly essential to differentiate between monetary and non-monetary benefits of the project.

4.2 Benefits of Monitoring and Evaluation in the Long Run

For the EbA projects submitted to UNFCCC so far, long-term monitoring and evaluation of project activities is not included as a component of most projects (Milman and Jagannathan 2017). The lack of awareness of ecosystem services provided by NbS may influence the decision-making process. An appropriate framework for evaluating long-term economic benefits of NbS is required to be employed for appraisals of projects addressing climate change (Seddon et al. 2020a). Evaluation of ecosystem services will not only reduce the risk of future mitigation projects but will make adaptation projects more likely to earn credits under market mechanisms of the Paris Agreement. It will no more be susceptible to the drawbacks for which Clean Development Mechanisms (CDM) was widely criticized. Paris Agreement also establishes Non-Market Approaches (NMAs). Additionally, the Paris Agreement also encourages countries to submit adaptation communications in which it would become imperative to outline adaptation needs, projects, and its effects.

For effective climate mitigation, it would be imperative to evaluate uncertainties in ecosystem service quantifications and develop framework for systematic analysis of effectiveness indicators. Use of ecosystem services evaluation will act as a decision making and monitoring approach for NMAs. Land management and planning also require models to map and analyze ecosystem services. Also, because ecosystem services do not vary independently, many operational issues ultimately relate to the mitigation of ecosystem services trade-offs, for which multiple services and their interactions need to be considered (Vannier et al. 2019).

M&E of ecosystem services for NbS will also be useful to report quantitative information in the biennial transparency reports under the enhanced transparency framework of the Paris Agreement. As per paragraph 112 of the Modalities, Procedures and Guidelines for the Transparency Framework for Action and Support referred to in Article 13 of the Paris Agreement (Decision 18/CMA.1), information needs to be reported on establishment or use of domestic systems to monitor and evaluate the implementation of adaptation actions. Parties will be required to report on approaches and systems for M&E, including those in place or under development (UNFCCC 2019).

Although information on ecosystems and their services, which is required to support policy making, is still not captured in economic statistics of most countries, evaluation of ecosystem services of NbS would be a powerful tool for decision-makers. Ecosystem services can be evaluated by: putting a label (a price tag); creating a basis for comparison; creating a basis for exchange; and creating a rallying point (a basis for negotiation) (Ganguly 2013). These evaluations can be a part of national policy and biodiversity strategies. This will also help a country in meeting its climate change and biodiversity targets by better accounting for net gains. For example, in place of fast growing tree plantations, the vibrant, balanced and natural ecosystems would support countries in meeting the targets of the Paris Agreement (Seddon et al. 2019). Evaluating ecosystem services of NbS would give an opportunity of bringing holistic perspective while implementing the Paris climate deal. For inclusion of NbS in NDCs, it has been suggested that measurable and time-bound numeric targets (e.g. hectares of ecosystems utilized, CO_2 sequestered and/or number of people benefiting from increased resilience) may be prioritized (WWF 2020). In the absence of monitoring, it would be difficult to make them a part of market mechanisms but non-market instruments. For finance, it could depend on crowd finding, green bonds, and payment of ecosystem services (IUCN French Committee 2019).

Some work has already been done on evaluating ecosystem services of NbS in different contexts. Raymond et al. (2017) proposed a framework for assessing and implementing the co-benefits of NbS in urban areas. Albert et al. (2016) applied ecosystem services indicators in landscape planning and management and proposed the Ecosystem Services-in-Planning framework. Anzaldua et al. (2018) came up with the DESSIN Ecosystem Services (ESS) evaluation framework specifically to address water management practices in the Ecosystem Services Approach. Baral et al. (2016) proposed a framework for assessing ecosystem goods and services from planted forests. Logsdon and Chaubey (2013) developed a quantitative approach to evaluating ecosystem services using SWAT model. However, none of these frameworks is an integrated and universally accepted one for NbS.

4.3 Need for a Universal Measurement, Reporting and Verification (MRV) Framework

A robust MRV framework of ecosystem services is required for full accounting and reflection of the range of benefits that is accrued from NbS. Conventionally, the 'do no harm' approach has been in practice in the carbon markets which relied on the statements that the project does not create harm to the environment, society, and economy. The mechanisms—such as CDM—relied on contribution of the projects to the sustainable development of the host country with a voluntary sustainable development tool being in place; however, there is no requirement to measure, report and verify these benefits in CDM. This lack of MRV led to plain statements being accepted as promises, but no mechanism being in place to ascertain the same. In the Paris Agreement, both the market and non-market approaches will benefit from an MRV framework for the potential ecosystem services of the NbS activities. This will instill confidence in the investors, buyers and financers who look for a real, measurable, long term and additional reduction in greenhouse gases and/or robust adaptation to the potential impacts of climate change while contributing to the other ecosystem services. While the tools and methods for such evaluation are a future area of research, a universally agreed framework that can be tested and used for the MRV of NbS by evaluating ecosystem benefits needs to be developed first. The framework should take into account both qualitative and quantitative evaluation methods and should be able to deliver the needs of both market and non-market approaches of the Paris Agreement.

5 Conclusion

NbS are gaining momentum in climate change negotiations in addition to biodiversity negotiations. The Paris Agreement offers significant opportunities for new solutions to replace the old and narrow market mechanisms of the Kyoto Protocol. This comes from Paris Agreements' equal inclination towards adaptation and mitigation. NbS, thus, became more important as they can not only help improve the resilience of the communities and improve overall adaptive capacity, but can also help advance mitigation by reducing the ecological degradation. However, these mechanisms face M&E challenges due to lack of appropriate indicators and baseline methodologies. Most EbA projects, which fall under the purview of NbS, lack methodologies to assess the effectiveness of the project due to the lack of quantification of ecosystem services. Although frameworks to develop such indicators are available, it would depend largely on local conditions and the type of ecosystem considered and its assets. Therefore, indicators based on system understanding and its linkages with the values and benefits to the people need to be developed. This would help in M&E of future NbS projects and reduce the uncertainty which makes it more feasible to be included in the market mechanisms of the Paris Agreement, while also being friendly to biodiversity and dependent communities.

It is interesting to know that there are already methods and models for calculation of risk reduction, livelihood index, climate vulnerability, social vulnerability, and ecosystem mapping. Each of these methods are different, but in the case of NbS, all of these parameters would come under one umbrella solution. Therefore, an overarching and encompassing method of calculating metrics with an integrated framework for NbS would also be needed. As of now, this is a major gap area and a hurdle for the successful implementation of NbS. In addition to this, selection criteria for NbS must be developed. This should include a checklist of negative and positive aspects of a project based on the criteria for NbS so that appropriate safeguards are available. This will significantly rule out the chances of maladaptation in case of adaptation projects, and double counting or business -as-usual in case of mitigation projects.

References

Albert C, Galler C, Hermes J et al (2016) Applying ecosystem services indicators in landscape planning and management: the ES-in-Planning framework. Ecol Indic 61:100–113. https://doi.org/10.1016/j.ecolind.2015.03.029

Anzaldua G, Gerner NV, Lago M et al (2018) Getting into the water with the Ecosystem Services Approach: the DESSIN ESS evaluation framework. Ecosyst Serv 30:318–326. https://doi.org/10.1016/j.ecoser.2017.12.004

Baral H, Guariguata MR, Keenan RJ (2016) A proposed framework for assessing ecosystem goods and services from planted forests. Ecosyst Serv 22:260–268. https://doi.org/10.1016/j.ecoser.2016.10.002

Barrow E, Wit P, Hawes P, Mainguy G (2012) IUCN commission on ecosystem management. SAPI EN Shar Knowl Prep Future 57

Bellard C, Bertelsmeier C, Leadley P et al (2012) Impacts of climate change on the future of biodiversity. Ecol Lett 15:365–377. https://doi.org/10.1111/j.1461-0248.2011.01736.x

Brown C, Reyers B, Ingwall-King L, et al (2014) Measuring ecosystem services: guidance on developing ecosystem service indicators. UNEP-WCMC, Cambridge

Chong J (2014) Ecosystem-based approaches to climate change adaptation: progress and challenges. Int Environ Agreem Polit Law Econ 14:391–405

Cohen-Shacham E, Andrade A, Dalton J et al (2019) Core principles for successfully implementing and upscaling Nature-based Solutions. Environ Sci Policy 98:20–29. https://doi.org/10.1016/j.envsci.2019.04.014

Cohen-Shacham E, Walters G, Janzen C, Maginnis S (eds) (2016) Nature-based solutions to address global societal challenges. IUCN International Union for Conservation of Nature

COP-10 (2010) Biodiversity and climate change: Decision X/33. Secretariat of the Convention on Biological Diversity, Nagoya, Japan

Doswald N, Osti M (2011) Ecosystem-based approaches to adaptation and mitigation: good practice examples and lessons learned in Europe. BfN, Federal Agency for Nature Conservation

Eggermont H, Balian E, Azevedo JMN, et al (2015) Nature-based solutions: new influence for environmental management and research in Europe. https://www.ingentaconnect.com/content/oekom/gaia/2015/00000024/00000004/art00010;jsessionid=9ldi8es2e0unf.x-ic-live-02. Accessed 25 Apr 2020

EU (2020) The EU and Nature-based solutions. European Commission. https://ec.europa.eu/info/research-and-innovation/research-area/environment/nature-based-solutions_en. Accessed 9 Jan 2021

Ford J, Berrang-Ford L, Lesnikowski A, et al (2013) How to track adaptation to climate change: a typology of approaches for national-level application. Ecol Soc 18. https://doi.org/10.5751/ES-05732-180340

Ganguly A (2013) Valuation of Ecosystem Services: Linking theory and practice. The Energy and Resources Institute. Presented at IGNFA Dehradun, India. https://www.slideserve.com/damia/valuation-of-ecosystem-services-linking-theory-and-practice

GIZ, UNEP-WCMC, FEBA (2020) Guidebook for monitoring and evaluating ecosystem-based adaptation interventions. Deutsche Gesellschaft für Internationale Zusammenarbeit (GIZ) GmbH, Bonn, Germany

Griscom BW, Busch J, Cook-Patton SC et al (2020) National mitigation potential from natural climate solutions in the tropics. Philos Trans R Soc B Biol Sci 375:20190126. https://doi.org/10.1098/rstb.2019.0126

Gupta H (2015) PhD Thesis: Gaps and linkages between climate change and biodiversity conventions: science, politics and policy. Panjab University, Chandigarh, India. http://hdl.handle.net/10603/199269

Gupta H, Dube LC (2018) Addressing biodiversity in climate change discourse: Paris mechanisms hold more promise. Int for Rev 20:104–114. https://doi.org/10.1505/146554818822824282

IISD (2018) ALIVE - adaptation, livelihoods and ecosystems planning tool: user manual. Version 1.0. United Nations Environment Programme – International Ecosystem Management Partnership & the International Institute for Sustainable Development
IPBES (2019) Summary for policymakers of the global assessment report on biodiversity and ecosystem services of the Intergovernmental Science-Policy Platform on Biodiversity and Ecosystem Services. IPBES Secretariat, Bonn, Germany
IPSI Secretariat CONCEPT | International Partnership for the Satoyama Initiative. https://satoyama-initiative.org/concept/. Accessed 31 Dec 2019
IUCN (2019) A global standard for nature-based solutions. In: IUCN. https://www.iucn.org/theme/ecosystem-management/our-work/a-global-standard-nature-based-solutions. Accessed 24 Apr 2020
IUCN (2016) WCC-2016-Res-069-EN. Defining nature-based solutions. https://portals.iucn.org/library/sites/library/files/resrecfiles/WCC_2016_RES_069_EN.pdf
IUCN French Committee (2019) Nature-based Solutions for climate change adaptation and disaster risk reduction. France, Paris
Kabisch N, Stadler J, Korn H, Bonn A (2016) Nature-based solutions to climate change mitigation and adaptation in urban areas. Federal Agency for Nature Conservation
Logsdon RA, Chaubey I (2013) A quantitative approach to evaluating ecosystem services. Ecol Model 257:57–65. https://doi.org/10.1016/j.ecolmodel.2013.02.009
Maes J, Jacobs S (2017) Nature-based solutions for Europe's sustainable development: Europe's sustainable development. Conserv Lett 10:121–124. https://doi.org/10.1111/conl.12216
Mead L (2019) Nature-based climate solutions support all SDGs: COP 25 side event. In: SDG Knowl Hub. https://sdg.iisd.org:443/news/nature-based-climate-solutions-support-all-sdgs-cop-25-side-event/. Accessed 14 May 2020
Millennium Ecosystem Assessment (2005) Ecosystems and Human Well-Being: Our Human Planet: Summary for Decision Makers. Island Press, p 129. ISBN: 978-1-55963-387-1
Milman A, Jagannathan K (2017) Conceptualization and implementation of ecosystems-based adaptation. Clim Change 142:113–127. https://doi.org/10.1007/s10584-017-1933-0
Nakanuku-Diggs L, Hesselink F (2014) Think nature: improving community resilience through nature-based solutions. International Union for Conservation of Nature
Nature based solutions what are nature-based solutions (NBS)? | Nature BASED SOLUTIons. https://www.nature-basedsolutions.com/. Accessed 21 Apr 2020
Nature4Climate (2019) Momentum for nature-based solutions continues to grow despite disappointing UN climate talks. In: Nature4Climate. https://nature4climate.org/news/momentum-for-nature-based-solutions-continues-to-grow-despite-disappointing-un-climate-talks/. Accessed 12 Apr 2020
NBS Coalition (2019) Compendium of contributions: nature based solutions. Climate Action Summit 2019. https://wedocs.unep.org/bitstream/handle/20.500.11822/29988/Compendium_NBS.pdf?sequence=1&isAllowed=y
Nesshöver C, Assmuth T, Irvine KN et al (2017) The science, policy and practice of nature-based solutions: an interdisciplinary perspective. Sci Total Environ 579:1215–1227. https://doi.org/10.1016/j.scitotenv.2016.11.106
Ortiz DA (2019) San Jose Principles for High Ambition and Integrity in International Carbon Markets. In: Dir. Cambio Climático. https://cambioclimatico.go.cr/press-release-leading-countries-set-benchmark-for-carbon-markets-with-san-jose-principles/. Accessed 13 Apr 2020
Raymond CM, Frantzeskaki N, Kabisch N et al (2017) A framework for assessing and implementing the co-benefits of nature-based solutions in urban areas. Environ Sci Policy 77:15–24. https://doi.org/10.1016/j.envsci.2017.07.008
Reid H, Bourne A, Muller H, et al (2018) A framework for assessing the effectiveness of ecosystem-based approaches to adaptation. In: Resilience. Elsevier, pp 207–216
Scarano FR (2017) Ecosystem-based adaptation to climate change: concept, scalability and a role for conservation science. Perspect Ecol Conserv 15:65–73. https://doi.org/10.1016/j.pecon.2017.05.003

Schirpke U, Kohler M, Leitinger G et al (2017) Future impacts of changing land-use and climate on ecosystem services of mountain grassland and their resilience. Ecosyst Serv 26:79–94. https://doi.org/10.1016/j.ecoser.2017.06.008

Schirpke U, Timmermann F, Tappeiner U, Tasser E (2016) Cultural ecosystem services of mountain regions: Modelling the aesthetic value. Ecol Indic 69:78–90. https://doi.org/10.1016/j.ecolind.2016.04.001

Secretariat of CBD (2009a) Connecting biodiversity and climate change mitigation and adaptation: report of the second ad hoc technical expert group on biodiversity and climate change. Secretariat of the Convention on Biological Diversity, Montreal, Quebec, Canada

Secretariat of CBD (2009b) Biodiversity and climate change action. Secretariat of the Convention on Biological Diversity

Seddon N, Chausson A, Berry P et al (2020) Understanding the value and limits of nature-based solutions to climate change and other global challenges. Philos Trans R Soc B Biol Sci 375:20190120. https://doi.org/10.1098/rstb.2019.0120

Seddon N, Daniels E, Davis R et al (2020) Global recognition of the importance of nature-based solutions to the impacts of climate change. Glob Sustain 3:e15. https://doi.org/10.1017/sus.2020.8

Seddon N, Turner B, Berry P et al (2019) Grounding nature-based climate solutions in sound biodiversity science. Nat Clim Change 9:84–87. https://doi.org/10.1038/s41558-019-0405-0

Seidl R, Albrich K, Erb K et al (2019) What drives the future supply of regulating ecosystem services in a mountain forest landscape? For Ecol Manag 445:37–47. https://doi.org/10.1016/j.foreco.2019.03.047

TEEB making nature's values visible. In: Econ Ecosyst Biodivers http://www.teebweb.org/about/. Accessed 25 Apr 2020

TEEB (2013) Guidance manual for TEEB country studies. United Nations Environment Programme

UNCCD - Ecosystem-based-adaptation. http://www.unccd.int/en/programmes/Event-and-campaigns/WDCD/Pages/What-is-Ecosystem-Based-Adaptation.aspx. Accessed 1 Nov 2014

UNDP (2019) Pathway for increasing nature-based solutions in NDCs: a seven-step approach for enhancing nationally determined contributions through nature-based solutions. United Nations Development Programme, New York, USA

UNFCCC (2017) Subsidiary body for scientific and technological advice. Forty-sixth session. Bonn, 8–18 May 2017. Nairobi work programme on impacts, vulnerability and adaptation to climate change. Adaptation planning, implementation and evaluation addressing ecosystems and areas such as water resources

UNFCCC (2015) Subsidiary Body for Scientific and Technological Advice. Forty-third session. Paris, 1–4 December 2015. Nairobi work programme on impacts, vulnerability and adaptation to climate change. Good practices and lessons learned in adaptation planning processes addressing ecosystems, human settlements, water resources and health, and in processes and structures for linking national and local adaptation planning: a synthesis of case studies

UNFCCC (2019) Decision 18/CMA.1: Modalities, procedures and guidelines for the transparency framework for action and support referred to in Article 13 of the Paris Agreement; Report of the Conference of the Parties serving as the meeting of the Parties to the Paris Agreement on the third part of its first session, held in Katowice from 2 to 15 December 2018

United Nations (2015) Paris agreement. United Nations Framework Convention on Climate Change. http://unfccc.int/files/essential_background/convention/application/pdf/english_paris_agreement.pdf

Vannier C, Lasseur R, Crouzat E et al (2019) Mapping ecosystem services bundles in a heterogeneous mountain region. Ecosyst People 15:74–88. https://doi.org/10.1080/26395916.2019.1570971

Vargas L, Willemen L, Hein L (2019) Assessing the capacity of ecosystems to supply ecosystem services using remote sensing and an ecosystem accounting approach. Environ Manage 63:1–15. https://doi.org/10.1007/s00267-018-1110-x

Wamsler C, Luederitz C, Brink E (2014) Local levers for change: mainstreaming ecosystem-based adaptation into municipal planning to foster sustainability transitions

World Bank (2017) Guidebook on ecosystem accounting. World Bank. https://elibrary.worldbank.org/doi/pdf/10.1596/29829

WWF (2020) Enhancing NDCs through nature-based solutions: 8 simple recommendations for integrating nature into NDCs. https://c402277.ssl.cf1.rackcdn.com/publications/1318/files/original/enhancing_ndcs_through_nature_based_solutions.pdf?1585149353

FEBA (Friends of Ecosystem-based Adaptation) (2017) Making Ecosystem-based Adaptation Effective: A Framework for Defining Qualification Criteria and Quality Standards (FEBA technical paper developed for UNFCCC-SBSTA 46). Bertram M, Barrow E, Blackwood K, Rizvi AR, Reid H, von Scheliha-Dawid S (authors). GIZ, Bonn, Germany, IIED, London, UK, and IUCN, Gland, Switzerland, p 14.

The Common Heritage of Humankind (CHH): Visions of Developed Countries vs. Developing Countries

Elhoucine Chougrani and Moulay Hicham El Amrani

Abstract The Common Heritage of Humankind (CHH) includes shared systems and resources such as the climate system, ozone layer, outer space, oceans, water, forests, biological diversity, etc. The use of CHH should aim at benefitting the human society across the globe, hence the vital need to preserve its sustainability and balance. However, there are currently certain problems and difficulties in recognizing a common vision about the concept and practice of CHH. Therefore, it seems very important to think and rethink the Commons in an era marked by globalizing dynamics and conflict of interests among key actors. As many developing countries support the view that CHH is a *cogens norm* in international law, the need to enforce a strict regulation has been advocated. However, several liberal countries prefer focusing on the free market mechanism (the invisible hand and market fundamentalism) which facilitates the over-exploitation of the Global South's natural resources, thus limiting the sovereignty of developing countries over their resources. This has been possible through many processes such as privatization, and deregulation. The transnational corporations, as influential tools, often use their financial and economic power to decrease the role of states in enforcing and monitoring compliance with social and environmental norms and regulations. Therefore, the major concern is how to manage and regulate the CHH while meeting the needs of human societies. In this chapter, we try to shed light on the conflict of interests between the Global North and the South with regard to CHH, before proposing a framework to share responsibilities and burdens. Finally, the analysis focuses on the needed shift to be made from the ultra-liberal *laisser faire, laisser passer model,* and the free trade ideology to the possibility to reduce or eliminate environmentally destructive activities, thus limiting the damage affecting the ecological balance and sustainability of natural resources worldwide.

E. Chougrani (✉)
College of Law, Economics and Social Sciences, Cadi Ayyad University, Marrakesh, Morocco
e-mail: chougranielhoucine1@gmail.com

M. H. El Amrani
College of Law, Economics and Social Sciences of Agadir, Ibn Zohr University, Agadir, Morocco
e-mail: elamrani@mondeannuaires.com

M. Behnassi et al. (eds.), *Social-Ecological Systems (SES)*,
https://doi.org/10.1007/978-3-030-76247-6_3

Keywords Common Heritage of Humankind (CHH) · Developed countries · Developing countries visions · Conflict of interests · Future generations

1 Introduction

The CHH should aim at benefiting the human society across the globe, hence the vital need to preserve its sustainability and balance. Every country should bear its responsibility regarding the use and exploitation of the current resources without jeopardizing the rights of present and future generations and the sustainability of the global ecosystem.

Since there are certain issues and problems in recognizing a shared vision about the concept and even the practice of the CHH, it is crucial to re-think the Commons in an era marked by globalizing dynamics and conflict of interests among key actors. In this regard, several developing countries, particularly many, African ones, support the view that CHH is a jus *cogens* norm in international law, since the concept has become part of customary international law, and therefore. It is necessary to enforce strict regulations. However, many other countries and actors, motivated by neo-liberal considerations, think only in terms of market mechanisms, in which (-the invisible hand and the absence of State intervention-) occupy a key place. While doing so, their objective is to exploit natural resources of the Global South, upon which developing countries can only exercise limited sovereignty. The capitalism has operated against the interests of the mass of a common humanity, and threatens the natural foundations of human society (Heller 2011: 239).

In the era of globalization, economic dynamics are characterized by the privatization of every process with a market value. In this regard, transnational corporations (TNCs) as influential tools often, use their capital and economic power to undermine the role of States in enforcing and monitoring compliance with social and environmental norms and regulations.

Therefore, the major concern in this Anthropocene era is how to manage and regulate the CHH while meeting the needs of human societies. In this chapter, we try to shed light on the conflict of interests between the Global North and the South with regard to CHH, before proposing a framework to share responsibilities and burdens. Finally, the analysis focuses on the needed shift to be made from the ultra-liberal *Laisser* faire *Laisser* passer model and the free trade ideology to the possibility to reduce or eliminate environmentally destructive activities, thus limiting the damage affecting the ecological balance and sustainability of natural resources worldwide.

2 CHH and Competing Paradigms

2.1 CHH: Scope of the Concept According to the Literature

The Common Heritage of Humankind (CHH) principle, *res communes humanitatus,* is a wholly new concept of property rights (Buck 1998: 28). It is a modern alternative to the traditional idea of exclusive ownership or free unlimited access. Some authors think there are very broad notions of CHH's components; Ian Brownlie, for example, enumerates the Earth as a whole and its constituent elements such as soil, water, plants and animals, human beings and societies, cultural heritage, energy resources, and information (Fitzmaurice 2002: 158). Accordingly, all stackholders have access to the information related to every environmental resource.

According to Baylis and Smith (2001: 474), the global Commons are the resources open for use by the entire international community, not falling under the jurisdiction of any particular State. The elements that belong to the CHH, especially, oceans, atmosphere, deep-sea, Antarctica, and outer space - are beyond the States' sovereignty and national jurisdiction.

The understanding of the global Commons began with the definition of the tragedy of the Commons. According to the American Ecologist Garrett Hardin 1968: the tragedy of the Commons is the over-exploitation (or over-utilization) of open access resources by users pursuing their individual and self-interests (Baylis and Smith 2001: 474. Indeed, the tragedy of the Commons is the outcome of the interaction of rational individuals pursuing their interests and the circumstances in which they find themselves (Grieco et al. 2015: 392). These circumstances are threefold if we take the example of a pastureland; first, there is a growing population which creates an incentive for the herders to operate in a way focusing on available resources. Then, the resources used in this Common become finite; the pastureland is big but has a predetermined carrying capacity in terms of the cattle it can support. Finally, the most difficult situation which should be managed is the absence of an authority with the power to regulate the use of the Common pastureland.

There are certain features of the concept of CHH which may also be applicable to the environment. Such as the common interest to preserve the Earth's environment; sharing benefits and burdens; the non-appropriation by individual States; and the possibility of an international mechanism to supervise the implementation of this principle (Fitzmaurice 2002: 159. The Garrett Hardin's argument of the Commons presumes that if a Pasture is open to all, it is probable that each herdsman will try to keep as many cattle as possible on the Commons; therefore, the freedom in Commons brings ruin to all (Hardin 1968: 1243–1244). Access to the Commons is multi-level; it can be individual, on the state level or international (Fig. 1).

In his classic work, *Progress and Poverty*, the American economist and social philosopher, Henry George, by refers to all-natural resources, makes a clarion call: we must make land common property (George 2006: 241). He admits that the common right to land does not require any shock or dispossession, it can be reached by a simple

Fig. 1 Levels of access to the commons *Source:* Grieco et al. (2015)

and easy method of taxing only land values (George 2006: 17). The principles of taxation show that this is the best means of raising revenue.

2.2 *CHH in International Law*

2.2.1 CHH in Multilateral Treaties

The 1967 Outer Space Treaty, including the moon and other celestial bodies, is not subject to national appropriation either through any claim of sovereignty or occupation. Yet, in reality, the superpowers dominate space and make it increasingly difficult for other countries to enjoy the same access. In this regard, the Treaty includes the following components: the use of Antarctica for peaceful purposes only; facilitation of scientific research in Antarctica; facilitation of international scientific cooperation in Antarctica; facilitation of the exercise of the rights of inspection provided for in Article VII of the Treaty; questions relating to the exercise of jurisdiction in Antarctica; and preservation and conservation of living resources in Antarctica. In this respect, Paolillo pointed out that there are four essential elements, which at this very moment clarify the idea of the CHH. These are: no State should appropriate any of these spaces; the necessary elaboration and application of international control and regulation regimes; peaceful use, which means that no State, collectively or individually, should use any of these spaces for war purposes (Faramiñán Gilber 2019: 8). Space should be utilized for the benefit and advantage of humankind. Based on this, the 1967 Treaty provides a fundamental vision of how the international community, the States and actors should use the Commons. It is then crucial to discuss how rationally all stakeholders use their resources.

The main innovation included in the 1982 United Nations Convention on the Law of the Sea (UNCLOS) is the concept of the Common Heritage of Mankind which has a revolutionary character (Scovazzi 2007: 11–13). Article 136 of the Convention explicitly provides that "The Area and its resources are the common heritage of mankind."

It presupposes a third kind of regime (intra and intergenerational equity) which is different from both the traditional regimes of sovereignty applicable to the territorial sea, and of freedom applicable to the high seas. Both the UNCLOS and the 1984 Moon Treaty attach to this concept four common attributes: non- appropriation; international management; sharing benefits; and the peaceful use (Fitzmaurice 2002: 153).

The 1992 United Nations Framework Convention on Climate Change (UNFCCC) acknowledges in its Preamble that changes in the Earth's climate and its adverse effects are a common concern of humankind. Accordingly, the Article 3 of the UNFCCC states that "The Parties should protect the climate system for the benefit of present and future generations of humankind, based on equity and following their common but differentiated responsibilities and respective capabilities". Subsequently, the 2015 Paris Agreement confirms the same by stating that climate change is a common concern of humankind.

2.2.2 CHH in International Soft Law

The General Assembly has made efforts since 1960s within the direction of protecting the Commons (such as seas, air, and water). In many resolutions-for example, Resolution 2340 (XXII) of 18 December 1967, Resolution 2467 (XXIII) of 21 December 1968, Resolution 2754 (XXIV) of 15 December 1969, Resolution 2749 (XXV) of 17 of December 1970, Resolutions 3029 (XXVII) of 18 December 1972 and 3067 (XXVIII) of 16 November 1973- using the expression 'Common Heritage of Mankind', the objective of the General Assembly was clear at the outset. It embodies the idea that the resources of the seabed and floor, beyond the limits of national jurisdiction, belong to all peoples and should be used for the benefit of all" (Egede 2011: 63).

2.3 CHH According to Countries Visions

2.3.1 Developed Countries' Vision

Several liberal countries are mainly guided by the free-market mechanism to use and exploit the natural resources of the Global South in which developing countries could only exercise limited sovereignty over their natural resources. In this context, and according to McCay et al. (2003: 353), changes in natural resources management systems and the privatization of the Commons can lead to the destruction of both social capital and trust.

Developed countries envisaged the implementation of the CHH as an extension of the concept of freedom of the high seas and the deep sea-bed. It follows that they should all enjoy the freedom of access to these areas such as sea–bed miners participating States on equal footing under the current system of licensing (Fitzmaurice 2002: 154).

2.3.2 Developing Countries' Vision

Developing countries insist on their right to reach economic development (Sands 2000: 380). However, there are disagreements regarding the common environmental concerns related to their economic development. These countries, often self-perceived as vulnerable, feel that they are in the margins of an industrial evolution which primarily benefits the industrial countries.

Developing countries consider that the CHH means sovereignty over or ownership of the concerned territories and their resources are common and indivisible (Fitzmaurice 2002: 154). The exploration and exploitation of the concerned territories and their resources should be collectively allocated through a mechanism of universal membership. Such a mechanism would plan and oversee all activities and distribute the benefits to all parties taking into account their needs.

The Common heritage principle has become the cornerstone of the contemporary law of the sea (White 1982: 516). It has also come to be a powerful inspiration for the increasingly vocal group of G77 + China, a powerful alliance of developing countries. For African States, the CHH has been included as a human right in the African Charter on Human and Peoples' Rights, which come into force in 1986. Such a particular treaty is based on the African interpretation of human rights and involves the communal nature of the African society. In article 22 (1), of the Charter it was stated that: "all peoples shall have the right to their economic, social and cultural development with due regard to their freedom and identity and in the equal enjoyment of the Common heritage of mankind".

In addition, several African States support the view that CHH is a *jus cogens* in international law; it has even become part of customary international law. Therefore, it is necessary to enforce a strict regulation.

In the African Union Commission 2015 report (Agenda 2063. *The Africa we want*), the Aspiration 5 states that: Africa has a strong cultural identity, Common heritage, values, and ethics. In addition, the aspiration 7 stipulates that: Africa is a strong, united, and influential global player and partner (A major social, political, and economic force in the world), with its share in the global Commons (land, oceans, and space).

2.4 The CHH and Intergenerational Equity

Firstly, sustainable development prioritizes the needs of the current generations in a manner that does not compromise the needs of future generations (WCED 1987). Secondly, the preferences and choices made for the benefit of the present generations significantly influence the welfare of future generations. The current generation (Wolf 1995: 791) has property rights over the Earth' resources, but these resources will also be needed by future generations, who can neither claim and defend their relevant rights nor ask us to protect their interests due to the non-existence of these generations.

In essence, socially sustainable development provides a sound argument for the protection and preservation of the rights of future generations. For this reason, the present generations do not have an absolute right to excessively exploit and indiscriminately abuse natural resources at the expense of future generations (Vibhute 1997: 287). It is legally as well as morally obligatory to preserve, conserve, and save the environment for posterity. This means that actors should take their fair share of the Commons that belong to different generations in terms of equitable access to environmental resources.

Certainly, the principle of equity is central to the achievement of sustainable development (Cordonier and Khalfan 2004: 99). Both of them refer to the intergenerational equity, that is, the rights of future generations to enjoy a fair level of the Common patrimony. Both of them also highlight the intra-generational equity that is the rights of all peoples, to fair access to the current generation's entitlement to the Earth's natural resources. In this perspective, Weiss (1992: 20) claims, that the intergenerational equity theory presumes that, human species should share common environmental resources of the planet with other species, and peoples, from past, present, and future generations. Such a vision promotes the links between fauna, flora, and humans. In addition, the principles of intergenerational equity form the basis of intergenerational rights and obligations (Weiss 1989), which derive from the temporal position of each generation with each other.

Among the most influential international law, scholars working on intergenerational equity, we can mention here Spijkers (2018), Weiss (1992, 1993, 1989, 2008), and Fitzmaurice (2002). In their research, one finds a great variety of approaches. For instance, under the concept CHH, Weiss (1989) covers outer space and ocean beds, water, weather, climate, the ozone layer, genetic and cultural diversity. Weiss (1992) believes that there are two relationships which should shape any theory of intergenerational equity in the environmental context: The first relates to our relationship with the natural system of which we are apart; and the second relates to our relationship with other generations. Besides, there are three normative principles of intergenerational equity. *First*, each generation must conserve options, that is the right to a diversity of the resource base hence the importance of conserving the diversity of the natural and cultural resource base. *Second*, each generation should be required to maintain the quality of the Earth. The latter should be passed on to the future generation in a condition which is not worse than that in which it was received by the present ones. It will then be necessary to develop a predictive awareness of natural resource diversity and quality. *Third*, each generation should provide its members with equitable rights of access to the legacy of previous generations and secure this access for future generations (Weiss 1992, 1989).

The conservation of the Commons requires a huge effort to understand the interaction between actors and generations. The Intergenerational equity compels the present generations to limit their use of natural resources in order to leave future generations with equal living conditions (Shelton 2013: 4). Its other aspects also imply the principle of solidarity which calls countries for sharing the obligation to use natural resources in a way that takes into account the needs of other countries. In this perspective, it is necessary according to Petrella (2001: 8) to start by recognizing water as a common global heritage of humanity, a source of life, and a fundamental resource for sustainable development.

2.5 The United Nations Efforts and the CHH

The Brundtland Report (WCED 1987) establishes an ethical collective responsibility for the CHH which is not synonymous with collective international rights to particular resources within nations. In this regard, such an approach should not interfere with the concept of national sovereignty. But, it certainly does not mean that individual nations would no longer be left to rely on their isolated efforts to protect species within their borders.

Theoretically, the CHH is a concept that represents a milestone in international cooperation between actors and stakeholders. According to the Report, *the Future we want* (United Nations 2012), the humanity should promote the global responsibility and intergenerational solidarity for the achievement of social sustainable development. For the Report *Transforming Our World* (United Nations 2015): a call has been made to strengthen efforts to protect and safeguard the world's cultural and natural heritage. In the 2030 Agenda for sustainable Development, all countries and stakeholders assert that the planet Earth and its ecosystems are a Common home, and that "Mother Earth" is a Common expression in several countries and regions.

Despite the existence of such universal frameworks, there are still certain problems in recognizing a common vision concerning the notion and the practice of the CHH. Some authors, such as- White (1982), consider the CHH as a philosophical and legal tool for equitable redistribution of the world's wealth including resources still untapped. According to Geldenhuys (2015: 8), the concept of CHH has become more difficult for countries to accept the current legal framework for the regulation of exploitation and use of the resources on the moon and other celestial bodies. For the authors of this chapter, it is very important to think and re-think the Commons in a world marked by globalizing dynamics and conflict of interests between actors and economic liberal strategies.

3 CHH Norms vs. Practices

3.1 Developed Countries: Individual Ownership of Property

In recent years, the concept of property has become more significant, both in legal and financial terms. Viewed as goods or objects inherited from ancestors that concern just one person or family, the property has acquired the status of a socio-cultural system. The heritage is no longer considered to be separable from the whole system as it is part of a common human property.

The UNESCO's World Heritage Program[1] promoted and popularized the concept in different parts of the world, thus challenging the old traditional perceptions and practices especially in a liberal system where private property is the key. Within such a new framework, some conflicts arose regarding a few World Heritage Sites such as the meaning of the 'heritage' as well as the rights that owners need to use and manage these sites. Issues linked to such approval situations get more complicated with the different cultural, social, economic, and political agendas' actors.

In developed countries, prosperity and property are intimately interrelated. Private property gives the owners the right to fully use and exploit its resources which are supposed to have positive impacts on economic development. Property rights have two essential elements, namely the exclusive right of individuals to freely use the resources and to willingly transfer the property rights. Both elements value the goods and services. Thus, wealth will be efficiently created, although the question of sharing the wealth fairly often remains unanswered.

Naturally, individuals have conflicts of interest (Thomas Hobbes rules). They will always try to get richer regardless of environmental or social impacts. Such a pursuit of wealth will create an imbalance between citizens. The authority[2] to solve these negative impacts defines the competition as a rule. In sum, the property rights' system in society settles the authorized forms of competition. Besides this principle, the authorities also add transparency dependence and an efficient judicial system.

---

[1]The World Heritage Convention, adopted in 1972, established two mechanisms of work: the World Heritage Committee and the World Heritage Fund, which have been in operation since 1976. The Convention is always updated, and the recent one adopted in the 43th session hold on July 23th, 2019 (available at: http://whc.unesco.org/en/sessions/43com/).

[2]United Nations Convention on the Law of the Sea (1982), Sect. 4. The authority subsection A. General Provisions Article 156 Establishment of the Authority "There is hereby established the International Seabed Authority, which shall function in accordance with this Part. 2. All States Parties are ipso facto members of the Authority. 3. Observers at the Third United Nations Conference on the Law of the Sea who have signed the Final Act and who are not referred to in article 305, paragraph 1(c), (d), (e) or (f), shall have the right to participate in the Authority as observers, in accordance with its rules, regulations and procedures. 4. The seat of the Authority shall be in Jamaica. 5. The Authority may establish such regional centres or offices as it deems necessary for the exercise of its functions".

3.2 Developing Countries: Common Property

The management of common resources inspires three main trends: a socialist, a capitalist, and a social scientists' paradigm. The socialist paradigm asserts that economic poverty is caused by the inequitable distribution of resources and power, which is bound to cause resource destruction. The capitalist paradigm considers that resources commonly held are currently subject to degradation; therefore, the privatization of public resources is the key to resolving this problem. The last paradigm supposes that the multiplicities of economic, political, and social dynamics are factors of degradation. Preserving the common property requires from the policy-makers the recognition support, and reinforcement of local control systems instead of replacing them with another new management strategy.

The common property resources, especially forests and pastures are decreasing and deteriorating in developing countries. This is mainly caused by the pressure of a growing population on nature for the sole objective to maximize individual benefits regardless of negative environmental impacts. The Neoclassical economists suggest privatization as a solution to overexploitation of common resources (Runge 1985). With private property, individuals will rationally exploit resources, will be competitive in the market, and eventually this competition will lead to appropriate management (Vernon 1988).

The socialist paradigm rejects the doctrine that the population is the main source of tragedy. Scientists of this stream consider the increasing rate of population in historical terms, not invoking simple aggregate population parameters. For example, the family size can reflect rational economic decisions. Indeed, the cost-benefits of having extra children, in economically poor societies or the case of inequitable resource distribution will have positive economic impacts. Children will work in agriculture or the informal economy and give the family additional income (Hecht 1985). Therefore, the analysis should not concentrate on the relationship between population and poverty but on historical, social, and political variables to understand the choices made by families. Accordingly, the socialist paradigm proposes the pooling or nationalizing of public resources instead of privatizing them.

For the sake of having the best framework to preserve natural resources, developing countries try to find the best model. The privatization model promotes only the wellbeing of a relatively few fractions of society at the expense of the rest. It deprives poor people from access to public resources and privileges those who already have this access. This system widens the gap between citizens and cause-increased inequality in wealth distribution.

The nationalization of public resources under the State control is not systematically the best solution. The State may become unable to ensure the continuity of preserving the environment and the equity of access for all citizens. For instance, the nationalization of the forest in Nepal in 1957 had caused an increase in deforestation instead of decelerating it (Kergoat 2007: 324). Therefore, many developing countries are in a dilemma as they do not know whether they conform to the capitalist or the socialist paradigms. However, their choice often depends on the economic balance of power, knowing that powerful countries have obligated them to follow the Western paradigm if they are to benefit from foreign financial aids and political support. More often, developing countries have no choice.

3.3 CHH Is a Jus Cogens *Norm*

According to the article 53 of the 1969 Vienna Convention on the Law of treaties "a peremptory norm of general international law (jus cogens) is a norm accepted and recognized by the international community of States as a whole as a norm from which no derogation is permitted and which can be modified only by a subsequent norm of general international law having the same character".

In this regard, the *jus cogen refers* to a compelling law, that is to say, norms of general international law considered superior in the hierarchy of international law architecture. Indeed, they are the rules, which cannot be derogated from and must be applied. *Jus cogens* will have an important impact on international relations and the way countries establish their contractual relations. Countries are required to respect international principles and rules, keep their relations peaceful and avoid the use of violence. For resort to war to be justified, all peaceful alternatives must have been exhausted; therefore, war would be an exceptional case to solve differences (i.e. the right to self-defense).

In consequence, *jus cogens norms* are considered as international constitutional rules. However, the concept of the *jus cogens* norms is not very clear, thus leading to the politicization of such a concept by countries. For instance, it is not clear whether the CHH is one of the peremptory norms or not. For example, African countries, along with some other developing countries in the Group of 77 + China, consider it as a peremptory norm. Conversely, the United States does not consider it a *jus cogens*.[3] Certainly, the Group of 77 + China is protesting against the unilateral legislation made by certain countries in the 1980s, considering it as illegal, because "the principal of the CHH is a customary rule which has the force of peremptory norm".[4] Therefore, several African countries endorsed the position according to which the CHH is a *jus cogens* norm.[5]

One more question should be raised regardless of the uncertainty of the application of the contents of *jus cogens*. In other words, in which terms the rules must be covered, either through over-inclusiveness or under-inclusiveness?

For instance, during the negotiations of the UNCLOS, developing countries had focused on the fact that they represent the interests of all humankind since they constitute its majority. As a consequence, they demand the respect of the *jus cogens*

[3]UNCLOS III, Official records, Vol. 17, p. 243.

[4]Letter 29 August 1980, Chairman of the group of 77, E.K. Wapenyi Of Uganda, to the President of the UNCLOS III, UNCLOS III, Official Records, Vol 17, pp. 111–114.

[5]The statements of the representatives of Burundi and Niger at UNCLOS III, UNCLOS III, Official records Vol. 6, Para. 42, p. 78 and Par. 6, p. 75.

norms as proclaimed in the 1970s (especially by the 1970 United Nations General Assembly resolution on the Seabed). However, some Western countries disagreed.[6]

The second example related to the legal nature of the principle of permanent sovereignty over natural resources resolution (UNGA RES 1803/1962) was raised at the Vienna Conference on the Succession of States concerning State Property, Archives, and Debts. The Draft Convention on Succession of States contained a rule requiring that agreements concluded between a predecessor state and a newly independent state concerning succession to state property of the predecessor state not "infringe the principle of the permanent sovereignty of every people over its wealth and natural resources" (Karmul 2005). Such a Conference was an occasion for developing countries to underline that the principle of permanent sovereignty over wealth and natural resources is a principle of *jus cogens*. Additionally, the Conference was used to discuss other principles such as the right of people to development, to information about their history and their cultural heritage.

In this regard, three categories of *jus cogens* found genuine support, as suggested by Karmul (2005: 84–85): first, the rules protecting the foundations of international order, i.e. the prohibition of genocide or the use of force in international relations except in self-defense; second, the rules concerning peaceful cooperation in the protection of common interests (i.e., freedom of the seas) and the rules protecting the most fundamental and basic human rights; and finally rules protecting civilians during wars.

3.4 CHH as a Means to Assess and Advance Equity

In the Preamble of the UNCLOS, the objective of the convention has been framed as follows: the UNCLOS bearing in mind that the achievement of these goals will contribute to the realization of a just and equitable international economic order which takes into account the interests and needs of humankind as a whole and, in particular, the special interests and needs of developing countries, whether coastal or land-locked". Therefore, the achievement of an international economic order is a key to creating the conditions for the assessment of this equality and not simply a matter of writing it down in the Convention.

The UNCLOS devised a new rule for regulating and managing the Area and its resources by creating an intergovernmental body designed to act and manage the common goods on behalf of mankind as a whole. To achieve this goal, the International Seabed Authority (ISA) has been created and all Member States of the UNCLOS are required to organize and control all seabed-mining activities. The ISA includes three sub-organizations: the Assembly, the Council, and the Secretariat. These sub-organizations are assisted by two specialized bodies: the Legal and Technical Commission, and the Finance Committee.[7] The ISA also incorporates

[6] Statement of the U.S Delegation, U.N Convention on Law of the Sea Part XVII at 243, Danielenko.

[7] 1994 Implementing Agreement, Annex, Sect. 1(4) and Sect. 9.

an Enterprise- which never existed- whose main role is to buy mining technology and achieve joint ventures with commercial operators in order to realize a profit, as part of CHH (Churchill and Lowe 1999: 228). Despite the adoption of the 1994 Implementing Agreement, its roles have been reviewed. It now includes some other tasks such as monitoring deep sea mining trends, development, assessments of the results of marine scientific research, and prospection and exploration in the area.[8] The mission of the Enterprise is currently assured by the Secretariat. Moreover, the ISA is competent in adopting appropriate rules for preserving human life, marine environment, and equitable benefit-sharing. It also has the power to ensure compliance in the Area and its provisions (art. 153 of the UNCLOS).

The full implementation by the ISA of the CHH comforts three main challenges. It requires the recognition of the pluralistic interests linked to humankind and such interests should go beyond classical stakeholders. Instead, it is crucial to establish a process where humankind would be effectively represented and to lift any ambiguity as to an equitable distribution of benefits. In this respect, the legislation tried to propose a solution by considering the principle of CHH through UNCLOS. Such legislation aims at identifying the world community as the owner of the sea bed (Rüdiger Wolfrum 1983: 315) which leads to the question: what kind of regime should be applied since it is considered as part of international law? Is access to the sea-bed is equal for all countries? In other words, does everybody have the right to participate in the debate, either through a political organization, civil society, or non-governmental organizations? The latter had been able to acquire the observer status in the ISA. They have neither right to vote nor any role in the decision-making process.

The last challenge, which the ISA faces, lies in its distributive dimension. This challenge is related to representation of mankind in ensuring the exploitation of the Common heritage. The beneficiaries are all countries, regardless of their geographic location or economic situation.

3.5 The Future of Humankind

The future of humankind is often considered as a contrast subject. Individual or collective decisions may have positive or negative consequences. Therefore, we should develop a realistic model of good practices for the sake of preserving human life and ensuring a sustainable future. The future is a reflection of both human fears and hopes; therefore, humankind must have a clear and realistic idea of what the future might bring in, in order to make the right decisions today including the preservation of humankind itself. Our ability to forecast the future possibilities and trajectories are reinforced every day, thanks to the expanded scientific and technological capacities and the increasingly complex human social organizations.

[8] 1994 Implementing Agreement, Annex, Sect. 2.

The current development models and social organization schemes have major impacts on the environment and ecological balance. The impacts of climate change, biodiversity decline, pollutions, and over-use of non-renewable natural resources, are among the emerging challenges to be dealt with by next generations. Therefore, all the measures to mitigate and manage such challenges should be taken in order to offer the next generations a healthy and viable environment. However, are we currently serious in asking the right questions and taking relevant bold decisions? Are we able to predict the future disruptions through the current science and technology?

The answer to this last question has been given by Crow and Sarewitz (2001: 89–101), who assert that the predictability is irrelevant: preparation for the future obviously does not require accurate prediction; rather, it requires a foundation of knowledge upon which to base action, a capacity to learn from experience, close attention to what is going on in the present, and healthy and resilient institutions that can effectively respond or adapt to change in a timely manner.

Unfortunately, humankind has abused its environment and natural resources, which already led to negative effects on human life, thus threatening the existence of human species. Human extinction is currently considered as a plausible scenario given our current excessive production and consumption patterns. Nowadays, environmental threats are somehow considered the same as nuclear threats. Also, the population dynamics growth influences sustainable development, thus challenging the well-being of future generations.

One of the solutions proposed and widely discussed is technological innovation, considered as a driver of long-term economic growth. Technology is perceived as a means to develop capital that will be used as an asset to implement a sustainable development model, which integrates the whole parameters of human conditions through a global vision. Therefore, as a result, technological innovations have the potential to influence human life quality, directly or indirectly, either individually or collectively, by changing current approaches of governance, entertainment, and human relationships.

The problem we face today is that, though we have agreed on the aim of sustainable development or more exactly, on reducing CO_2 emissions, policymakers have not come far on the road to developing effective policies. Companies recognize the scope for improvement, but most of them still cause huge environmental impacts (Seiler-Hausmann 2002: 10).

4 Conclusion

It is very important to shift from the ultra-liberal *laisser faire, laisser passer model* and the free trade ideology to the possibility to reduce or eliminate environmentally destructive activities, thus limiting the damage affecting the ecological balance and sustainability of natural resources worldwide, including the Commons. The current situation reveals that it is increasingly difficult to preserve the Commons in international relations despite the fact that while doing so, we are at the same time protecting our own survival. This is a huge dual challenge. Among the solutions currently proposed, advocating for climate justice, protecting the environment and the common natural resources while ensuring the social development and equitable access for vulnerable nations are pathways to be followed.

This chapter highlights the efforts made in international community to reach a consensus about CHH which confirms the key role that the international cooperation is playing in this area. The analysis has also focused on the ways through which thinking and re-thinking about the vital importance of the CHH can influence the conflicts of interests between the Global North and the South, thus enhancing a collective action by all relevant actors. We believe that this area of research necessitates an interdisciplinary thinking involving geography, sociology, economics, political science, anthropology, international relations, international law, natural science, etc.

An important difficulty currently facing the global North–South dialogue is the development of a global paradigm in order to save the common concern of humanity and to modify the situation about the degradation of the Commons. The international community must control and save our common future and prevent the degradation and destruction of natural common resources. The ultimate objective is to sustain and maintain the CHH according to the intra and intergenerational equity principles. In this perspective, the present analysis attempts to criticize the liberal market triumphalism and privatization common environmental resources. Many dynamics can reduce the destruction of the Commons, for example, a global policy can be developed comprising an effective economic instrument (such as the balance between adaptation and mitigation), the enforcement of strict regulations (norms, enforcement mechanisms, legal framework of the commons resources, etc.), and making explicit reference to ethical dimensions in order to facilitate public debate and assess societies' capabilities and limitations to save the future we want.

In fact, the North and South coalitions need to identify the special characteristics of the Commons, and use some of the constituent soft norms of international environmental law while dealing with the threats to the Commons. To do this successfully, all actors need to make the necessary efforts that benefit the humankind.

Finally, the protection of the CHH depends on the assessment of our behaviour and the recognition of the limitations of the free market mechanisms. This is crucial to sustain our common future and avoid the conflict over the Commons. Additionally, the highest priority must be granted to the integration of CHH into the process multilateral climate negotiations.

References

African Charter on Human and Peoples Rights Adopted in Nairobi June 27 1981

Baylis J, Smith S (2001) The globalization of world politics. An introduction to international relations. Oxford University Press, Oxford

Churchill RR, Lowe AV (1999) The law of the sea, 3rd edn. University Press, Manchester, p 228

Cordonier MC, Khalfan A (2004) Sustainable development. Principles, practices, and prospects. Oxford University Press, Oxford

Crow MM, Sarewitz D (2001) Nanotechnology and societal transformation. In: Teich AH, Nelson SD, McEnaney C, Lita SJ (eds.) AAAS Science and Technology Policy Yearbook (Washington, DC: American Association for the Advancement of Science), pp. 89–101

Shelton D (2013) The Oxford handbook of international human rights law. Oxford University Press, Oxford

Faramiñán Gilber JM (2019) The Common Heritage of Mankind Principle: The Moon and Lunar Resources. http://www.unoosa.org/pdf/pres/lsc2009/symp03.pdf. Accessed 9 Nov 2019

Fitzmaurice MA (2002) International protection of the environment. Collected courses of the Hague academy of international law. Martinus Nijhoff Publishers, The Hague

Garrett H (1968) The tragedy of the commons. Science, Dec. 13, 1968, 162(3859)

Geldenhuys A (2015) The legal status of the concept of Common heritage in the exploitation of resources on the moon and other celestial bodies: Is now the time for a legal regime? LLM, (University of Pretoria)

General Assembly resolution, Permanent sovereignty over natural resources, 1803 (XVII), 14 December 1962. https://www.ohchr.org/EN/ProfessionalInterest/Pages/NaturalResources.aspx. Accessed 29 Sep 2019

Hardin G (1968) The tragedy of the Commons. Am Assoc Adv Sci 162(N):3859

George H (2006) Progress and Poverty, 1st edn. Robert Schalkendbach Foundation, New York

Grieco J, John Ikenberry G, Mastanduno M (2015) Introduction to international relations. Enduring questions & contemporary perspective. Macmillan Education Palgrave

Heller H (2011) Capitalism and world history. Pluto Press, London

http://whc.unesco.org/en/sessions/43com/. Accessed 22 Dec 2019

https://repository.up.ac.za/bitstream/handle/2263/53130/Geldenhuys_Legal_2016.pdf?sequence=1&isAllowed=y

https://sustainabledevelopment.un.org/content/documents/21252030%20Agenda%20for%20Sustainable%20Development%20web.pdf. Accessed 10 Jan 2020

Karmul H (2005) The concept of Jus Cogens and the obligation under the U.N Charter. Santa Clara J Int Law 3(1):72–98

Seiler-Hausmann J-D (2002) Innovation & Sustainable Development, Institute for Global Environmental Strategies

Kergoat M (2007) Histoire politique du Népal - Aux origines de l'insurrection maoïste. Karthala, Paris

Letter 29 August 1980, Chairman of the group of 77, E.K. Wapenyi Of Uganda, to the President of the UNCLOS III, UNCLOS III, Official Records, Vol 17

Wolfrum R (1983) Max Planck Encyclopedia of public international law. The Common heritage of mankind

McCay BJ, Dolšak N, Ostrom E (2003) The Commons in the New Millennium. Challenges and Adaptation. Massachusetts Institute of Technology Press

MULTILATERAL Vienna Convention on the law of treaties (1969) vol 1155, 1–18232. https://treaties.un.org › Publication › UNTS ›. Accessed 25 Sep 2019 at 07h18

Multilateral Vienna Convention on the law of treaties, Article 53, vol 1155, 1–18232

Petrella R (2001) The water manifesto. Arguments for a world water contract, trans. Patrick Camiller, p 8. Zed Books

Runge CF (1985) Common property and collective action in economic development. In: Proceedings of the conference in common property resource management. National Academy Press, Washington, D. C

Sands P (2000) Environmental protection in the twenty-first century: sustainable development and international law. In: Richard L, Revesz P, Sands P, Stewart RB (eds) Environmental law, the economy and sustainable development. The United States, the European Union and the international community, pp 369–409. Cambridge University Press

Scovazzi T (2007) The concept of Common heritage of Humankind and the genetic resources of the Seabed beyond the limits of national jurisdiction, Agenda Internacional, Año XIV, vol 25, p 11+

Spijkers O (2018) Intergenerational equity and the sustainable development goals. Sustainability. MDPI. 10.3390/su10113836. Accessed 18 Dec 2019
Hecht S (1985) Environment, Development, and Politics: Capital accumulation and the Livestock. Sector in Eastern Amazonia. World development
Buck SJ (1998) The global Commons. An introduction. Island Press, Washington, D.C
The statements of the representatives of Burundi and Niger at UNCLOS III, UNCLOS III, Official records vol 6, Para. 42, P 78 and Par. 6 p 75
UN Transforming Our World: The 2030 Agenda for Sustainable Development Sustainable Development.un.org A/RES/70/1 https://sustainabledevelopment.un.org/content/documents/21252030%20Agenda%20for%20Sustainable%20Development%20web.pdf. Accessed 10 Jan 2020
UNCLOS III, Official records, vol 17, p 243
UNCLOS, art. 153
United Nations Convention on the Law of the Sea, November 16 1973
United Nations (2012) The Future We Want. https://sustainabledevelopment.un.org/content/documents/733FutureWeWant.pdf
United Nations (2015) Transforming Our World: The 2030 Agenda for Sustainable Development
Vernon R (1988) The promise of privatization, 1st edn. Council on Foreign Relations, New York
Vibhute KI (1997) Environment, present and future generations. Intergenerational equity, justice and responsibility. J Indian Law Inst 39(2/4):281–288
Vienna Convention on the law of treaties, Vienna, May, 23th, 1969, United Nations, Treaty Series, vol 1155. https://legal.un.org/ilc/texts/instruments/english/conventions/1_1_1969.pdf. Accessed 29 Sep 2019
World Commission on Environment and Development (WCED) (1987) Our Common future, A/42/427
Weiss EB (1993) Intergenerational fairness and water resources. The national Academies Press. https://www.nap.edu/read/2217/chapter/3. Accessed 12 Oct 2019
Weiss EB (2008) Climate change, intergenerational equity, and international law. Vermont J Environ Law 9. http://vjel.vermontlaw.edu/files/2013/06/Climate-Change-Intergenerational-Equity-and-International-Law.pdf. Accessed 1 Mar 2020
Weiss EB (1989) Fairness to future generations, common patrimony and intergenerational equity. United Nations University, Tokyo
Weiss EB (1992) In fairness to Future generations and sustainable development. Am Univ Int Law Rev 8(1)
Weiss EB (1989) Fairness to future generations, Common patrimony and intergenerational equity. United Nations University, Tokyo
White MV (1982) The common heritage of mankind: an assessment. 509 Case Western Reserve J Int Law 14(3). https://scholarlycommons.law.case.edu/jil/vol14/iss3/6
Wolf C. (1995) Contemporary Property Rights. Lockean provisos, and the Interests of Future Generations, vol. 105, N. 4
World Heritage Convention (1972) Paris, 16 November 1972, UNESCO. http://whc.unesco.org/en/conventiontext

Climate Change Induced Disaster, Gender Conflicts, Risks, and Insecurity in Lower-Gweru Community of Zimbabwe: Towards Capacity Building and Resilience in the Face of Social-Ecological Threats

Winniefridah Matsa

Abstract The ability of our planet Earth to sustain humanity is increasingly threatened as a result of climate change. This challenge is severely hitting the poorest and the most vulnerable resulting in changing patterns of interactions between societal and ecological entities. This chapter is premised on the socio-ecological theoretical framework. The research seeks to determine the unequal impact of climate change disasters on men and women in Lower Gweru community and explore the nexus between climate change, gender-conflicts, risks, and insecurities. It also assesses existing response mechanisms to foster resilience as well as capacity building to enable especially women to cope with challenges posed by climate change, risks, and insecurities. The research is a qualitative exploratory study of four villages of Lower Gweru. The sample consists of four village heads, two councilors, twenty household heads, two police officers, and two Community Health Workers, giving a total of 30 participants. The results revealed that climate change has unequal impact on male and females and cause gender-conflicts, risks, and insecurities. In addition, there are no viable response mechanisms or frameworks to foster resilience; therefore, the research recommends the design of a joint, coordinated and viable model for resilience that merges both gender analysis and climate change disaster resilience frameworks.

Keywords Socio-ecological systems · Gender-conflicts · Risks · Insecurity · Resilience

1 Background

Climate change is not an occurrence in the distant future but a current phenomenon that has recently become a key concern in development and is felt in all corners of the world (Brody et al. 2008; Dankelmam 2002; UNDP 2016). There has been little

---

W. Matsa (✉)
Gender Institute, Midlands State University, Gweru, Zimbabwe
e-mail: matsaw@staff.msu.ac.zw

M. Behnassi et al. (eds.), *Social-Ecological Systems (SES)*,
https://doi.org/10.1007/978-3-030-76247-6_4

research on climate change disasters and related gendered experiences, conflicts, risks, and insecurities in communities. The United Nations Development Fund for Women (UNIFEM) also notes that climate change is one of the most urgent global challenges facing the world today (UNIFEM 2015). The United Nations Secretary General points out that climate-induced disasters have increased in frequency and inflict costs at an average of $250–300 billion per year (UNDP 2016). The IPCC (2014) confirms that human influence on climate change system is increasing. The year 2015 was the hottest year on record globally.

Wedeman and Petraney (2016) claimed that climate change and environmental degradation are a great threat to poverty reduction and achievement of sustainable development goals (SDGs). Climate change disasters, poverty, and gender conflicts are triple evils deeply affecting development. In 2015, the 17 SDGs were launched to put much needed spotlight on poverty, inequality, and violence against women. The Hyogo Framework for Action 2005–2015 was replaced by the Sendai Framework for Disaster Risk Reduction 2015–2030. The latter incorporated gender-sensitive disaster risk reduction to promote a stronger role for women in building resilience. The 2015 Paris Agreement also set in motion the United Nations Framework Convention on Climate Change (UNFCCC) (UNDP 2016).

Dhliwayo (2019) highlighted that the impact of disasters is determined by the cultural and economic circumstances of communities in the context of poverty and social inequalities. Adverse impacts of climate change continue to burden the poorest and the most vulnerable, especially poor women (UNDP 2016). Women's historic disadvantages, limited access to resources, and restricted rights enhance their vulnerability to climate change disasters (KARAT Coalition 2013). Differences in vulnerability and risks emanate also from such non-climatic factors as multidimensional inequalities often produced by uneven development processes (IPPC 2014). Therefore, gender inequalities intersect with climate change risks and vulnerabilities.

The UNFCCC (2010) defines climate change as a change of climate attributed directly or indirectly to human activity that alters the composition of the global atmosphere. According to the United Nations Office for Disaster Risk Reduction (UNISDR now UNDRR), a disaster is a serious disruption of the functioning of a community or society due to hazardous events interacting with conditions of exposure, vulnerability, and capacity leading to human, material, economic and environmental losses (UNISDR 2017). On the other hand, climate change disasters refer to changes in global climatic conditions exacerbating climatic hazards. Effects are harsher on those located in marginal areas worsened by deforestation, water scarcity, and land degradation among others (UNDP 2016).

Disaster risks are defined by the UNDRR in Peters et al. (2019) as "the potential loss of life, injury, or destroyed or damaged assets which could occur to a system, society or a community in a specific period of time determined probabilistically as a function of hazard, exposure, and capacity". According to Bradshaw (2013), disaster risks can be climatological (drought, extreme heat, storms and floods), geomorphologic (earthquakes, landslides, volcanoes and erosion), bio-ecological (epidemic, plant diseases, plagues and depletion of species) as well as astronomical (hazards from space). Climate change increases disaster by altering the frequency and intensity of hazardous events and changing exposure patterns while at the same time decreasing resilience of households and communities (UNISDR 2017). Bradshaw (2013) points out that the world is becoming prone to disasters causing risks to those with low capacity to adapt. Climate change disasters cause a threat to economic livelihoods of the poor men and women making them vulnerable to conflicts and security risk (Brown et al. 2012).

Gender conflicts, risks, and insecurity are reported to be increasing during and after disasters; this occurs in all countries, at all stages of development, and at all levels (Amnesty International 2011). In Japan, in 1997 and 2010 earthquake, sexual abuse and physical violence increased on girls, women, evacuees, and volunteers causing risks and insecurities. Women and the poor live in fragile ecological zones with arid land, slopes, poor soils or land. Their lives are endangered with because of they are highly sensitive to climate variability (UNDP 2016).

Climate change disasters also create windows of opportunity for social structures to change (Masson et al. 2016). Unfortunately, traditional gender roles are re-emphasized and gender inequalities often worsen during and after climate disaster emergencies. They further note that different forms of violence increase encompassing physical, sexual, verbal, emotional and economic forms as shocks and stresses change gender norms and power relations. Responses to a changing climate require men and women to take different roles and responsibilities, thereby changing their livelihoods and lifestyles while exchanging masculine and feminine gender roles (ALPA 2013). Changes in livelihoods strategies create new spaces for men and women, thus resulting in creating shifts in expectations and perceptions of their roles. Men, to a greater extent, find it difficult to take up feminine roles, resulting in gender conflicts.

The United Nations (2015) recognise that climate change presents urgent and potentially irreversible threat to humankind and that action should be taken to enhance capacities of especially poor women to foster resilience. The European Union (2016) echoes that there is a need to enhance the adaptive capacity, strengthen resilience, and reduce vulnerabilities and risks linked to climate change disasters. The FAO (2013) shows that rural women are in the frontline battle to cope with climate change disasters to improve security. Unfortunately, women have less capacity to prepare for adaptation. There has been few or no response mechanisms to foster resilience and capacity building in climate change induced disasters. This is because the disaster, risks, and vulnerability discourse does not seem to inform development discourse despite the recognition that disaster affects development. Gender has also not adequately entered the disaster and development discourses. Attempts to engender disasters appear to be a number of decades behind (Bradshaw 2013). There are researches on gender and disasters but gender concerns are less evident as part of current disaster planning and policy. The concern has focused much on the impact of disasters on women and nothing beyond.

There have not been significant efforts in the direction of building capacity in communities, especially of women, with respect to their training, education, communication in science, and technology for risk reduction as well as their inclusion in post-disaster activities. Disaster Risk Reduction (DRR) is a systematic approach to identifying, assessing, and reducing the risk of disasters to curb socio-economic vulnerabilities and dealing with environmental factors that trigger risks (UNISDR 2017). The perspectives guiding incorporation of gender in natural disasters is based on the social construct of women as victims so that they are rescued to be active agents to fight against risks and vulnerability for resilience (Bradshaw 2013). The UNFCCC (2010) defines climate resilience as the capacity for the socio-economic system to absorb stresses and maintain function in the face of external stresses imposed upon it by climate change in order to adapt and reorganise to improve sustainability of livelihoods.

The study area Lower Gweru is situated in the Midlands province of Zimbabwe in the agro-ecological region four with occasional mid-season persistent droughts. The agro-features are rain-fed cultivation, natural livestock grazing, fruit gathering, market gardening, and irrigation in isolated places. Temperatures range between 11 °C in dry winters and 26 °C in wet summers that stretch from October to March. There are isolated wet places: Insukamini, Mambanjeni and Shagari irrigation schemes as well as Dufuya and Madigane wetlands. Wetlands allow people to irrigate or do market gardening using water from wells. Generally, droughts limit hectares put under cropping, and there are no alternative economic activities besides agro ones. Dry-land farming results in low food productivity, which enhances food insecurity. The common disasters are persistent droughts, and community members lack adaptive capacity and resilience, especially women. Hence, they become vulnerable to risks of poverty, food insecurity, and gender conflicts. It is against this background that this chapter seeks to determine the unequal impact of climate change disasters on men and women, explore the nexus of climate change disasters, gender conflicts, risks, and insecurities in Lower Gweru community as well as assess and proffer response mechanisms to foster resilience and capacity building.

2 Literature Review

2.1 Theoretical Framework

The paper uses the socio-ecological model as a theoretical framework to explain the unequal impact of climate change disasters on men and women and the manner it leads to gender conflicts, risks, and insecurities. Koontz et al. (2015) define social-ecological systems as an interplay of both natural, ecological, and human components that interact to affect systems' dynamics of multiple factors governing institutions. A social-ecological system is linked to people and nature, and emphasises that humans must be seen as part, not apart from nature (Berkes and Folke 1998).

The ecological model explains human behaviour, that is, gender conflicts, risks, and insecurities in terms of complex and multidimensional environmental and social factors (Bronfenbrenner in Doku 2012). As such, the unequal impact of climate change on men and women is a result of structural socio-economic factors; so is their resilience to climate change disasters, risks, and insecurities. The interplay of socio-cultural and environmental factors determines the extent of disaster impact, vulnerability to risks, and the extent of resilience. Heise (1998) echoes that causes of vulnerability risks lie in the ecological, social, and economic structures like culture, socialisation and environmental factors. Folke (2006) notes that bio-physical factors interact with social factors to influence resilience. Vulnerability and risks are rarely due to a single cause but result from processes of inequality in income, socio-economic status, and exposure (IPCC 2014).

These socio-cultural factors are located in the micro-sphere (family situation), mesosphere (community settings) or the exosphere (macro-level policies by government or development agencies) (Heise 1998). The macro and micro levels combine to determine the extent of the impact of climate change disaster risks on men and women, influence gender conflicts as well as resilience. Yet response mechanisms in most cases ignore social-ecological systems. There is a need for frameworks that takes note of ecological systems in resilience mechanisms.

2.2 *Empirical Findings on Unequal Impact of Climate Change Disasters on Men and Women*

Climate change impact is any change in the physical character of the environment reflected by cumulative interacting social and economic impacts. The changes also relate to human well-beings, disruption of infrastructure, networks, access to goods and services as well as adaptive capacities. However, climate change impacts are not evenly distributed between genders, the poor and even regions. Dankelmam (2002) also notes that climate change impacts are differently distributed among regions, generations, age, classes, income groups, and gender. In Zimbabwe, Ngangu Chimanimani area, all of the survivors found on roof and tree tops were men. The majority of those hard hit were women and children as a result of social constructions about men and women. Women lacked defensive mechanisms like swimming, fleeing, climbing or information about the approaching of cyclone Idai.

A majority of women live in rural areas while men seek jobs in urban centers and mines. Women also live in poor houses made of mud and thatch (Dhliwayo 2019). Out of a total of 250 000, 52% translating to 130 000 women were affected in both Chimanimani and Chipinge in Zimbabwe. About 75 000 women were pregnant, hence at risk of health related and other complications at delivery. Some who spoke to humanitarian agents expressed feelings of insecurity as a result of exposure to sexual attacks in public shelters like schools, clinics, and churches that lacked privacy. There is a need to understand the nexus between climate change disasters and their human impacts as an intertwined phenomenon (Habtzion 2013). Vulnerability and risks are a product of intense social processes that result in inequalities in socio-economic status as well as gender, class, and age.

The poor and the marginalised are structurally vulnerable and at risk (UNDP 2014). Poverty alongside socio-economic and political marginalisation cumulatively puts women into disadvantaged position in coping with adverse impacts of climate change; therefore, at risk. Masika (2002) had earlier noted that different power relations and levels of access to resources are key to understanding men and women's vulnerability, their exposure to risks, their coping capacity, and their ability to recover from disasters. In South East Asia, life expectancy for women and children was affected negatively, especially among the poor. Usually, the poor people settle on hazardous and flood prone areas. These together, with the type of residential space and building, the casualties were higher for women and children. In Indonesia and Sri Lanka, poor women could not buy or build safe houses on suitable land for settlement. Most of their homes were flooded (Murillo and Tan 2017).

Habtzion (2013) pointed out that Bengali women and children perished in large numbers waiting for male relatives to take them to safer places because of cultural restrictions on their movement. Bengali women had never learnt swimming and this reduced their chances of survival. In the 2003 European heat wave, about 80 000 deaths were recorded especially the elderly. Female deaths increased by 21% every month causing distortions in gender and age structure (Robie et al. 2007). In the Katrina disaster in USA, African American women were the ones who were trapped because of poverty, cultural restrictions and lack of survival skills. African Americans of 65–80 years were the first and hardest hit in the first year of the storm, Katrina. This was because they were deprived of basic, social and material necessities. About 65% perished because they had no capacity to evacuate on their own or to get on roof tops especially women (Adams 2011). In 1991in Bangladesh cyclone Gorky had 140 000 deaths in 1991, 14 women for every man and the 2004 Oceanic Tsunami had 70% of 250 000 fatalities being women (Sander et al. 2018).

2.3 The Nexus Between Climate Change Disasters, Gender, Risks, and Insecurities

Climate change leads to gender conflicts over resources in short supply (UNFCCC 2010). Those with least resources often have the least capacity to adapt and are vulnerable and conflicted. Gender and poverty intersect to produce vulnerabilities, risks, and conflicts. Women lack power over resources while men often have this power, even in times of shortages during post disaster situations. After Hurricane Mitchell in Honduras and Nicaragua, women could not return to work because of the increased domestic work and care. This disadvantaged women in terms of outside work. Males migrated in search of employment since domestic work made it difficult for many men to engage in income generating activities, thus causing conflict between men and women as men are expected to secure source income for bread winning. Some decisions and responsibilities were imposed on men and women.

The Swedish International Development Cooperation Agency (SIDA) discovered that there was correlation between climate change disasters and eruption of conflicts between nations, households, and genders (SIDA 2018). Climate related changes influence the factors that exacerbate conflict such as reduced water supply, food shortage, and changes affecting the distribution of roles and responsibilities. The World Health Organization (WHO) (2010) noted that an extra burden is placed on fathers as providers in the aftermath of disasters despite that the economic activities they depend on would have been affected by such disasters. Men end up taking risky behaviours and drowning trying to fish or cross flooded rivers. Failure to provide for the family results in conflicts at household level and psychological problems. The UNFCCC (2010) found that changing livelihood patterns result in anger, frustrations, conflicts, and violence from both men and women. In India, suicide increased amongst male farmers following droughts and failure to assure their livelihoods. Stephens (2013) highlights that in the Caribbean productive roles changed for men and women as a result of disasters; men had to help in the provision and prepara tion of food to ensure rehabilitation. Women gathered and provided food; a situation which had forced them to take work outside the home and travelling at night, hereby upsetting cultural expectations. The situation had potential for conflicts and, in some instances, misunderstandings that resulted in gender conflicts. The UNDP (2014) echoed instances of gender conflicts related to limited options of livelihoods and social and cultural discrimination in distribution of resources or food. Women sacrifices to gather foods for children, but culturally men are fed first. Women ended up going without food for the sake of their children.

In developing countries, FAO (2013) reported increased restrictions on women's rights with respect to access to land use, freedom to plant crops of their choice, and access to other productive resources. Women's historic disadvantages are enhanced; for example, the lack of land ownership of and decisions regarding its use. Rohr (2008) also noted that climate change resulted in greater scarcity of resources, water, arable land, thus leading to unequal distribution for men and women which worsens conflicts between them and between communities. Holden et al. (2016) highlight that in most rural societies, women access land rights through customary systems. The authority to decide on land allocation is ascribed to patrilineal lineage with most decisions on land taken by elderly men on behalf of the family. Women's land rights are determined by marital status, inheritance, and divorce. Such secondary rights are of an uncertain duration, not well-defined, subject to change, and usually depend on the maintenance of good relations between the involved parties. Women cannot decide on management of the plot or long-term plans about it.

In Mozambique, Niger, Kenya, and Ghana, men controlled the proceeds from women's small-scale projects and decided on the manner proceeds are used. Men controlled money from both big and expensive assets like livestock and fish. As a result of lack of employment opportunities and food scarcity, men married off their daughters for low bride price (GGCA 2016). In India, women were forced to eat less due to food scarcity. In Bangladesh, women and girls are first to skip meals as a result of food shortages.

Heise (1998) noted that poverty resulting from disasters threatens family structures as men and women seek alternative means of livelihood. Homes break up or separate because of migration and wage labour as families struggle to make ends meet. In Ghana, alternative livelihoods – like migration and wage labour by men in Saamini region resulted in breakdown of families. Drought-induced poverty increased rates of divorce, forcing men to have fewer wives in East Africa. The ALPA (2013) discovered that climate change reduced livestock herds, thus leading to divorce or inability to support many wives and children. When men especially fail to fulfill their gender roles, conflicts manifest.

The UNIFEM (2015) highlights that the impacts of climate change challenge models of feminity and traditional gender roles. Change of men's roles and increasing participation of women in the public arena and paid work challenge male's identity as bread winners. Challenging notions of masculinity results in conflicts between males and females. The Care International (2011) discovered that, in Kenya, droughts reduced livestock, thereby striking at the core of people's life styles and culture in Shanta Abqq and Nanighi regions. Sometimes, women started selling their produce to male middle men. This resulted in increase in women's negotiating power, thereby defying patriarchal and cultural expectations, which results in conflict. Men wanted to continue to decide on women's income. The roles had been changed and space is opened for men and women to the extent that both of them can transgress gender roles, norms, and gender division of labour. The ALPA (2013) discovered that due to climate change impacts, gender division of labour blurred in Mozambique: men were forced to work alongside women or women were doing men's work. This included activities such as weeding, harvesting, and processing agricultural produce which are traditionally women's work. In Kenya, men were forced to help women collect scarce fuel and water. In Kenya and Mozambique, crops grown are gendered in so far as men grow mostly cash crops and women traditional crops for consumption. Climate change sometimes force shifts from cash to traditional crops specific to women, thus forcing men to work together with women in the production and sale of such crops. In Azagar in Niger, women and men collaborated on almost everything including child care and food preparation. Working together sometimes result in conflict and violence.

In Ghana and Mozambique, the ALPA (2013) reported that although men and women had new gender roles, men still refused to give women control and remained the sole decision makers concerning assets, prices, information, and markets. It was further noted that interaction of men and women in work and markets led to backlash and violence as well as clashing cultural norms. Women in Kenya and Niger began to have freedom of movement, participated in the public sphere, and had voices or decision-making rights. This contradicted the resurging conservative interpretation of Islam. New roles violated social customs and values causing backlash, stigma, violence and marginalisation of one gender (ALPA 2013). Women entry into markets in Kenya and Ghana was met with violent resistance. Women who took up these roles were viewed with disdain. They were given derogatory language '*po a gandooo*' (female giants), yet the culture viewed the empowered women with a negative eye.

Loss of herds in Kenya increased drought-induced poverty, thus making it difficult for men and women to provide for their households. This further increased rates of gender conflicts and divorce. Men who felt marginalised or demeaned turned to drugs which exacerbated conflicts (ALPA 2013). This is echoed by Masson et al. (2016) who noted that, in Uganda, males faced 'masculinity crisis' linked to the livestock and other livelihoods. This is because agriculture was considered women's work. Failure to get alternative work led to a crisis which caused high levels of idleness, alcoholism, and gender-based violence. In Philippines, the farm roles for females changed as societies struggled to adapt to floods. Women were forced to challenge traditional gender roles and levels of education to take greater managerial responsibilities in projects designed to reduce water scarcity. In Sri Lanka, the gendered norms were challenged by involving women in water governance and management activities to ensure fair distribution of water to men and women (GGCA 2016). Upsetting power imbalances causes gender conflicts.

3 Methodology

This chapter is based on a qualitative exploratory study of four villages in Lower Gweru, Zimbabwe. It was composed of 4 village heads (one female and 3 males), 2 councilors (both males), 2 police officers (a male and a female), 20 household heads (12 females and 8 males), and 2 Community Health Workers (CHWs) (both females) giving a total of 30 participants all of who were purposively sampled. Village heads and councilors are knowledgeable about the impacts of climate change on the community and are the ones who represent people and report to the district and provincial administrators about the situation in their constituencies. Police officers and village heads also handle issues of gender conflicts. Councilors worked with development partners and the government in the area. Community health workers are also involved in issues that have to do with risks, vulnerabilities, and insecurities related to health, food, peoples' rights, malnutrition, gender conflicts, and in general gender harassment.

Data was collected through unstructured interviews with male and female members of households found working in their fields and gardens. Direct observation of people working on their market gardens, fields, transportation and delivery of goods to markets was made. Vendors outside shops, clinics, and schools were visited.

4 Findings and Discussions

4.1 Unequal Impact of Climate Change Disasters on Men and Women in Lower Gweru Community

The unequal impact of climate change disasters on Lower Gweru communities relates to gender and ecological aspects of this study. To understand the gendered impact of climate change, disaster risks, and insecurities in the Lower Gweru community, we need to explore the characteristics of the study area. Lower Gweru has an agro-based economy which is rain-fed. Results from direct observation show that the dominant agricultural activities are market gardening and fruit production from mainly exotic trees: bananas, guava, mango, and a few indigenous ones including Flacourtiaindica *(umtshwankela/tsubvu),* monkey orange *(umkhemeswane/mazhumwi),* snot apple *(uxakuxaku/matowe)* and wild medlar *(umviyo/matufu).* The market gardeners and fruit growers depend on water from streams, rivers, and wells. The place is semi-arid with low fertility status of the soils favourable to their variety of fruits and vegetables. Crop production and livestock heading is also practiced for subsistence. Commercial agriculture is skewed towards fruit production and market gardening.

The Lower Gweru community makes up the largest fruit and vegetable supplier in Gweru, the third largest city in Zimbabwe. The fruits and vegetables flood the Gweru Kudzanai bus termini, Kombai agricultural market, and Jamela vegetable market throughout the year. They produce varieties of green and yellow vegetables namely: rape, covo, cabbages, spinach, green beans, onions, butter nuts, cucumbers, carrots, green and red pepper, peas, okra, tomatoes, and beetroot among others. During the summer season they beef up their products with indigenous vegetables; for example, indigenous okra *(derere/idelele),* bean leaves *(munyemba/isanyonkwane,* pumpkin leaves *(muboora/ibhobola),* and spider wisp *(munyovhi/ulude).* The traditional fruits sold at town markets are bird plum *(nyii)*, snot apple *(matohwe/xakuxaku,* and monkey orange *(matamba/umhlali).* Other commercialised crops are sugar cane, watermelon, green and yellow pumpkins, fresh nuts, ground nuts, and green mealies.

Both men and women are supposed to equally participate in the production, processing, transportation, and marketing of products. However, because of gender division of labour and gender stereotypical roles, most of the heavy work is done by males. The clearing of the land, ploughing, planting, and watering through canes is for both, but more men than women meet the demands of this kind of work. Fruits and vegetables are perishable and need to be delivered to the Gweru markets almost on a daily basis; a distance of more than 60 km from Sogwala, 70 km from Ntabamhlophe-Shagari, and 65 km from Cross roads-Dimbamhiwa roads.

With climate change and disasters' impacts, women and men risk losses out of droughts, incessant rains, and floods. Women are relegated to poor pieces of land which are swampy, sloppy with poor arid soils, degraded lands, and stressed forests. At the same time, women lack resources to face and adapt to climate change impacts even if their livelihoods depend mostly on natural resources. Lack of resources affects even the kind of crops they grow, houses they build (poor thatched mud), and huts that cannot resist floods and storms. Hence, the lack of resources reduces their resilience capacity. Women, because of their work in the private sphere (household chores, child bearing, rearing, care work), cannot be on the move every day transporting goods to the markets in Gweru. As a result, transportation is mostly done by men. Crops, fruits, and vegetables are packed in buckets, crates or sacks. Sugar cane and indigenous canes are bundled and loaded on top of buses or commuter bus carriers. This is a work which demands masculine strength according to gender role expectation in this community. However, some brave women defy the prevailing cultural norms.

As a result of the perishable nature of the products, harvesting, especially during the hot season, is done in the evenings around 5–7 pm, while transportation and loading is done early in the morning over 2–5 am so that fresh goods reach the Gweru markets between 6 and 8 am. The nature, weight, and time of the activities contradict with gender stereotypical roles and cultural role expectations of men and women. Consequently, more men than women are involved in large-scale transportation of crops, fruits, and vegetables to commercial markets in town. More male farmers end up producing for large-scale markets while women produce for subsistence local markets and commercial markets to a lesser extent.

Respondents highlighted that agricultural activities were the back bone of their livelihood before they were hard hit by climate change disasters like drought, incessant rains, and floods. Fruit and vegetable production were vibrant to the extent that the area experienced little labour migration by the educated to work in towns. A sizeable number of civil servants like teachers, police, agritex personnel and nurses commute from their rural homes to town for work. They are also engaged in fruit and vegetable farming to supplement their incomes. Men own and control land, livestock, and water use. Women have user rights which depend on the liberty of men. Lower Gweru is largely made of communal reserves whose land is solely registered in male names and a few resettlement schemes with land in the hands of male and females. Such socio-cultural, economic, and ecological arrangement of society lays the basis for unequal impact of climate change disasters, risks, and insecurities on men and women in the area. The socio-ecological systems in this area also influence gender conflicts, vulnerability, and resilience by each gender.

Vegetable production is an all-seasons' activity. However, it was noted that for the past 10–15 years, the area has been hard hit by climate change disasters, especially droughts. With droughts, streams dry around July much before the onset of rains in October–November. This causes low and poor yields. Most exotic vegetables are watered using watering canes. Distance of water bodies, rivers and wells, the slippery, muddy and sloppy terrain are huge barriers, especially for the women. During droughts, fruits too are affected. The indigenous ones are found far away on hills and on tree tops. Gathering in such cases becomes difficult for women. The quality and quantity of products go down, which results in low supplies and low profits. Women are also particularly hard hit because of gender role expectations and the fact that their activities are largely skewed towards crops and vegetables that are rain-fed. This includes melons, pumpkins, leafy indigenous vegetables, and varieties of nuts. The exotic vegetables demand a lot of labour from women, watering, in addition to problems of loading and transportation to Gweru markets. Women end up taking their products to vending markets in rural bus stops, shopping centres, and institutions like schools and clinics. The disadvantage here is that town markets order products in bulk. A Dufuya woman bemoaned losses made by women through this vending mode:

> *In rural markets, our vegetables take hours or days and lose freshness. The wilted or dried products do not attract buyers.*

Women also produce for consumption purposes. If crops for subsistence are affected by droughts or floods, they end up using their commercial products for consumption or profits from commercial products to buy subsistence needs. Women reported that low profits also meant they fail to meet their gender role expectation. This covers providing food for the family and daily needs like relish, sugar, salt, soap, petroleum jelly, books, and pens for especially children in schools. Their families risk vulnerability to poverty and hunger. Women also risk insults, humiliation, and disdain from the community, extended family, and even husbands. One woman made this comment:

> *Umuzi ngumfazi* (it is the woman who is the backbone of the home).
>
> *Intandane enhle ngumakhothwa ngunina (A better orphan is the one under the care of a mother).*

This puts pressure on women to provide and care for families, especially children and husbands, the failure of which their marriages are at risks and insecure. Disasters indirectly plunge women into economic, cultural, and social risks and insecurities. Men in Lower Gweru community cited that market gardening and fruit marketing have been their only decent sources of employment and livelihood since time immemorial. They reported that the two have been viable and enough for their family livelihoods with supplements from crop production. Males reported that they themselves were fed, clothed, educated, and well catered for by their parents through these activities and have done the same with their own children. They moaned recent climate change disasters that devastated their livelihoods leaving them at the mercy of hunger, malnutrition, poverty, and unemployment. They felt they were now viewed by the community, wives, and children as unproductive and useless. This has resulted in gender conflicts and loss of respect, control of families, and dignity as fathers and bread winners.

Drought also disrupted women's market gardening and crop production. They lose time fetching water for household use. They travel long distances to boreholes at the expense of market gardening. If there are torrential rains, vegetables such as tomatoes, muboora, nuts and fresh mealies rot easier. They cannot dry their products or travel to markets every day because of incessant rains and household work.

Floods were reported to destroy crops or make it difficult to transport products to markets. Droughts lowered their crop, vegetable, fruit yields, and livestock production. Livestock should be taken to distant boreholes where one fetches water for them to drink. As a farmer, one is forced to reduce the number of livestock. Fetching water and grazing for herds is done at the expense of agricultural activities; otherwise, one risks losing everything from livestock rearing. In the process of diverting attention to livestock, one then neglects market gardening that brings almost daily earnings. Varying economic activities is a way of fostering resilience; but climate change disasters limit that opportunity.

4.2 Nexus Between Climate Change Disasters, Gender, Risks and Insecurities

Some participants highlighted that poverty and lack of resources lead to misunderstandings between couples. This insight emerged through in-depth interviews and questionnaire-based interviews about the link between climate change and gender conflicts. Men cited that the combination of climate change impacts that have negatively affected their agricultural activities and their unemployment has enhanced their life's misery. The Lower Gweru community still strongly believes that it is the duty of men as the head of the family to fend and support families as bread winners. Those who fail to meet this societal expectation are looked down upon not only by women and their children, but also by relatives and the community at large.

Unemployed males compared their lives to the agony of getting crushed between two grinding stones (*amatshe okuchola*). If they stayed at home, women (mothers and wives) would complain and insult them as useless men who had no plans for life. If they tried to look for alternative jobs and brought nothing or very little, they would be insulted for coming home empty handed. Most of the piece jobs available in the neighbourhood were pre-dominantly attributed to females. Women who did piece jobs complained about men as useless. Below are some of the verbatim quotes from men.

> *Umfazi uyakubalisela (*Women complain and there won't be any peace).
>
> *Ufuna ukugcinwa yimi (*You want me a woman to look after you)
>
> *With no money and food, there is no love and peace but with money, there is love even in the bedroom*

One of the female health workers pointed out that society views males as providers of families and that males who failed to live to this expectation lost respect and were viewed as useless. She mentioned that in one of their health counselling lessons, one woman openly declared that if she fends for the family, no one should restrict her in terms of time and movement. The angry woman went on to say that men had become useless to the extent of eating food brought by women without questioning its source.

> If I bring food, *uphongukukhamisa umlomo* (he just opens his mouth and eat without asking where the food came from)

In years when there is serious drought or incessant rains, women and men are forced to turn to other economic activities as *maricho* (short-term employment) workers for richer families. Women work as till operators in shops or as house cleaners. Some women engage in cross-border trading or sell goods to illegal miners.

This is usually shunned by the community and resisted by males. This makes women vulnerable to illegal gold miners, business people, and professionals who employ them. The illegal gold miners were reported to be abusive; they could refuse to pay, take goods by force, beat, stab, or kill anyone who crosses their paths. Women risked their lives and those of their husbands. Such working environments were insecure.

Men were also forced to take up alternative paths for livelihood even if these were risky. These were illegal gold mining and illegal migration where they risked their lives and those of families. Circumstances were forcing them to allow their wives to also take up these activities. Denying the woman employment in these circumstances resulted in conflicts, and refusing to take up risky alternatives was not the best option either. It is the equivalent of men caught between a rock and a hard surface. One middle-aged man from *Ntabamhlophe* shook his head and gave this comment:

> *To allow my wife to work for another man, a business person, hmmm. . . . nci... nci . . . nci its difficult. At the same time denying her leads to domestic conflicts.*

New alternative economic activities tend to separate the two genders. Their economic activities differ and the profits are not transparent. This creates suspicions and, consequently, conflicts. In illegal mining and piece jobs, men and women do not know about the quantum of money they respectively receive or spend. Women also sold second hand clothes (*mabhelo*) and craftwork (*madoyili*); men are unable to know about such profits. They work in separate places, carry out their own respective activities, and make separate decisions. This sets notions that the other spouse had money but did not want to use it to support the family. Community leaders were also of the idea that climate change impacts seem to have liberated men from agricultural activities that forced them to work with women around the home. It has given men independence to work away from home earning proceeds unknown to each other as couples. Community leaders also confirm that most of the conflicts they dealt with revolved around money, resources, and times of returning from work, suspicions related to work environments, and failure to fulfill duties of bread winning. Some women were day house cleaners in teachers' cottages or nurse aids to the neighbouring sick. Their husbands suspected them of having extra marital affairs with teachers, business people, or male neighbours who hired them as nurse aids or domestic workers.

Shifting gender roles as a result of climate change impacts worsened gender conflicts and risks in Lower Gweru. More women shifted gender roles than men. Women left husbands at home for work, and returned back home to do all household chores. Women expected to find their husbands having fetched water, firewood, and prepared meals for school children. Men and the community are still strict with gender division of labour, especially reproductive work. Women who suggested or questioned men on such issues were viewed as rude and unrespectful. The feelings of males were captured below:

> *Umfazi angasebenza uyakudelela* (If a woman works, she becomes disrespectful)
>
> *Ufuna ukukuhlala ekhanda akugezisise imiganu* (Such women sit on your head and make you clean the dishes).

A health worker also felt that unemployed men staying with working women were unhelpful, citing that climate change impacts have left males with few roles to play in their homes or the community. In her words, fathers do not do their roles.

> *Imizi ithwelwe ngabafazi* (It is women who have families on their shoulders)
>
> *To tell the truth; omama babhensile* (women are carrying heavy burdens)
>
> *Bahlalele abantwababo* (endure for the sake of their children)
>
> *There is no love, no peace of mind. Couples have different beds and different blankets.*

Respondents around *Maboleni* area cited that climate change induced shortages worsened gender conflicts. It was pointed out that firewood was available only at places far away and hard to come by. Water sources were now fewer; therefore, hours taken in search of these resources were longer. Some women and girls were insulted or beaten up for coming late or taking longer hours. One 40-year-old woman mentioned that her husband counts minutes and hours. In *Ntabamhlophe*, grazing lands, firewood, and water sources were not a serious problem, but men and women used shortages as excuses to get time for unfaithful errands.

In times of shortages, power inequalities work in favour of males. Women reported that in the face of climate disasters and threatened livelihoods, men decided on crops to be grown, places, quantity, times of planting, and purposes of crops. Sometimes, males would plant their crops or vegetables first when the weather conditions were still favourable at the expense of women's crops and vegetables. Gender conflicts would then ensue. Most resilient crops grown by women are for subsistence, for example, drought-resistant small grains, melons, water melons, green and yellow pumpkins, sugar cane, and indigenous vegetables among others. Either women would fail to get land to plough these or they are planted late into the season. These were sources of conflict between couples during the cultivation season. Traditional leaders confirmed that most of the cases they attended to, had to do with land ownership and use, money for inputs, failure to care for resources especially seeds, changing gender roles, and failure to meet gender role expectation as well as conflicts over decision making.

4.3 Response Mechanisms to Foster Resilience Against Climate Change Disasters, Risks, and Insecurities of Men and Women in Lower Gweru

Participants cited that there were no response mechanisms to assist Lower Gweru community to cope and adapt or to foster resilience to climate change disasters. This was true of all agencies including government, non-governmental organisations or community leadership, and other development partners. Households or individuals had their own coping or resilience strategies against risks and insecurities but these did not go very far because of environmental, socio-cultural, and economic constraints.

They had not received any assistance in terms of building dams or sinking of boreholes. The latter were sunk in institutions like schools and clinics for community and household water use not for projects. One traditional leader noted that at some point they had been promised that they would be assisted with transport to ferry their fruits to town or with the setting up of a canning industry, but the idea never materialized. Individual households had their own strategies of resilience. Women practiced intercropping where varieties of seeds are planted on one piece of land to avert detrimental effects of droughts or floods. Drought-resistant crops gave better yields during dry seasons and those with high affinity for water thrived during the wet seasons. However, women sometimes do not have decision-making power over land use, types of crops grown, and the seeding plans. It depends on the generosity of the male owner of land.

One woman from *Maboleni* related an incident where she planned to use organic fertilizer for her crops but was over-powered by the husband who used inorganic fertilizer everywhere. She consequently lost all her leafy crops, melons, and varieties of pumpkins. Preparation of land by ploughing immediately after harvest to trap water is a useful method for resilience, but all this can be decided by males when they see the need. Women do not make decisions over reduction of livestock, for example, selling some to remain with a few to be given food supplement. Resilience for women is linked to gender relations. Some men dig trenches to conserve water and to make watering lighter, but for women digging trenches is a hard labour. Limited resource base makes it difficult for people to practice crop rotation or to get technology for water conservation. They also try to settle and cultivate on higher ground to avoid swampy areas, but land shortage is a barrier. Women try to dry fruits and vegetables, but quantities are reduced and the market value of the products goes down.

Councillors pointed out that food aid from the government, and earlier from ADRA and Care International, has assisted in coping with especially food risks. However, these institutions have never trained or educated the community on climate disaster risks management. That knowledge was made available to large-scale commercial farmers. Local leadership, together with development partners, dealt with manifestations of climate change risks, starvation, malnutrition, domestic and gender-based violence as well as poverty, thereby treating symptoms rather than causes. Modern or indigenous knowledge of weather forecasting has not been availed to the community. Only a few benefit from indigenous knowledge systems. Some felt indigenous knowledge non-scientific, and thus unreliable. Shifting settlements away from disasters was hampered by land shortage. Poverty makes it difficult for community members to fortify themselves against risks and vulnerability.

5 Conclusions

The study concludes that men and women in Lower Gweru are hard hit by climate change disaster risks, especially through drought and incessant rains from cyclones. However, women feel the greatest impact because of unequal power relations and gender role expectations. Climate change disasters accentuate gender conflicts as both try to negotiate their space for resilience in the face of environmental threats. The structural factors of socio-ecological systems, that is, lack of policies on disaster risks management and cultural and economic arrangements of society are a barrier to resilience building. Both men and women find it difficult to foster resilience against vulnerability, risks, and insecurity. The latter however are more vulnerable. Currently, there are no direct response mechanisms or frameworks for resilience developed by the government, non-governmental organisations, and development partners dealing with climate change, risks, vulnerabilities, and insecurities. These tend to focus on warnings and evacuations during floods or food relief in times of hunger from drought. Police officers, lawyers and community leaders treat symptoms of climate change disasters like hunger, gender violence and conflicts, and quarrels over land disputes. Vulnerabilities, risks, gender conflicts, and insecurities are structurally linked to climate change disasters. Resilience strategies depend on economic, cultural patterns, and values of the socio-ecological systems of the community. Response mechanisms that lack structural transformation cannot reverse the undesired trends perpetuated by climate change disasters. There is an urgent need to incorporate socio-ecological systems and gender analysis into climate change disaster frameworks. New and expanded frameworks that consider interactive systems are quite necessary.

Way Forward

Frameworks dealing with climate change and gender should mainstream climate change disaster risk reduction and resilience. Disaster risk resilience should take into consideration environmental, socio-economic, and cultural factors. There should be a participatory approach where all stakeholders contribute to resilience building. These include government, non-governmental organisations, development partners, community leaders, men and women. Both genders should be empowered with information on climate change disasters, preparedness, gendered impacts, and approaches to minimise risks and insecurities associated with disasters. There is also a need to be gender-conscientised on differential impacts and socio-cultural constraints on resilience. Economically, men and women need to be empowered to access capital and other resources to enable them to be resilient. Government and other development partners, like agricultural extension officers and community leaders, should be educated on climate change disaster risks and gender frameworks in order to be able to assist men and women in community resilience strategies.

Below is suggested a coordinated model or framework for gender resilience to climate change disaster risks (Fig. 1). It merges frameworks pertained to gender, climate change disaster risks reduction, and resilience. It could be more robust than the Hyogo and Sendai frameworks because it considers gender analysis frameworks and environmental, economic, socio-cultural, and stakeholder participatory approaches.

The strategies should focus on structural factors ranging from gender mainstreaming policies, climate change disaster and gender resilience frameworks. Environmental, socio-cultural and economic factors should be taken into consideration so that the end receivers become resilient men and women with resilient livelihood.

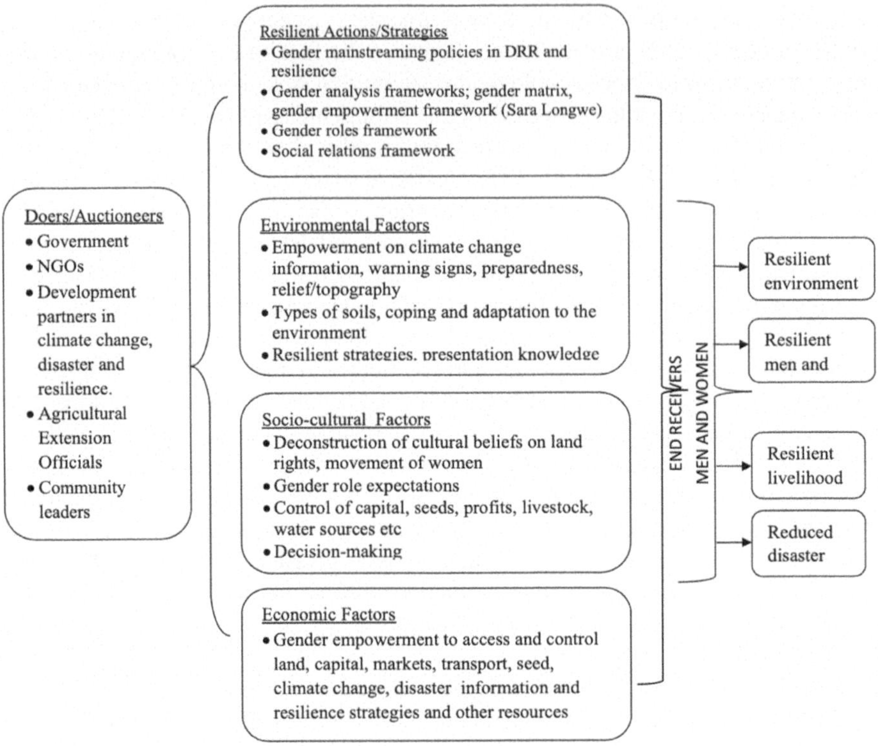

Fig. 1 Blended gender analysis frameworks and climate change risk reduction and resilience framework. *Source* Developed by the author

They should be able to manage the environment and to cope with impact of climate change disaster risks and insecurities.

References

Adams V (2011) Aging disasters: mortality, vulnerability and long-term recovery among Katrina survivors. Med Anthropol 30(3):247–270

Adaptation Learning Programme for Africa (ALPA) (2013) Gender Dynamics in a changing climate: how gender and adaptative capcity affects resilience. Care

Amnesty International (2011) The gender trap, violence and poverty. Amnesty International, London

Bradshaw S (2013) Gender, development and disasters. Edward Elgar Publishing Ltd, Massachusetts

Berkes F, Folke C (1998) Linking social and ecological systems: managing practices and social mechanisms for building resilience. Cambridge University Press, Cambridge

Brody A, Demetriade S, Esplen E (2008) Gender and Climate Change: lMapping the linkages. A Scoping study on knowledge gaps. University of Sussex. DAID, Bridge

Brown D, Chanakira RR, Chatiza K, Dhliwayo M, Dodman D, Masiwa M, Muchadenyika D, Mugabe P, Zvigadza S (2012) Climate change impacts, vulnerability and adaptation in Zimbabwe. IIEDCC. Workings paper no 3

Care International (2011) Climate change, vulnerability and adaptative capacity in Garisacounty. https://insights.careinternational.org

Dankelmam I (2002) Climate change: learning from gender analysis and experiences of organising for sustainable development in J. Masika, Gender, development and climate change. Oxfam, Oxford

Dhliwayo O (2019) Social and economic dimensions of cyclone Idai on women. Zimsentinel News. Accessed 29 Mar 2019

Doku PN (2012) The Mental Health of orphans and vulnerable children within the context of HIV/AIDS in Ghana. University of Glasgow: Mental Health and Well Being Institute

European Union (2016) Implementing the Paris agreement: Issues at stake in view of the COP 22 climate change conference. European Union, Paris

Food and Agriculture Organization (FAO) (2013) Gender and climate change research in agriculture and food security for rural development. Training guide. FAO. www.fao.org

Folke C (2006) Resilience: The emergence of a perspective for socio-ecological systems analysis. Glob Environ Chang 16(3):253–267

GGCA (2016) Gender and climate change in Africa, Facts from gender and climate: a closer look at existing evidence. GGCA. Factsheets

Habtzion S (2013) Gender and disaster reduction. Policy brief Asia and the Pacific. UNDP

Heise LL (1998) Violence against women: an integrated, ecological framework. Violence against Woman 4(3):262–290. https://doi.org/10.1177/1077801298004003002

Holden S, Ali DA, Hilhorst T (2016) A land tenure module for LSMS. Paper prepared for presentation at the 2016 World conference on land and property. The World Bank. Washington DC, 14–18 March 2016

IPCC (2014) Gender and Climate Change in Africa: Facts from Gender and Climate Change: A closer look at existing evidence. Fact sheet

KARAT Coalition (2013) Climate change: Women in developing countries, the hardest hit, Policy Brief. https://www.karat.org/wp-content/uploads/2012/02/Brief-on-Gender-and-climate-change.pdf

Koontz TM, Gupta D, Mudliar P, Ranjan P (2015) Adaptative institutions in Socio-Ecological Systems, Governance: A Synthesis Framework. Environmental Science and Policy. https://daneshyari.com/article/preview/10504581.pdf

Masika R (2002) Gender, development and climate change. Oxfam, Oxford

Masson VL, Lim S, Budimir M, Podboj JS (2016) Disasters and violence against women and girls: can disasters shake social norms and power relations. Overseas Development Institute, London

Murillo MQ, Tan S (2017) Discovering the differentiated and gendered consequences of natural disasters on the gender gap and life expectancy in SE Asia. journalnet. Hazardsearth system

Peters K, Peters LER, Twigg J, Walch C (2019) Disaster risk reduction strategies. Working paper

Stephens C (2013) Disaster risk management and hurricanes in the Caribbean: A gendered analysis. Working paper series no 7

Robie JM, Cheung SL, Roy S, Van H, Harmann OFR (2007) Report on excess mortality in Europe during summer 2003. EU community Action progress for Public Health. http://ec.europa.eu

Rohr U (2008) Gender aspects of climate induced conflicts in environment and conflict. Special edition news letter on Gender, environment and conflict. Adelphi research. http://www.eco-platform.org

Swedish International Development Cooperation Agency (SIDA) (2018) The relationship between climate change and violent conflicts. SIDA. Working paper

United Nations (2015) Framework on climate change. UN, Paris

United Nations Development Programme (UNDP) (2016) Gender climate change adaptation and disaster risk reduction. Training module 2

United Nations Framework Convention on Climate Change (UNFCCC) (2010) The need for adaptation. Fact sheet
United Nations Development Fund for Women (UNIFEM) (2015) Masculinity and GBV. Fact sheet. Unifemsheet
United Nations Office for Disaster Risk Reduction (UNISDR) (2017) Terminology-UNISDR. www.unisdr-org/owe/infor/terminology
UNDP (2014) Human Development Report 2014, Sustaining human progress: Reducing vulnerabilities and building resilience. http://hdr.undp.org
Wedeman N, Petraney T (2016) Invest in women to tackle climate change and conserve the environment. womendeliver.org
World Health Organization (WHO) (2010) Gender, climate change and health. WHO. www.un.org
Sander N, Lejano R, Aliposa J, Casas EJ, Rahman MS, Hossain L, Kabir L (2018) Challenge Fund Report on Gender and Extreme Weather Events. World Bank Global facility for Disaster Reduction and Recovery. https://www.gfdrr.org/sites/default/files/publication/Addendum%20D%20-%20Report%20on%20Gender%20and%20Extreme%20Weather%20Events.pdf

Resilience of Social-Ecological Systems: At the Limits of Hegemonic Masculinity

Mukoni Manuku

Abstract A key challenge of the Anthropocene era is to advance human development without crossing ecological thresholds or undermining critical ecological services. The purpose of this chapter is to show that we cannot hope for social-ecological resilience, viability, and stasis if the underlying value systems of society, especially hegemonic masculinity tendencies, remain untouched. The chapter underscores that hegemonic masculinity tendencies have a larger ecological footprint that is responsible for the emergence of many social-ecological risks and emergencies, which cause certain dynamics of vulnerability, insecurity, and crises for both social and ecological systems. The chapter observes that although attempts have been made to integrate a gender perspective into social-ecological resilience analysis, much of this research has focused on the understanding of the interdependent relations between social-ecological systems and gender. This analysis transcends such paradigms to argue that it is not enough. What is needed is not a mere incorporation of gender analysis but the transformation of hegemonic masculinity value systems given its significant influence on socio-ecological resilience. Furthermore, the analysis shows that it is the same hegemonic masculinity tendencies that contribute to the incapacity of society to manage risk dynamics and reduce vulnerability. The analysis refers to Raewyn Connell's (1987) theory to show how hegemonic masculinity acts as a structural and systemic driver of social-ecological systems insecurity, vulnerability, and risks. The chapter proposes a shift to a gender transformative paradigm in social-ecological resilience that targets hegemonic masculinity.

Keywords Hegemonic masculinity · Socio-ecological resilience · Ecological footprint · Ecological services · Carbon footprint

M. Manuku (✉)
Midlands State University Gender Institute, Gweru, Zimbabwe
e-mail: mukonim@staff.msu.ac.zw; mukonim@msu.ac.zw

M. Behnassi et al. (eds.), *Social-Ecological Systems (SES)*,
https://doi.org/10.1007/978-3-030-76247-6_5

1 Introduction

This chapter explores the intersections of hegemonic masculinity with the resilience of social-ecological systems. It seeks to contribute to the debate on structural drivers of vulnerability, risks, crises, and insecurity in such systems. The chapter acknowledges that while several researches in the past tried to link social-ecological systems with gender, the focus has been largely on the understanding of the interdependent relations between the social-ecological systems and gender. The impact of hegemonic masculinity on social-ecological systems has been given in most cases a cursive and anecdotal attention. Although, this analysis is largely based on a literature review, its style of writing invokes some empirical data gathered during my PhD research between 2014 and 2016 in the rural communities of Beitbridge district in Zimbabwe. Employing qualitative methods, the study focused on women's participation in community environmental education. Data generation was through individual in-depth semi-structured interviews, focus group discussions with women and men in the communities as well as participant observation of live community environmental education sessions facilitated by an organization based in Beitbridge. The data reported in this chapter did not directly address the issue of women's participation in community environmental education per se, but it emerged during the process of the study. The analysis presented here is motivated by this data.

It is widely acknowledged in literature that while women have played key roles in organizing against the causes of environmental and climate change, there have been little interrogation of how different forms of masculinity shape and influence men's contributions to environmental challenges and their response mechanisms (MacGregor and Seymour 2017; Pease 2016). In support of this observation, Pease (2016) further argues that most masculinity theorists have neglected a critical interrogation of men and the natural world, hence it is high time for critical masculinity scholars to seriously interrogate the environmental consequences of masculinity and men's practices. In concurrence, MacGregor and Seymour (2017) argue that what remains under-addressed are the myriad ways in which masculine roles, identities, and practices shape the human relationships with the more than world-men and nature. It is against this backdrop that this analysis seeks to bridge the gap by considering and challenging the ways in which hegemonic masculinity exacerbates social-ecological risks, crises, and insecurities.

To provide an entry point to the theoretical debates and frameworks that the chapter seeks to contribute to, the concept of hegemonic masculinity will be firstly unpacked. This is followed by an analysis of the linkages between hegemonic masculinity and social-ecological systems risks and insecurity. The latter section presents an analysis of the existing response mechanisms and assesses their ability and inability in fostering the resilience and viability of social-ecological systems. It was observed that these mechanisms are underpinned by the same value systems that promulgate and propagate the attitudes and practices which increase the vulnerability of social-ecological systems, therefore warranting the need for transformative approaches to address the effects of hegemonic masculinity. Following this, the usefulness of the transformation of hegemonic masculinity as a paradigm shift for moving social-ecological systems to higher levels of viability and resilience is mapped. After presenting these response mechanisms, the chapter demonstrates how hegemonic masculinity as a value system can be transformed to move social-ecological systems to viability and resilience. It is underscored that hegemonic masculinity tendencies have a larger ecological footprint that is responsible for the emergence of many social-ecological risks and emergencies, which cause many of the dynamics of vulnerability, insecurity, and crises for both social and the ecological systems.

2 Theoretical Framework: Raewyn Connell's Theory of Masculinity

To understand how the power dynamics emanating from hegemonic masculinity value systems limit the resilience of social-ecological systems, the chapter adopts Raewyn Connell's (1987) theory of hegemonic masculinity as a theoretical lens. Raewyn Connell is an Australian sociologist who first coined the concept of hegemonic masculinity. According to Connell (1995), patriarchal norms of masculinity can also be called hegemonic masculinity. She claims that hegemonic masculinity refers to the configuration of gender practice. This embodies the currently accepted answer to the problem of the legitimation of patriarchy which guarantees the dominant position of men and the subordination of women (1995). Connell (2005) and Schleef (2010) define hegemonic masculinity as referring to the current dominant form of masculinity which is a culturally idealized form of behaviour that creates and reinforces institutions of male dominance over women and other men. According to Connell, masculinity therefore refers to simultaneously a place in gender relations, the practices through which women and men engage that place in gender, and the effects of these practices in bodily experience, personality, and culture. While, Ratele in Porter (2013) conceptualises hegemonic masculinity as an amalgamation of social behaviours and actions that produce gender-based hierarchies and provide the impetus to further reinforce such hierarchies. The notion of hegemonic masculinity held that a specific configuration of social practices – not just a set of expectations or an identity – allows men to perpetuate power over and subjugate women (Porter 2013). It is clear that hegemonic masculinity is not biological, but an ascendancy achieved through culture, institutions, and persuasion (Hacker in Connell and Messerschidt 2005). This notion shows that hegemonic masculinity is a social construct that can be contested and deconstructed such that new and alternative, less oppressive and un-hierarchical or egalitarian forms of masculinity could perhaps arise and become hegemonic (Porter 2013). Although most men do not measure up to it, they rank themselves against it, and enact and play out what Connell termed complicit masculinities. The conception further upheld that those men who do not overtly demonstrate masculine dominance, but nonetheless benefitted from its patriarchal dividend, supported hegemonic masculinity despite the fact that they enact complicit masculinity. It is assumed that not all men acted out hegemonic masculinity; indeed it maintained that perhaps only a minority did so but that all men aspired to its standards which included competitiveness, independence, and bravery that involve taking risks for a successful life.

These archetypes of an ideal man as upheld in the conceptualization of hegemonic masculinity are the underlying causes of today's social-ecological systems risks, crisis, and insecurities. For example, today ecology is under threat because men and women are no longer exercising a felt need but greed to police their valued competitive standard. Schwalbe cited in Pease (2016) argues that hegemonic forms of masculinity promote an expectation that men enact control over themselves, other men, women, and the environment. Connell's (1987) theory of hegemonic masculinity conceptualised masculinity as plural, contested, and changing over time. Informed by this notion in this chapter, I consider and challenge the ways in which hegemonic masculinity limits the achievement of social-ecological resilience as well as how it can be contested for the viability and security of social-ecological systems. Having outlined the theory of hegemonic masculinity, the below section examines how enacting and playing out hegemonic masculinity impacts social-ecological systems.

3 Impacts of Hegemonic Masculinity on Social-Ecological Systems

This section presents and discusses the interactions of hegemonic masculinity with social-ecological systems. It attempts to tease out the impacts of hegemonic masculinity on social-ecological systems and show how such impacts hinder the resilience of such systems. Four themes emerged from the data and the analysis of literature on these interactions as discussed later under this section. A conceptual clarification of the concepts that underscored this analysis is presented as a forerunner.

3.1 Definition of Key Concepts and Literature Review

The concept of 'social-ecological systems' was first coined by Berkes and Folke in 1998 to connote the interactions between humans and nature. It links social and ecological systems (Janssen 2011). This concept recognises that people are part of nature and depend on ecosystems for their survival (Folke et al. 2016) while impacting ecosystems. For example, the current global food system as a social-ecological system is highly impacted by biodiversity loss and climate change through changes in temperature, precipitation, extreme weather events and sea level rise. The global food system, in turn, is a major driver of biodiversity loss and climate change given its higher carbon emissions, depletion of freshwater and land resources, increased pollution of aquatic and terrestrial ecosystems, deterioration of public health, increased mortality and animal suffering. Similarly, many risks affecting social systems such as poverty, hunger, conflicts, displacements, injustices, inequalities are increasingly induced by ecological system dynamics such as climate change. According to Marshall (2015), the social-ecological system framework was developed to support communication across multiple disciplines concerned with sustainable provision and/or appropriation of common-pool resources (CPRs).

Resilience and vulnerability characterize social-ecological systems. Emanating within natural science research, the concept of 'resilience' was first coined by Holling (1973) who viewed resilience as the amount of disturbance a system can absorb and yet remain within the same state or domain of attraction (Holling in Leslie and McGabe 2013). Since this groundbreaking conceptualization of resilience, it has been defined differently in available literature and its definitions are evolve and vary according to the discipline involved. For example, from a social science perspective, Adger in Leslie and McCabe (2013) conceptualized social resilience as the ability of human communities to withstand shocks to their social structure arising from environmental variability and economic or political upheaval. Walker and Salt (2006) define resilience as the capacity of a system to absorb disturbance and re-organize while undergoing change in a manner that allows for the persistence of system function, structure, and feedbacks. Broadening this definition of resilience, Folke (2006) adds a more explicit concern with flexibility and adaptive capacity by upholding that resilience is taken to mean more than simply the persistence of ecological relationships or of social structure and identity, but also the adaptive capacity to respond to opportunities and constraints that are presented by perturbations. This definition invokes the notion that resilience is synonymous to sustainability. Scoones (2016) claims that sustainability, or alternatively resilience, is the ability of a system to bounce back from shocks and stress to return to stable equilibrium states.

Although the original conceptualizations of the terms were scientific, conservative and technical focusing on a return to a status quo (Scoones 2016), the realization that the effects of the social and environmental factors mutual influence each other and are mediated through social, cultural, and economic structures and processes had given the impetus for social analyses of these concepts. Their analysis in relation to social policy for example has become increasingly recognized and gained prominence. This provided the basis for such frameworks of analyses as the social-ecological systems approach and the social-ecological system resilience approach. For, example, the latter with its origin in systems ecology, offers a field of scholarship of relevance to planning at a time when global ecological challenge requires urgent attention.

In the context of social-ecological systems, 'resilience' is the capacity of a social-ecological system to absorb or withstand perturbations and other stressors such that the system remains within the same regime, essentially maintaining its structure and functions (Holling 1973; Gunderson and Holling 2002) capable of offering its ecological services to humankind. Walker et al. in Kawarazuka et al. (2017) define the 'resilience of social-ecological systems' as the capacity for interrelated ecological and social systems to absorb or adapt to shocks or stressors without changing state.

It describes the degree to which the system is capable of self-organization, learning, and adaptation (Leslie and McCabe 2013). When the resilience is enhanced, the system is more likely to tolerate disturbing events without collapsing into a qualitatively different state that is controlled by a different set of processes. This shows that resilience is closely linked to sustainability. The latter involves maintaining the functionality of a system when it is perturbed, or maintaining the elements needed to renew or reorganize if a large perturbation radically alters structure and function. The ability to do this is termed 'resilience' (Walker et al. 2002). On the other hand, the vulnerability of social-ecological systems refers to both a system's exposure to disruptive shocks and trends and to its ability to prepare for, cope with, and recover from the impacts of such shocks and trends. Thus, adaptation has a temporal aspect: human systems can be adjusted in anticipation of change or as a reaction to its consequences.

Goal 15 of Sustainable Development Goals (SDGs) seeks to protect, restore, and promote the sustainable use of terrestrial ecosystems, sustainably manage forests, combat desertification, and halt and reverse land degradation and biodiversity loss (UN 2015). However, as long as the underlying systemic and structural value systems such as hegemonic masculinity are not transformed, this goal remains distant and unachievable. Many studies on gender and the environment have focussed on the situation of women (Arora-Johnson 2014; FAO 1998) and much less interest has been on the male aspect of environmental politics (Gaard 2014). For those who tried to address the issue of men and environmental politics, the concern was on the gender-technology-environment nexus (Alaimo 2010). A few researches also focus on the gender roles and gender inequity in relation to the environment, technology, and development (Dankelman 2002; Gupta 2015). Much concern has been on the construction of hegemonic masculinity to examine how it is embedded in environmental policy (Anshelm and Hultman 2014; Gaard 2014). The studies on how hegemonic masculinity ideologies influence and impact social-ecological resilience remain marginal.

The few studies that have tried to bring analysis of gender and social-ecological resilience together did not link hegemonic masculinity ideologies with social-ecological resilience. For example, some of them focussed on social ecological resilience in the context of gender violence. Aregu et al. (2016) looked at the impact of gender blindness on social-ecological resilience. Kawarazuka et al. (2017) focussed on the challenges and opportunities for bringing a gender analysis together with social ecological resilience analysis. Delgado-Serrano (2018) focused on gender and cross-scale differences in the perception of social-ecological systems. This chapter argues that bringing a gender analysis perspective is not enough; instead it calls for the unsettling of hegemonic masculinity as a deep-seated value system of inequities for the social-ecological system to move towards viability and resilience.

3.2 Hegemonic Masculinity and Ecological Services

The concept of 'ecosystem services' refers to the transformation of natural assets into things that humans value. For example, fungi, worms, and bacteria transform sunlight, carbon, and nitrogen into fertile soil, keeping a sustainable nutrient cycle for the benefit of humankind (EJOLT) within ecological thresholds. However, contemporary human development practices informed by hegemonic masculinity value systems have in this era crossed many of the ecological thresholds, undermining these critical ecological services. For example, gender role expectations and the concomitant pursuit of wealth as a measure of the idealized hegemonic masculinity force men to engage in an array of practices. These, in turn, drive risk for both social and ecological systems in order to affirm their valued hegemonic masculinity, such as overproduction for commercialisation and overconsumption as they are viewed and expected to provide for their families. The way we produce and consume food, especially in northern and emerging countries, is considered as a human and environmental disaster (Pease 2016). For example, during participant observation 2, when the communities were being educated on sustainable harvesting of Mopani worms, the following excerpts were captured:

> *Traditional Mopani worm harvesting was done by women but nowadays, because it has been commercialised, men are interested and play a bigger role using unsustainable methods such as having to cut down the Mopani trees for easy access to the worms.*

One male participant argued in the session that:

> *These resources were given to us by God to sustain our livelihoods why do we need to conserve them when our lives are at stake.*

The argument raised in the excerpt is that the ecosystem is balanced in such a way that there are fewer high life forms which include humankind on the upper end of the food chain who are supposed to benefit from the ecological services at all cost. The excerpt confirms the forms of masculinity or masculinism as characterised earlier by Brittain cited in Pease (2016). These pertain to manhood in terms of the capacity to dominate others and the natural environment to achieve eco-balance. Similarly, Cornall in Pease (2016) refers to the masculine mindset as a dominator model which he regards as the cause of environmental problems and ecological disasters. Such attitudes and notions increase the risks and insecurity which threaten social-ecological systems, thus leading to many challenges for the humanity. These include food and health security and related systems in addition to the biodiversity loss due to overconsumption. These, in turn, impact socio-economic development and political integration with many implications for livelihoods, wellbeing, productivity, and regional and domestic security. This implies that, to address environmental destruction, we must challenge the institutional power of culturally powerful men in addition to the internalization by men of the dominator model of human consciousness. Moreover, scientists predict that the hegemonic masculinity also drives the carbon footprint as illustrated below.

3.3 Hegemonic Masculinity and Carbon/Ecological Footprint

Carbon footprint is a measure of the total carbon dioxide emissions that directly or indirectly arise from anthropogenic activities. Hegemonic masculinity gives men more privileges and privileged people have larger carbon footprints (MacGregor and Seymour 2017). The assumed characteristics associated with hegemonic masculinity have negative implication on the social-ecological system resilience. For example, during participant observation 8, a group of community members were being taught about land degradation and mitigation strategies. When the facilitator challenged the members to consider selling some of their cattle and goats as a way of reducing overgrazing which was causing land degradation, the following sentiments were heard:

> Ndidopfi ndimunna de kana ndisina danga, uyakundwa ngavhanwe vhanna, dzingatori dzitshiya dipeni dzidzothe ngauanda ndihone uri vhathu vhauofhe vhauthonifhe ngauvhani ukhovha umuna vhukuma kana unasambi lihulwane *(How can I be considered a man without a kraal.*[1] *You will be outcompeted by other men. As a man you must have a large herd such that during the days when they go for dipping you are reserved a day for your herd alone due to numbers. This will make people fear and respect you because you will be a real man symbolised by a large herd of cattle.*

The excerpt above reflects the hegemonic value systems which entail competition to earn respect due to the accumulation of wealth at the expense of environmental protection. Men take what it costs to protect their masculinity and uphold their valued masculinity at the expense of the resilience of social-ecological systems. Similarly, Anshelm and Hultman (2014) argue that some forms of masculinity are so fragile that they have to be affirmed and reproduced. This affirmation makes men skeptical about climate science as a threat to their masculinity. A similar observation was made by Pease (2016) who argues that men are thus more concerned with the protection of a particular form of industrialised and patriarchal society that affirms their masculinity and their class privilege than the protection of environment. With such mindsets, it becomes very difficult for society to benefit from ecological services without undermining the critical ecological thresholds. Keeping large herds of cattle for example increases the carbon footprint for these communities due to emissions of carbon dioxide. This concurs with an earlier observation made by Daly in Amuson and Olutula (2017) who draws particular attention to the destructive nature of men as the main cause of many problems that confront humanity today including ecological crises – such as global warming and mineral extraction – and their attendant socioeconomic issues. Dankelman (2002) noted earlier that climate change was framed as a technical problem that ignored the social and political dimensions of human induced changes in the climatic systems. Yet, Masterson et al. (2019) have argued that responding to sustainability issues requires understanding their roots in intertwined social and ecological system. There is a need to better understand women's situations in the context of their interactions with social-ecological systems. They are often responsible for collecting firewood, fodder, and water in many rural areas of the

[1] Term borrowed from the shelter of cattle in Zimbabwe used here to connote a bigger herd of cattle.

world in addition to their partial role as caretaker of home and children. This is why women are more disadvantaged during natural disasters (Fordham 2003) and bear the real brunt of a changing climate (Gupta 2015). Global discussions are now stressing women engagement in decision making and climate negotiations (UNFCCC 2012; WEDO 2012) which may challenge hegemonic masculinity in the long run.

'Ecological footprint' simply means the impact of human activities on the globe and the amount of resources necessary to produce the goods and services necessary to support a particular lifestyle, in terms of the area of biologically productive land and water. It also measures resources needed to produce what we need, what we consume, and dispose of. The concept of 'ecological footprint' was first defined and developed by Mathis Wackernagel and William Rees in the mid-1990s and it has since been implemented as an indicator for sustainability. It was defined as the ecologically productive territory needed to produce the resources used and to assimilate the waste and emissions produced by a population with a specific life mode indefinitely. The footprint concept highlights the need for a balance between the consumption and use of resources. It also represents the demands made on the ecosystem and natural resources. Researches show that women consume comparatively less resources, are less polluting, and usually have a lower GHG footprint than men (Johnsson-Latham 2006; Toro et al. 2019). Additionally, throat-cut competition emanating from hegemonic masculinity in society today has resulted in large ecological footprints on the planet. For example, if one person buys a land rover today due to the competition enshrined in hegemonic masculinity value systems, it triggers competition and everyone would want to compete, thus leading to more land rovers, resulting in a larger ecological footprint (Mukoni 2015). The same hegemonic value systems hinder the effectiveness of sustainability interventions.

3.4 *Hegemonic Masculinity and Sustainability Interventions*

Pease (2016) notes that for men to be vulnerable to external forces is a challenge to their masculinity and their sense of self as a man. This is because being in control and taking risks are the two of the key dimensions of hegemonic masculinity. Upholding such notions has serious implications for social-ecological systems. Franz-Balsen (2014) argues that hegemonic forms of masculinity are at odds with ecological sustainability. If this is so, then the promotion of environmental awareness and knowledge about the social impact of global warming will pose a threat to this form of masculinity. For example, during participant observation 5 on wildlife conservation for ecotourism, when facilitators were teaching the community on the need to conserve biodiversity for ecotourism, two male participants were heard commenting that:

Vhusina tshamuno avhuleyi (Without salt it's not edible).[2]

[2]This comment was a counter-argument to conservation practices meaning that *without meat food becomes inedible.*

Ndingasivhuthomi, ngamuroho zwinyagavho nama nga thungo (I can't start eating sadza[3] *with vegetables only. Vegetables must be accompanied with some meat*

The excerpts confirm Hultman's observation that men are a big part of ecological problems because they travel a lot, eat too much meat, and live in energy-intensive buildings (Hultman in MacGregor and Seymour 2017). Such practices are informed by hegemonic masculinity and have negative consequences for both social and ecological systems. Leduc in Pease (2016) argues that gender inequality is a part of unsustainable development which, in turn, drives climate change and increases natural disasters. I concur with Pease's (2016) argument that there are common structural causes of environmental degradation and gender discrimination, hence the realization of greater gender equity will most likely parallel more progressive environmental politics. However, as Pease (2016) argues, while this is often cited to promote the improvement in the status of women, less attention is given to what it means by progressive gender and environmental politics for men as discussed in the next section.

4 Current Strategies to Manage Hegemonic Masculinity Impacts on Social-Ecological Resilience

Strategies have been advocated within the perspective to neutralise the impacts of hegemonic masculinity on social-ecological resilience. For example, 'Add women and stir' strategies are a set of some relevant strategies. That has been proffered. For example, given that energy policies and programmes, fossil fuel companies, and funding mechanisms addressing climate change are all male-dominated, it is important to ensure that women have access to the key decision-making processes related to climate change (Dankelman 2002). In support of this view, Gonda (2015) explains that working towards gender equality in the context of climate change requires an investigation of how gender inequalities play out in the climate change policies and interventions. Research has, therefore, focused more on making policies and policy discourses gender-responsive by bringing women to the negotiation table. This has resulted in an increase in the number of women participating in climate change policy formulation forums, for example, in the hope that their ideas and preferences will be taken into cognisance. However, Torri (2010) and Naganag (2014) have long argued that getting women to the policy making table is not enough. Magnusdottir and Kronsell (2015) also argue that gender balance alone is not sufficient to ensure gender-sensitive policy making as masculine norms are deeply institutionalized in climate institutions.

This sidelining of women has negative implications for both social and ecological systems. Excluding women's ideas increases their vulnerability to socio-ecological risks such as floods and bush fires and exacerbates inequalities, including gender

[3] Is a local term referring to the staple food of these communities, usually prepared from mealie meal and is thicker than porridge.

inequality (Kameri-Mbote 2013). Similarly, when women are not involved or do not fully and equally participate, they resort to ineffective farming methods in an effort to support their families with food, a role that solely lies with them in many cases and for which they are primarily responsible (Mitra and Rao 2016). Studies by the FAO also highlight the role and conditions of women with respect to agriculture and farming (FAO 1998; FAO and Doss 2011) mentioning that the labour burden of rural women exceeds that of men, and includes a higher proportion of unpaid household responsibilities related to preparing food and collecting fuel and water.

Climate change adds to women's existing burden as it threatens households' food security, making women who are the providers of food in the household a crucial target group when it comes to dissemination of knowledge about adaptation and mitigation strategies. The lack of knowledge on climate change, for example, makes women resort to poor farming methods, thus deepening poverty levels. This is further exacerbated by the fact that environmental disasters impact more negatively women than their male counterparts. This is attributed largely to the traditional gender roles imposed by the patriarchal system of the society, which emphasizes masculine privilege (Mukoni 2018).

Aregu et al. (2016) in Ethiopia studied how the failure to take into account gendered roles in the management of communal pasture can affect the resilience of this social-ecological system. The study established that women are excluded from the informal institutions that define access and use rules which guide the management of the communal pasture. Consequently, women's knowledge, preferences, and needs are not taken into account. This negatively affects the resilience of the communal pasture because: firstly, the exclusion of women's knowledge leads to future adaptation options being overlooked; and secondly, as a result of the failure to address women's needs, they start to question the legitimacy of the informal institution. The case study thus shows how exclusion of women, by sidelining their knowledge and needs, weakens social learning and adaptiveness of the management rules. Being blind to gender related issues may thus undermine the resilience of a social-ecological system. Similarly, Stoddard and Tindall (2011) observed that notions of hegemonic masculinity are becoming salient as part of an interpretive framework.

The same sentiments were raised by Fleming (2017) who argues that intellectual deliberations on climate change are dominated by white Northern male voices. Yet, Raworth in MacGregor and Seymour (2017) argue that it is impossible to fully grasp the contours of the Anthropocene without understanding power relations within human societies and how particular cultural values and practices have become barriers to pro-environmental knowledge as the pro-environmental knowledge and action can itself replicate unequal power relations. The above analysis reinforces the view articulated in the early 1990s that the current patriarchal gender regime will need to change if we are to pursue a more sustainable world (Johnsson-Latham 2006). Gender inequality is accordingly part of unsustainable development, which in turn drives climate change and increases natural disasters (Leduc in Pease 2016). This warrants transformation of structural and systemic systems that tackle hegemonic masculinity values, attitudes, and practices. This calls for gender-transformative approaches in social-ecological systems resilience.

5 'Gender-Transformative Paradigm': A Flagship to Social-Ecological System Viability and Resilience

The previous sections show how hegemonic masculinity limits the realization of social-ecological resilience by being a driver of risks and insecurities that undermine such resilience. This section highlights and proposes the need for gender-transformative approaches to secure the resilience of interdependent social-ecological systems. While such approaches have been widely used in health and sexuality (Cornwall et al. 2011; Peacock and Barker 2014) as well as gender-based violence research and initiatives (Jewkes et al. 2015; Peacock and Barker 2014), little has been done in the area of resilience of social-ecological systems (Connell 2017). This literature shows how transformation of hegemonic masculinity can be operationalized for the improvement of social-ecological systems. Connell (2017) suggests that knowledge about masculinities will be valuable for environmental activism too. A gender-transformative approach seeks to engage men of all ages to purportedly transform masculinities (Porter 2013). Peacock and Barker in Jewkes et al. (2015) argue that women can also negatively reinforce gender norms for hegemonic masculinity. Building up on this argument, I suggest the use of collective action and community mobilization with both genders to explore and become conscious of their respective conditioning together about the social construction of hegemonic masculinity and its impacts on social-ecological resilience. This invokes the use of social learning in the communities to conscientise men and women on the effects of power-based gender performances. The pre-condition is a change in the mindsets of men and women about the notion that masculinity is a state to be achieved in addition to the change of current standard measures of hegemonic masculinity (Porter 2013). This can go a long way in exposing men and women to alternative masculinities that are egalitarian.

Jewkes et al. (2015) further notes that most gender transformative strategies in place are short term and technical. In the context of this analysis, the approach will entail the engagement of men and women in the communities in dialogue through participatory learning approaches that facilitate critical consciousness raising (Gibbs et al. 2014). This will in turn provide spaces that allow men and women to reflect on the negative impacts of hegemonic masculinity on social-ecological systems. Women and men need to be exposed to alternative definition of hegemonic gender performance that are egalitarian.

6 Conclusion

This analysis started with the exposition of hegemonic masculinity as espoused by Raewyn Connell, and then moved on to discuss the concept of 'social-ecological system' and the related concept of 'social-ecological resilience'. It further analysed how hegemonic masculinity acts as a structural driver that limits the realization of the resilience of social-ecological systems as it informs the values, actions, and practices that fuel the risks and insecurities affecting such systems. The analysis culminated in the conclusion that if the underlying value structures and systems of hegemonic masculinity are left untouched, we cannot hope for the resilience, viability, and security of social-ecological systems. The analysis finally examined how hegemonic masculinity can be transformed, and then concludes with a discussion on the importance of consciousness raising of men and women about egalitarian masculinities to move social-ecological systems from risks and insecurity to viability and resilience.

References

Alaimo S (2010) Bodily Nature: Science, Environment and the Material. Indiana University Press, Bloomington

Amuson L, Olutula O (2017) Contextualizing African women's empowerment in agriculture: challenges from climate change and mineral extraction perspectives. J Women's Stud 18(4):117–130. https://vc.bridgewedu/jiws/vol18/iss4/y

Anshelm J, Hultman M (2014) Discourses of Global Climate Change. Routledge, London

Aregu L, Darnhofer I, Tegegne A, Hoekstra D, Wurzinger M (2016) The impact of gender- blindness on social-ecological resilience: the case of a communal pasture in the highlands of Ethiopia. Ambio 45(s3):287–296

Arora-Johnson S (2014) Forty years of gender research and environmental policy where do we stand? Women Stud International Forum 47:295–308

Connell R (1987). Gender and Power, Society, the Person and Sexual Politics. Cambridge, Polity Press

Connell R (1995) Masculinities. Polity Press, Sydney

Connell R (2017) Foreword: masculinities in the sociocene. In: MacGregor S, Seymour N (eds) Men and Nature: Hegemonic Masculinities and Environmental Change, pp 1–5. Rachel Carson Centre

Connell R, Messerschmidt J (2005) Hegemonic masculinity: rethinking the concept. Gend Soc 19(6):829–859. https://doi.org/10.1177/0891243205278639

Cornwall A, Edstrom J, Greig A (2011) Men and Development: Politicizing Masculinities. Zed Books Limited, London

Dankelman I (2002) Climate change: learning from gender analysis and women's experiences of organizing for sustainable development. Gend Dev 10(2):21–29

Delgado-Serrano MM, Semerena RE (2018).Gender and cross-scale differences in the Perception of social-ecological systems. Sustainability 10:2983, 1–21.

EJOLT Ecosystem Services. In: Environ. Justice Organ. Liabilities Trade. http://www.ejolt.org/2013/05/ecosystem-services/. Accessed 3 Oct 2020

FAO (1998) Rural Women and Food Security: Current Situation and Perspectives. Food and Agriculture Organization of the United Nations, Rome

FAO and Doss C (2011) The role of women in agriculture. Food and Agriculture Organization

Fleming J (2017) Excuse us while we fix the sky: WEIRD Supermen and climate engineering. In: MacGregor S, Seymour S (eds) Men and Nature: Hegemonic Masculinities and Environmental Change, pp 23–28. Rachel Carson Centre

Folke C (2006) Resilience: the emergence of a perspective for social-ecological systems analyses. Glob Environ Change 16:253–267. http://dx.doi.org/10.1016/j.gloenvcha.2006.04.002

Folke C, Biggs R, Norström A, et al (2016) Social-ecological resilience and biosphere-based sustainability science. Ecol Soc 21https://doi.org/10.5751/ES-08748-210341

Fordham M (2003) Gender, development and disaster: the necessity for integration. In: Pelling M (ed) Natural Disasters and Development in a Globalizing World. Routledge, New York, pp 57–74
Franz-Balsen A (2014) Gender and (un)sustainability: can communication solve a conflict of norms? Sustainability 6:1974–1991
Gaard G (2014) Towards new ecomasculinities, ecogenders and ecosexualities. In: Adams CJ, Gruen L, (eds) Ecofeminism, Feminist Intersections with other Animals and the Earth. Bloomsbury Publishing, pp 225–239, New York
Gibbs A, Jewkes R, Sikweyiya Y, Willan S (2014) Reconstructing masculinity? A qualitative evaluation of the stepping stones and creating futures intervention in urban informal settlements in South Africa. Cult Health Sex 17(2):110–122
Gonda N (2015) Discursive analysis of environmental policy? Lessons learnt and new Perspectives revealing the patriarchal side of climate change through feminist critical discourse analysis: A case study from Nicaragua. In: Paper presented at the 10th International conference in interpretive policy analysis, pp 8–10 July 2015
Gunderson LH, Holling CS (2002) Panarchy: Understanding Transformations in Human and Natural Systems. Island Press, Washington, D.C.
Gupta H (2015) Women and climate change: linking ground perspectives to the global scenario. Indian J Gend Stud 22:408–420. https://doi.org/10.1177/0971521515594278
Holling CS (1973) Resilience and stability of ecological systems. Ann Rev Ecol Syst 4:1–23. https://doi.org/10.1146/annurev.es.04.110173.000245
Janssen MA (2011) Resilience and adaptation in the governance of social-ecological systems. Int J Commons 5(2):340–345
Jewkes R, Flood M, Lang J (2015) From work with men and boys to changes of social norms and reduction of inequities in gender relations: a conceptual shift in prevention of violence against women and girls. Lancet 385(9977):1580–1589
Jewkes R, Morrel R, Hearn J, Lundqvist E, Blackbeard D, Lindegger G, Quayle M, Sikweyiya Y, Gottzen L (2015) Hegemonic masculinity: combining theory and practice in gender intervention. Cult Health Sex 17(S2):S112–S127
Johnsson-Latham G (2006) Initial study of lifestyles, consumption patterns, sustainable development and gender: Do women leave a smaller ecological footprint than men?
Kameri-Mbote P (2013). Climate change and gender justice. International policy and legal responses. In: Ruppel OC, Roschmann C, Ruppel-Schlichting K (eds) International Climate law and Global Governance: Legal responses to a changing environment. Nomos Publishers, pp 323–348, Baden-Baden
Kawarazuka N, Locke C, McDougall C, Kantor P, Morgan M (2017) Bringing analysis of gender and social-ecological resilience together in small-scale fisheries research: challenges and opportunities. Ambio 46:201–213
Leslie P, McCabe JT (2013) Response diversity and resilience in social-ecological systems. Curr Anthropol 54(2):114–214
MacGregor S, Seymour N (2017). Introduction. In: MacGregor S, Seymour S (eds.) Men and Nature: Hegemonic Masculinities and Environmental Change, pp 9–14. Rachel Carson Centre
Magnusdottir GL, Kronsell A (2015). The (in)visibility of gender in Scandinavian climate Policy-making. Int Feminist J Polit 17:1–19
Marshall GR (2015). A social-ecological systems framework for food systems research: accommodating transformation systems and their products. Int J Commons 9(2):881–908. http://doi.org/https://doi.org/10.18352/ijc.587
Masterson VA, Enqvist JP, Stedman RC, Tengo M (2019) Sense of place in social-ecological systems: from theory to empirics. Sustain Sci 14:555–564
McCright AM, Dunlap RE (2011) "Cool Dudes": the denial of climate change among conservative white males in the United States. Glob Environ Chang 21(4):1163–1172
Mitra A, Rao N (2016) Families, farms and changing gender relations in Asia. In: FAO and MSSRF (eds) Family Farming Meeting the Zero-Hunger Challenge. Academic Foundation, pp 41–124, New Dehli

Morrell R, Jewkes R, Lindegger G (2012) Hegemonic masculinity/masculinities in South Africa: culture, power, and gender politics. Men Masculinities 15(1):11–30
Mukoni M (2015) Ecological sustainability: reinvigorate and emulate African principles of life or cease to exist? Int J Asian Soc Sci 5(9):514–521
Mukoni M (2018) Women's participation in community environmental education: a qualitative study of one community organization in Zimbabwe. Unpublished PhD thesis, University of Kwazulu-Natal, South Africa
Naganag EM (2014) The role of indigenous women in forest conservation in Upland Kalinga Province, Northern Phillipines. Int J Adv Res Manage Soc Sci 3(6):75
Peacock D, Barker G (2014) Working with men and boys to prevent gender-based violence principles, lessons learned, and ways forward. Men Masculinities 17(5):578–599
Pease B (2016) Masculinism, climate change and man-made disasters: towards an environmental pro-feminist response. In: Enarson E, Pease B (eds) Men, Masculinities and Disaster. Routledge, London, pp 21–33
Porter A (2013) What is constructed can be transformed': masculinities in post conflict societies in Africa. Int Peacekeeping 20(4):486–501
Schleef D (2010) Identity transformation, hegemonic masculinity and research in professionals. Sociol Compass 4(2):122–135
Scoones I (2016) The politics of sustainability and development. Annu Rev Environ Resour 41:293–319
Stoddart MCJ, Tindall D (2011) Ecofeminism, hegemonic masculinity and environmental movement, participation in British Columbia, Canada, 1998–2007: Women always clean up the mess. Sociol Spectrum 31(3):342–368
Toro F, Serrano M, Guillen M (2019) Who pollutes more? Gender differences in consumptions patterns
Torri MC (2010) Power, structure, gender relations and community-based conservation: the Cawswe study of the Sariska region Rajasta India. J Int Women's Stud 11(4):1–18
UN (2015) A/RES/70/1 - Transforming our world: the 2030 Agenda for Sustainable Development. Resolution adopted by the General Assembly on 25 September 2015. https://sustainabledevelopment.un.org/topics/mountains/decisions. Accessed 4 Apr 2020
UNFCCC (2012) Promoting gender balance and improving the participation of women in UNFCCC negotiations and in the representation of parties in bodies established pursuant to the convention or the Kyoto Protocol
Walker B, Carpenter S, Anderies J, Abel N., Cumming G., Janssen M. Pritchard R (2002). Resilience management in social-ecological systems: a working hypothesis for a participatory approach. Ecol Soc 6(1)
Walker B, Salt D (2006) Resilience Thinking: Sustaining Ecosystems and People in a Changing World. Island Press, Washington
WEDO (2012) Women's participation in UN climate Negotiations 2008–2012. Women's Environment and Development Organization, New York

Gendered Perspective: Climate Change Adaptation Strategy in Malawi

Linus Agbleze

Abstract Recently, the Least Developed Countries Expert Group (LEG) reiterated the need to stress the understanding of climate change adaptation from a gender perspective. This does not only call for integrating the gender dimension into existing conceptual frameworks, but also mainstreaming it to capture existing research gaps and drawing linkages on the themes of characteristics of climatic shock, attributes of affected systems and response characteristics, hence the focus of this chapter. Accordingly, this research set-out to contextualize the gender as a conceptual framework, test the hypothesis that gender is independent from an adaptation strategy and investigate the factors influencing the adaptation strategy. Chi-squared test of independence and multiple logistic regression analysis were employed. The chi squared test produced p-value as 0.216, hence no statistically significant association between gender and type of adaptation strategy was concluded. The regression analysis revealed that: women with a basic and advanced education are more likely to adopt autonomous adaptation strategy compared to women without education; access to information resulted in adoption of autonomous adaptation; and women with access to credit are less likely to adapt autonomously. Moreover, poor women are more likely to adopt autonomous adaptation strategy compared to rich women. Therefore, it is recommended that the adaptive capacity of women should be focused on in improving adaptation rather than external support.

Keywords Climate shocks · Adaptation strategy · Gender perspective · Malawi

1 Introduction

Proceeding from an unequivocal scientific evidence on climate change occurrence (Urry 2015), the impacts of climatic shocks are expected to be differentiated at different scales and by type of system (Smit and Wandel 2006). Countries in the

L. Agbleze (✉)
Institute of Environment and Human Security, United Nations University, Bonn, Germany
e-mail: agbleze@student.unu.edu

University of Bonn, Bonn, Germany

M. Behnassi et al. (eds.), *Social-Ecological Systems (SES)*,
https://doi.org/10.1007/978-3-030-76247-6_6

Global South, mainly sub-Saharan Africa, are expected to be the most impacted (Kotir 2011). This feeds into the argument that adaptation strategies will vary not only as a reflection of the climatic hazard concern but also of the available capacity to cope in an affected region. In an attempt to understand the disparity in the various manifestations of climate change, many researches have assessed both climatic shocks and responses and in recent times within a gendered perspective, despite the fact that this trend is still plagued with enormous research gaps. Rightly identifying this loop in research and bringing it to international recognition, the Least Developed Countries Expert Group (LEG) of the United Nations Framework Convention on Climate Change (UNFCCC) stressed in 2015 the need to understand climate change adaptation using gendered lenses.

Irrespective of the system or individual affected, climate mitigation and adaptation have been identified as the main actions that define the climate profile of a region, country, household or individual. With the realization that mitigation alone will not reverse climatic impacts in the foreseeable term (Althor et al. 2016), adaptation to climatic shocks and hazards is currently gaining an international recognition. In support, many relevant researches have concluded that adaptation is no more an option but a necessary strategy in response to environmental and climatic stimuli (Parry et al. 2007). Contributions have included theoretical studies of adaptation and development of assessment tools in addition to empirical researches which attempt to document and report adaptation responses and practices (McCarthy et al. 2001). While such studies have identified adaptation to be hazard-specific, critical understanding of how characteristics of climatic shocks relate to the type of adaptation strategy engaged remains a missed opportunity.

Early works in responding to this gap includes that of Smithers and Smit (1997) who synthesized available literature as conceptual input to guide adaptation research. By this, Smithers and Smit (1997) proposed studying adaptation by addressing the nature of climatic shock, attributes of system affected, and response characteristics as key themes. Existing literature has usually addressed these themes from many angles; however, the interpretation of existing conceptual frameworks reveals that the inter-linkages between the various themes are still unexplored. Moreover, this literature remains too broad, hence the need for a nuanced analysis to unravel the possible intricacies and feedback loops. Thus, this chapter attempts to fill this gap and aims at providing a precise contextualization through an empirical research.

Addressing the need for further conceptual development in the area of adaptation with a policy relevance implies that adaptation is conceptualized in a manner that yields understanding the roles of actors. In an attempt to analyze the adaptive response to inform policy and action, the criteria of type and role of actor is central, hence it is the focus of this research. This is discussed in the context where some adaptation measures are initiated by the central government and other actors in addition to the victims of climate change impacts (external adaptation). Measures implemented as response to these impacts by individuals directly affected are described as autonomous adaptation. As an external adaptation, Malawian women have received micro-finance and climate-smart agricultural support from the government as part of National Adaptation and Programmes of Action (NAPA), Climate Adaptation for

Rural Livelihood and Agriculture (CARLA), and Smallholder Crop Production and Marketing Project (AU 2012). These initiatives did not exclude men. Autonomous adaptation includes the use of personal savings and assets in adaption. The nature of climate shock is assessed based on suddenness; in this chapter, slow onset and sudden onset climatic shocks are distinguished. Proceeding from this approach, a good base is provided to draw a linkage and understand the relationship between the nature of climate shock and the type of adaptation strategy engaged. Further, a critical missing link identified in both conceptual and empirical understanding of adaptation is the factors influencing autonomous and external adaptation. This research seeks to fill this gap while mainstreaming the call by LEG, hence the gendered perspective particularly focusing on women.

Comparative assessment of climate change impacts on men compared to women – as in the works of Guloba (2014) and Huynh and Resurrección (2014) – has conclusively established that women are more impacted and less adapted. In essence, progress is made in the direction of focusing on women for a nuanced analysis in this chapter. For this purpose, the underlying factors that account for their adaptation level are investigated to understand how they influence and relate to their choice of adaptation strategy. This analysis is justifiable as Webb (2015) rightly indicates that women are not only unequally disadvantaged but have a unique and dynamic status. This contributes to differences in response to climate change impacts. Gender analysis focusing on women rather than a comparison between women and men will create a potent ground for realizing improved results from community-based adaptation interventions. Comparative assessment approach to gender analysis is likely to exacerbate inequality and provide fewer effective inputs when used as a basis for an adaptation intervention project. This is because it strives on an unwarranted assumption of uniform inequality within one gender group and generalization of women as the most marginalized (Webb 2015; Demetriades and Esplen 2010). For instance, while women are generalized to be engaged in unpaid jobs that limit them to undertake less expensive and simple adaptation measures, exceptions exist where some women with better education and control of assets have a better financial backing for adaptation.

An indepth analysis on how variance in income level or other factors explain women's choice of adaptation strategy will serve as a good basis to empirically understand how support and intervention programs for women will influence their adaptation strategy and pathways. To achieve precise and concrete understanding, research needs to be contextualized by clearly defining the scale and the hazard type considered and focussing on a gender group to properly intersect characteristics of gendered adaptation to climate change impacts. Such contextualization is duly expressed in this chapter with the climatic hazards identified as droughts and floods, and the scale of analysis as households across Malawi with a focus on women.

In an attempt to fill the identified conceptual gaps while satisfying the need for empirical evidence, this chapter set-out to test a number of hypotheses as part of its objectives. These hypotheses include the following: (i) there is a significant relationship between the type of climatic shock or hazard and the type of adaptation strategy adopted, and (ii) the gender is independent of adaptation strategy. In addition to testing these hypotheses, the objectives of this chapter include: devising and contextualizing a framework that captures the hazard and adaptation strategy nexus; and assessing how various socio-economic factors influence the type of adaptation strategy engaged.

Responding to the call by Smithers and Smit (1997) for further research using their framework as an organizational tool, their generic adaptation framework has been comprehended to tease-out aspects that allow linkages to be drawn. Such linkages were based on the nature of climatic hazard, characteristics of affected system, and the type of adaptation strategy. Moreover, available literature on conceptualization of adaptation has been carefully reviewed, and the analysis of their strength and weakness has been carried out. In this chapter, the human–environment interaction where households are impacted and responding to a changing climate is explored. With Malawi, identified as a hotspot in need of household and gendered assessment of adaptation, this chapter provides the empirical evidence of identified gaps. Females (nationwide), who experienced drought and flood, were selected for this analysis. Drought and floods were conceptualized as slow and sudden onset climatic shocks, respectively. With a focus on the human system, socio-economic variables such as educational level, subjective well-being, and access to means of information and credit were analyzed to understand their influence on the choice of adaptation strategy.

This chapter starts with weaving through the conceptual fabrics of climate change hazards and adaptation. It proceeds through the undertaking of a climate change hazard and adaptation profile analysis of Malawi and throws a spotlight on gender debate. Then, this debate is put into the proper perspective while providing empirical evidence which captures adaptation in Malawi. A critical contribution is the testing of a number of hypotheses that reflect key theoretical questions.

2 Conceptualizing Climate Change Related Hazards in Relation to Adaptation Response

Noting the inconsistent and conflicting use of terms in the literature in understanding adaptation to climate change, Smithers and Smit (1997) provided a conceptual framework on adaptation that synthesized the various perspectives on adaptation strategy. This was intended as an organizational tool to guide adaptation research while removing the identified conceptual challenges. Three key themes are distinguished from the framework and recommended for adaptation research: climatic shock or hazard characteristics; characteristics of affected systems; and the different shades of adaptation strategy (Kelly and Adger 2000; Ravankhah et al. 2019). This section focuses on the theme of climatic shock properties with discussions on its conceptualization.

Conceptualizing hazards based on their characteristics aims at understanding the level of impact experienced by victims and their response (Few 2007). Salient properties of climate change related hazards or impacts have consequential effects on human response. The magnitude, frequency, duration, suddenness, areal extent, and expected impact of a hazard have been the main criteria for differentiating between hazards by researchers and defines victims' response strategy (Parry et al. 1996; Wilhite 2000; Melching and Pilon 2006). Based on magnitude, climatic hazards of high magnitude produce impacts that are beyond the absorptive capacity of a system, hence the need for adaptation at a fast rate and large scale (Parry et al. 2007). Thus, understanding hazard based on magnitude offers the utility of undertaking an adaptation analysis which seeks to evaluate the amount of resources required and assess the adequacy of adaptive capacity. Moreover, the spatial scale of a hazard, being either localized or wide-spread, defines the type of adaptation strategy to be engaged (Thornton et al. 2019). Suggested by the Special Report on Managing the Risks of Extreme Events and Disasters to Advance Climate Change Adaptation (SREX 2012), localized hazard will require adaptation different from other regions.

Understanding climatic hazard based on frequency, duration, and suddenness explores the temporal dimension of hazards and provides a clue to decision makers' awareness (Smithers and Smit 1997; Dilley et al. 2005). Climatic hazards of high frequency are expected to be better expressed and easily observed; hence they are often considered by decision makers. This also feeds into the notion that adaptation responses to high frequency climatic hazards need to be prioritized and made readily available. The duration of a hazard event is such that long-lived hazards are expected to create higher impacts on social-ecological systems compared to short-lived climatic hazards (De Nys et al. 2016). Long-lived climatic hazards present an opportunity for concurrent adaptation where impacts of climatic hazards are being experienced (Smit et al. 2000). However, short-lived climatic events usually require reactive adaptation (Ranger et al. 2010).

Related to the duration of a climatic hazard, the suddenness of hazard defines the rate at which impact is initiated and fully realized (Boon et al. 2012). Slow onset events generally take a longer time for their full impacts to manifest, hence they require incremental adaptation strategy (Wilhelmi and Morss 2013). As an example of slow onset climatic hazard, droughts usually begin with less visible impacts in early stages and extreme low soil moisture in later stages (Smakhtin and Schipper 2008). Prior to drought attaining their extremes, less capacity may be required to adapt while the burden of adapting increases with time towards the worst impact. The impact of sudden or fast-paced climatic events manifests in the short term, hence it requires higher sense of urgency in adapting. Flooding from rainstorm or heavy rainfall is considered a sudden climatic event (Wilhelmi and Morss 2013). The amount of readily available adaptive capacity and timeline for implementation define how well a system will adapt to a sudden climatic hazard (Agrawal 2008).

Sudden climatic hazards are usually more difficult to predict due to their shorter time span for indepth analysis, hence predispose to reactive adaptation. Based on the property of suddenness of a climatic hazard as assessed by Glade and Alexander (2013) and Smith and Petley (2009), drought and floods are conceptualized as slow onset and sudden climatic shock, respectively in this research. Understanding hazards based on suddenness provides insights on adaptation preparedness which feeds into identifying the forerunner in initiating adaptation response (Agrawal 2008). In developing countries such as Malawi, sudden onset climatic hazards with dire consequences are not adequately addressed by the central government in the short term (Agrawal 2008). This is because government institutions are not only under-resourced, but also challenged in their logistical ability to deploy relief and support services to remote places (UNICEF 2020). Thus, delay in government support for adaptation to sudden climatic shocks makes reliance on public authorities' efforts an irrelevant option. Rather, building individuals adaptive capacity as the first call for adaptation to sudden onset hazard provides the benefit of matching a shorter response time interval with a short-lived hazard. The reverse for slow onset climatic hazard is equally true though may be seen as a simplification. This line of argument suggests that individuals in attempt to make rational decision for effective and efficient adaptation will employ their own adaptive capacity for sudden onset climatic hazards where urgency is required. Following the same logic, external adaptation will be used as a response to slow onset climatic hazard where delay in support will be well accommodated without experiencing the worst impact. Such an understanding feeds into the hypothesis that there is a significant relationship between the type of climatic hazard and adaptation strategy as tested in this research.

2.1 Climate Change Adaptation Strategy

Adaptation, an adjustment by a socio-ecological system in anticipation of or response to actual or expected climatic stimuli in attempt to reduce adverse impacts or harness an opportunity, has been conceptualized differently (Smit et al. 2000; Smit and Wandel 2006; Pielke 1998). To discern these differences, a review of how adaptation has been conceptualized in the literature is undertaken in this section. The aim is to assess the scientific and policy relevance of each concept and its application through a focus on adaptation response characteristics-oriented research.

Adaptation entails identification of the hazard and assessment of available adaptive capacity (Smit and Wandel 2006). This suggests that adaptation measures and adaptive capacity is hazard specific (Brooks et al. 2005). The decision of whether an affected individual will adapt or not depends mainly on available adaptive capacity, while the type of adaptation measure implemented will primarily depend on the hazard characteristics, hence a foreseeable relation. Reflecting on the available literature, Smit and Wandel (2006:282) stated that "the concepts of adaptation, adaptive capacity, vulnerability, resilience, exposure and sensitivity are interrelated", and this remains an undeniable fact irrespective of the type of relation which is not being debated in this chapter. The relationship between the various concepts have been discussed without relating them to the understanding of hazard. This is identified as a missing gap. It is suggested in this chapter that a relationship between hazard and adaptation strategy is discernible. Hinted by Brooks et al. (2005), the need to assess the relation between adaptation options and the type of climatic hazard is well expressed. Evaluating the relationship between adaptation options and hazards will help in assessing the extent to which available adaptive capacity is appropriate in attaining the best mitigation of a disaster impact. Supported with assessment of factors influencing adaptation strategy, a baseline for practical adaptation implementation is provided in this chapter.

As an overview of human adaptive response, adaptation strategy can be analyzed based on intent, role of government, decision making process, spatial and social scales, mechanism and form among others (Smit and Wandel 2006; Handmer 2003). Adaptation based on intent questions whether the process was undertaken purposefully, or happened incidentally (Waggoner 1994). Adaptation is mostly planned with deliberate efforts in direct response to a climate change impact which is usually well assessed and understood (Grothmann et al. 2013). Situations where measures not primarily implemented against climate change are found to be effective against its impacts have also been cited. These are described as incidental adaptation (Polsky and Easterling 2001). Evaluating adaptation strategy based on intent has the utility of assessing and measuring the readiness of a community, organization or individual (collectively described as system) and the appropriateness of the measure for a climate change impact. However, this approach does not undertake indepth assessment of the actual adaptation measures in place nor analyze their effectiveness. Thus, while planned adaptation strategy may be assessed to be a positive action of readiness, the element of possible maladaptation is not accounted for. Moreover, no insightful meaning for prioritizing adaptation measures is deductible.

Assessing the duration, short-lived measures are considered as an adjustment or a coping strategy while long-term adaptation measures evolve as part of a system (Burton et al. 1993). This is also understood as tactical and strategic adaptation strategy for short- and long-term measures respectively (Smit and Pilifosova 2003; Holland and Smit 2014). This approach to understanding adaptation when juxtaposed with hazard duration allows the comprehensiveness and efficiency of a response to be tested. Thus, a tactical adaptation strategy may be effective as a response to sudden and short-lived hazard while a strategic adaptation may be rightly paired with a slow onset or long-term hazard impact. For a slow onset climatic hazard, a strategic adaptation is a more comprehensive approach in comparison to a tactical adaptation. Nonetheless, understanding adaptation strategy based on duration seems less insightful in undertaking comparative analysis of adaptation given that its effectiveness is not accounted for.

Based on the time of implementing adaptation relative to the occurrence of a climate change impact, adaptation is classified as anticipatory or reactive (Bosello et al. 2010; Glantz 1992). Anticipatory adaptation includes measures that are implemented in expectation or projection of an adverse climate change impact or opportunity while reactive adaptation is implemented in response to a manifested climate change impact or opportunity (Smit and Wandel 2006). This method of framing adaptation is highly dependent on the hazard rather than adaptation strategy itself, hence this does not allow indepth assessment of the adaptation options available to an affected system. However, some amount of salient information, such as the predictive power of the affected system, is deductible. It is likely that a community that is able to predict a climatic hazard, risk or opportunity adequately will undertake anticipatory adaptation while a system that lacks such capacities may rather react to manifested climate change impacts. Argued by Chen and Kates (1994), developing countries are highly constrained in adequately responding to manifested climate change impacts, hence they expected such impacts to be beyond their means of adaptation. The shortcoming of assessing adaptation based on time is evident in its disregard for the effectiveness of adaptation activities.

Analyzing adaptation to be representative of available adaptive capacity requires assessing the form of adaptation, which is classified to be technological, financial, informational, behavourial, and institutional among others (Smit et al. 2000; Murtinho and Hayes 2008). The criteria of conceptualizing adaptation based on form expresses the fact that adaptation is hazard-specific. Thus, while adaptation form specifies the exact measure being implemented, it is not appropriate for comparative assessment. However, it allows indepth assessment of available adaptive capacity and appropriateness for the climate change impact. Framing adaptation based on form provides important clues in determining the relevant actor to initiate and champion the path and measures for adaptation. Behaviourial adaptation strategies, which include changes in activities being implemented, are usually initiated by individuals affected by the hazard (Tompkins et al. 2010). Even in situations where policy incentives and instruments are used at the central level, their main intention is to stimulate changes in behaviour and lifestyle at the individual scale. Technological adaptation is mostly engineering and structural in approach (Klein et al. 2001); hence it requires technical and specialized capacities that are more likely to be possessed by central governments and large organizations rather than an individual. This approach to evaluating adaptation provides the utility of indicating the level of diversity in adaptive capacity.

Moreover, conceptualizing adaptation based on the role of actor allows the distinction between adaptation measures initiated and led by government or organizations, and those by individuals directly affected by climate change (Tobey 1992; Agrawal 2008, 2010; Ivey et al. 2004). Affected victims acting on their own efforts and resources are analyzed to be adapting autonomously while dependence on government and other organizations' support to adapt is regarded as external adaptation strategy (Smit and Pilifosova 2003; Monnereau and Abraham 2013; Agrawal 2010). Attempts to investigate adaptation with this framing provides insights into existing power relations in a country and how that influences response to environmental change impacts. This offers an agency perspective on adaptation strategy by assessing the role of various actors implementing adaptation strategies. A critical shortcoming in the use of this framing remains that it does not evaluate the effectiveness of the actual adaptation measures implemented but remains superficial as to who implemented it. Moreover, this assessment approach does not provide any analytical guide to undertake adaptation research that estimates how much climate change impacts can be mitigated by implementing a set of adaptation strategies. Nonetheless, this conceptualization of adaptation provides the most potent framing for assessing some policy provisions. For instance, at the high political level, countries are expected to "formulate and implement measures to facilitate adequate adaptation to climate change" as indicated in article 4.1 of UNFCCC (Smit and Wandel 2006). This explains the role of governments in providing resources to enable adaptation to climate change by communities, hence a call for external adaptation. The responsibility of adapting is not entirely borne by governments, but also victims impacted who depend on their own resources to undertake autonomous adaptation. By assessing a number of supports provided by government to victims affected by climate-induced disasters, the extent to which a government is achieving its task of facilitating adequate adaptation can be evaluated.

The role of government in adaptation is both direct and indirect (Smith et al. 2009). Using the case of drought, direct adaptation measures will include provision of drought resistant seeds and crop varieties to farmers while an indirect adaptation role by government will be to undertake drought monitoring research and disseminate information to farmers to implement their own adaptation measures. This has led some researchers to question autonomous adaptation by indicating that victims' strategies are not entirely self-initiated. For instance, affected farmers, while implementing their own measures, may receive external support in the form of information and weather forecast and policies and programs from government agencies (Malik et al. 2010; Burton 1994).

2.2 *Adaptive Capacity as a Factor Influencing the Type of Adaptation Strategy*

The assessment of adaptive capacity reveals that it varies both spatially and temporally with underlying factors in the environment. Argued by Smit and Wandel (2006), adaptation is the result of adaptive capacity which is an existing potential rather than the actual adaptation (Brook 2003). Prevailing factors in an environment influence adaptation (Uddin et al. 2014). Depending on the scale of analysis, different factors determine adaptation. At the household and community level, education and income levels, access to financial resources, social networks, and health status among others are determinants of adaptation (Shikuku et al. 2017; Piya et al. 2013). These factors do not only determine the level of adaptive capacity, but also influence the path and type of adaptation strategy adopted. Argued by Brook (2003), micro-scale factors which define individuals' adaptive capacity are not enough to guarantee a successful adaptation. The adaptive capacity of individuals is a reflection of remote and underlying political, economic, and social environment (Change 2016; Wise et al. 2014).

At a larger scale such as country level, underlying factors definitive of adaptive capacity are noted such as the rate of economic growth, regulatory environment, policy directives, effectiveness of relevant institutions, international relations, trade, and policies among others (Bowen et al. 2012; Brooks and Adger 2005; Berman et al. 2012). Conditions identified at the local, national and international level interact to define adaptive capacity at any given scale (Brown et al. 2010). For instance, a trade embargo or high tariffs between two countries may have the local consequence of reducing farmers ability to export their farm produce and translate into a low-income level. Low income means low adaptive capacity to purchase a high yielding or drought resistant seedling to adapt to a localized drought hazard. Thus, adaptive capacity is not entirely inherent in an individual but also accounted for by large-scale processes and conditions (Nelson et al. 2008). Curtailment of government's role in adaptation implies that victims of climate change impacts are left with lesser supports, hence they have to adapt autonomously. Constituting a basis for the hypothesis being tested in this chapter, Brook (2003) noted that the nature and duration of a hazard determines the nature of adaptive capacity and adaptation strategies.

2.3 Understanding the Role of Gender in Adaptation Narratives (Gendered Adaptation)

Gender inequality is considered an important issue in adaptation research yet remains a missing link in the mainstream, particularly in developing countries, and Malawi is no different (Pratiwi et al. 2017). At the international level, discussions on climate change impacts, causes, mitigation, and adaptation are increasingly held along the lines of gender and advocacy for integrating this approach in climate change policies is growing. Reference can be made to the United Nations Development Program (UNDP 2011). Unfortunately, such gender mainstreaming exercises at the political level have turn-out to be technocratic. By this, women are included on board discussions and projects while the underlying factors that disempowered them from building high adaptive capacity continue to strive (Van Aelst 2016). In explaining limited gendered assessment of climate change, the literature is dominated by technical assessment using models and neoclassical economic approaches among others, which are inherently configured as masculine (Terry 2009). Moreover, while social scientists have addressed climate change discussions through the perspectives of development and sustainability among others, this has not led to copious capture of gender analysis until recent years. Understanding gendered roles and their dynamics in adaptation remains focal in properly comprehending and streamlining adaptation strategies and programmes, hence the review and discussion of literature in this section.

Existing narratives on gendered climate change research are quite clear with the conclusion that women are the most vulnerable yet active agents for effective adaptation (Tschakert and Machado 2012). In resource management and access discourse, gender has been understood as a key factor in explaining current trends in inequality (Smit and Wandel 2006). While remaining underprivileged to access resources, women's role in resource management at the household level has endowed them with extensive knowledge that proves very beneficial in adaptation (Habtezion 2013). Gender inequality influences the level of adaptive capacity which translates into adaptation; the relation is thus crystal clear (Pratiwi et al. 2017; Cohen et al. 2016; Phan et al. 2019).

Gender difference in adaptation is the result of gendered climate change impacts where inequality in workloads, roles, and opportunities are exacerbated (Nelson 2015; Van Aelst 2016). Gendered roles contribute to the difference in concern and priorities over climate change impact. In support of this, Pratiwi et al. (2017) indicated that women may be more concerned over getting clean and affordable water while men express grievances over low livelihoods despite being affected by the same climatic hazard (drought). Intensification of roles has also been witnessed with women travelling longer distance in search of water in areas of prolonged and more intense drought (Goh 2012). The difference in concern for climate change impact among men and women is also explained by varied levels of dependence on natural resources (Denton 2002). Men usually have the option of easily diverting to a sector not directly stricken by climate change. This is usually not the case for women; hence their adaptation options are usually directed towards adapting their productive natural resource to a changing climate rather than abandoning it altogether. Supported by empirical evidences from Thelaphi, Dube, and Mkhwanazi rural communities in South Africa, women have also resorted to diversifying their livelihoods in response to climatic threats (Babugura et al. 2010). Moreover, women have assummed greater role for food security in households making innovation and diversification a necessary part of their roles (Babugura et al. 2010). Changes in livelihood strategies are expected to result in changes in adaptation strategies irrespective of gender (Webb 2015).

With the assumption of equal level of climate change awareness, as argued by Pratiwi et al. (2017), explanation of the difference in adaptation strategy between men and women needs to be sought from other socio-economic factors. Factors that have been tested as influential on the level of adaptive capacities and adaptation strategy include education level, access to resources, information, and credit, ability to innovate, social networks, and livelihoods among others (Asante et al. 2012; Onta and Resurreccion 2011). With women identified as having lower access to all these factors and being poor among the poorest, their adaptive capacity has been generalized as being low (Denton 2002; Alhassan et al. 2018). With regards to the form of adaptation, women favour behavourial change as response to climatic hazards compared to technical, structural, and technological fixes by men (Roehr 2007).

The education level, as a factor influencing adaptation, has been copiously captured in the literature (Gebrehiwot and Van Der Veen 2013; Deressa et al. 2009; Obayelu et al. 2014). Evidence from communities in part of Niger and Kenya indicates high illiteracy rate among women impedes their access to information which is helpful for adaptation (Webb 2015). While such studies have been useful in understanding the possible impact of education level on whether or not to adapt, there is a dire need to push the frontiers to unravel the adaptation pathway to be realized. Hence the importance of identifying the type of adaptation strategy women are likely to adopt with the increase in education level; a gap which is covered by this chapter.

Access to information as a factor influencing adaptive capacity has been assessed by Asfaw and Admassie (2004) to be gendered. They argued that male households have greater access to information which influences their adoption of new technologies or practices. Within an adaptation context, this is likely to translate into a gendered adaptation strategy. Substantiated by Tenge De Graffe and Heller (2004), limited access to information by women constrains the adoption of soil and water conservation measures as adaptation to drought. An opposition has been raised by Deressa et al. (2009) who indicate the contextuality of the gendered access to information with evidence suggesting that women have greater access to information and experience, especially in the agriculture sector, hence they are more likely able to adapt to climate change. Irrespective of the narrative as to who benefits from access to information, a clear line of argument remains that access to information influences adaptation to climatic shocks. The new twist being provided in this research is to understand the type of adaptation to be adopted as a result of access to information.

Deduced from the work of Deressa et al. (2009), access to credit builds people's capacity to adapt to climate change impacts. A positive relationship between access to credit and the level of adaptation has been established (Tazeze et al. 2012; Hassan and Nhemachena 2008). Availability of credit constitutes a ground for external adaptation strategy, hence the people with access to credit are less likely to adopt autonomous adaptation strategy irrespective of their gender. This proposition is being tested in this chapter.

3 Materials and Methods

3.1 Case Study: Climate Change Hazards in Malawi and Its Adaptation Profile

Located in the south eastern region of Africa, Malawi is not different from other sub-Saharan countries that are identified among the most vulnerable to climate change. Indeed, the country is vulnerable to droughts, floods, storms, dry spells, and intense rainfall among others (Murray et al. 2016). The implications are rather diverse ranging from farm losses that threaten livelihoods, health impacts in the form of malaria and cholera, and water and food insecurity among others (Lowe et al. 2013; Stevens and Madani 2016). Droughts and floods have been cited as the main climatic hazards that threaten natural resource base with the poor being the most exposed and vulnerable (Asfaw et al. 2015). Considering spatial assessment of climatic hazards, flooding is localized to the south while drought manifests across the entire country (Pauw et al. 2010; Glad 2010). A rather worrying trend over the past two decades is the increase in severity, frequency, magnitude, and intensity of droughts and floods which have produced adverse impacts mostly on rural dwellers (Coulibaly et al. 2015; Adeloye et al. 2015). Floods and drought are the main drivers of persistent food insecurity which has been assessed to be a national burden (Nangoma 2007). This is demonstrated by floods causing 12% annual loss of maize as compared to 4.6% by drought.

Socio-economic conditions in Malawi are such that about 70% of the populace as of 2017 lived below the poverty line of US$1.90 daily coupled with low per capita gross national income of US$320 (World Bank 2018b). With about 35% of poor households headed by women, one can imagine the extent to which poverty driven gendered adaptation strategy is expressed (Chigwada 2004). The fact that about 78% of Malawians are engaged in agriculture implies that the severity of flooding and drought will directly affect most peoples' means of livelihood (Wood and Mayer 2006; Garrity et al. 2010). A more bewildering situation is the impact of climatic hazards on the economy. Floods are estimated to cost 1% reduction in Gross Domestic Product (GDP) and this figure is higher for droughts (Nangoma 2007). A linkage between drought and other socio-economic characteristics of households is evident. Drought contributes to 1.3% poverty increase and extreme drought results in poverty increase by 17% (Nangoma 2007). The level of drought impact is gendered with female headed households suffering from acute malnutrition and health conditions (Nangoma 2007). Food insecurity is expected to worsen with projection of shorter rainy seasons, hence the rising food prices which equally affect non-farm workers (Tadross et al. 2007; Pauw and Thurlow 2009). This justifies the rational of not only focusing on farmers but all households who reported drought in this study.

Adaptation measures for drought have varied depending on the actor involved. Malawi government have preferred construction of dams, deep wells and rain harvest structures for drought adaptation while households in Malawi focus on storing grains in granaries and relocation among a number of behaviourial changes (Syroka and Nucifora 2010; Fisher et al. 2010; Panganga et al. 2012). Increasing frequency and magnitude of droughts and floods are rendering such measures ineffective, and Jørstad and Webersik (2016) claim that livelihood diversification is a more appropriate adaptation measure.

3.2 Method

In proposing a conceptual framework adequately contextualized as a tool for empirical research, discussions made in earlier sections of this chapter and the work of Smithers and Smit (1997) were chiefly drawn upon. A deliberate attempt to establish a linkage between the various themes of conceptualization led to: the assemblage of slow onset and sudden onset climatic hazard (theme of climatic hazard characteristics); external and autonomous adaptation strategy (theme of adaptive response characteristics); and low and high adaptive capacity (attributes of the system affected). Proceeding from this categorization, a number of questions were asked as a guide to deduce how key characteristics of the various concepts will interrelate. This includes, (i) deductively, which adaptation strategy people with high or low adaptive capacity are likely to adopted, and (ii) which type of adaptation strategy will individuals use for the various hazard? Base on the characteristics associated with a concept, the supposed behaviour of an individual is deduced and summarized into a framework. Aspects of this framework are tested with an empirical research.

The study area for the empirical research is defined to be Malawi. Living Standard Measurement Survey (LSMS) from the World Bank was used as a secondary data (World Bank 2018a). The Fourth Integrated Household Survey which covers the period April, 2016 to April, 2017 was used. The sample size used in analyzing the relationship between gender and adaptation strategy was 3,026, constituting of 1,467 men and 1,558 women who experienced droughts and floods. Nonetheless, this sample size reduced after data cleaning for some variables. Focusing on how various factors influence adaptation strategy of women, a sample size of 1,216 was used. Droughts and floods are conceptualized as slow onset and sudden climatic shocks, respectively. Adaptation strategy is categorized as autonomous and external. Chi-squared test of independence and multiple logistic regression analysis were undertaken.

With a focus on the human–environment interaction, socio-economic factors such as educational level, access to information and credit, and subjective well-being were analyzed as independent variables to understand their influence on the choice of adaptation strategy which was treated as a dependent variable.

Taking a clue from the education system of Malawi, which is reflected in the dataset, Junior Certificate of Education (JCE) and Primary School Leaving Certificate of Education (PSLC) were classified as a basic education, and the Malawi School Certificate of Education (MSCE), non-university diploma, and university diploma degree were classified as an advanced education. Thus, the education level was classified into three categories, namely no education, basic education, and advanced education, hence a dummy was created for each of them for the multiple logistic regression. ‘No education’ was used as the reference category.

Subjective well-being was classified as binary variable with categories as poor and rich. Subjective well-being, step 1 to 3 – as captured in the dataset – were categorized as poor, and step 4 to 6 were classified as rich. Subjective well-being step 1 is understood as poor and varies in that manner to step 6 which translates as rich.

Access to credit was a binary variable with the categories as females who acquired any form of credit and those who did not.

Access to information in this research has been best identified as the possession of devices such as radio and television (TV). This was the best available variable from the dataset that could be used as a surrogate based on the assumption that women who possess radio or TV will use it to acquire information on drought and flood prediction and the availability of government or institutional support and relief. This can influence their adaptation strategy. Access to radio was treated as a dichotomous variable with categories as females who had radio and those without it. Access to TV was classified in the same manner.

Adaptation measures were analyzed to identify the actors that initiated the process. In analyzing adaptation strategy, the question of who is offering support as a response to climatic shocks was considered. Adaptation measures analyzed to be initiated by the government and NGOs or other entities were classified as external adaptation strategy. All other measures wholly initiated and implemented by victims impacted by droughts and floods were classified as autonomous adaptation strategy.

4 Results

4.1 Suggested Conceptual Framework for Analyzing Hazard-Adaptation Nexus/Interrelation

(See Fig. 1).

4.2 Results on Adaptation Strategy Is Gendered

To test the hypothesis that adaptation strategy is gendered, chi square test of independence was undertaken based on the following formulations:

- Null hypothesis (N0): Type of adaptation strategy is independent of gender.
- Alternate hypothesis (NA): Type of adaptation strategy is related to gender.

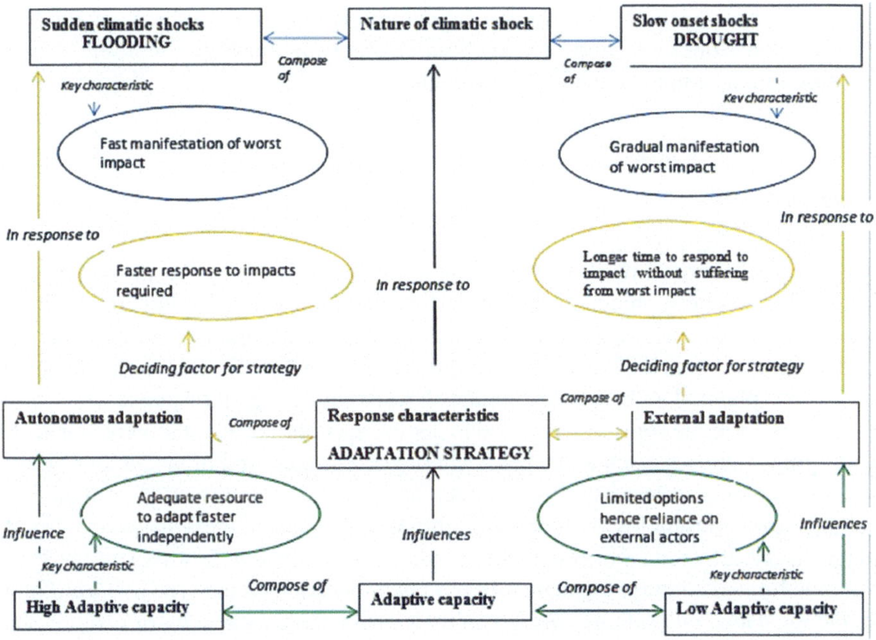

Fig. 1 Conceptual framework for analyzing climatic hazard-adaptation nexus (Source: Author based on Smithers and Smit 1997)

4.3 Result on the Type of Adaptation Strategy Is Related to the Type of Climatic Shock Among Women

The result of the second objective, which is to test whether there is a significant relationship between the type of climatic shock and the type of adaptation strategy, was undertaken with chi-square test of independence and is captured here. The hypothesis was formulated for testing as follows:

- Null hypothesis (N0): There is no significant relationship between the type of adaptation strategy and the type of climatic shock.
- Alternative hypothesis (NA): There is a significant relationship between the type of adaptation strategy and the type of climatic shock.

4.4 Result on How Various Factors Influence the Type of Adaptation Strategy Among Women

The result on assessing how various socio-economic factors are influencing the type of adaptation strategy in response to climatic shocks is illustrated in Table 3. This was based on a sample size of 1,216 women.

5 Discussion

The proposed framework allows relations to be drawn between the various themes in a clear and concise manner relevant to the case studies starting with an empirical test of the framework in Malawi. Evaluating climate change impact as slow and sudden onset event is indicative of the required rate of adaptation response. Societies and victims plagued with sudden climatic shocks require urgent response while slow onset events provide some grace period to deploy appropriate adaptation response. In developing countries, such as Malawi, where governments and organizations tasked with managing disasters are highly under-resourced, the deployment of measures in support of households to adapt to a climate change impact is usually delayed. Suggestively, affected victims in making rational decision as to how they utilize their available limited adaptive capacity will base their decision on the type of hazard they are facing. Thus, individuals will fall on their own adaptive capacity or resources as an autonomous adaptation strategy in response to sudden climatic events due to the urgency required; while they will wait and rely on government support as external adaptation strategy in response to slow onset events where worst impacts are not usually experienced in the short term.

Empirical testing of hypothesis suggested from the conceptual framework for climate change adaptation as devised in this chapter provides early insights in the

Table 1 Chi-square test of independence between gender and adaptation strategy

Case processing summary

	Cases					
	Valid		Missing		Total	
	N	Percent	N	Percent	N	Percent
Gender * Type of adaptation strategy	3025	100.0%	0	0.0%	3025	100.0%

Chi-Square tests

	Value	df	Asymptotic significance (2-sided)	Exact sig. (2-sided)	Exact sig. (1-sided)
Pearson Chi-Square	1.532[a]	1	.216		
Continuity correction[b]	1.434	1	.231		
Likelihood ratio	1.532	1	.216		
Fisher's exact test				.225	.116
N of valid cases	3025				

a. 0 cells (0.0%) have expected count less than 5. The minimum expected count is 412.70.
b. Computed only for a 2 × 2 table
Source: Author

context of Malawi. Illustrated in Table 1 is the test on the hypothesis that adaptation strategy is gendered. The p-value calculated was 0.216 which is greater than significance level (α) of 0.05, hence the acceptance of the null hypothesis. It is thus concluded that there is no statistically significant association between gender and the type of adaptation strategy in Malawi. With the conclusion that the difference in preference for a particular response is not clearly discernible between men and women, a comparison between the two groups is of little analytical relevance. Thus, this chapter focusses on women for further analysis.

Illustrated in Table 2 is a p-value of 0.511 which is greater than 0.05, hence the acceptance of the null hypothesis that there is no statistically significant relationship between the type of adaptation strategy and the type of climatic shock at a significant level of 5%. This result implies that Malawians do not take the time interval that a climatic shock is expected to have full impact on their well-being as main consideration for adaptation response. Irrespective of whether a climatic shock will manifest its impact suddenly or slowly, Malawians do not hesitate in deciding how to act. Both Government and individual supports available to them will be used to address shocks indiscriminately.

Depicted in Table 3 is the results on understanding how each of the independent variables influences adaptation strategy while controlling the other independent variable. For the education level, no education was used as the reference category. The results indicate that women with basic education (Odd ratio (OR) = 1.581) are more

Table 2 Chi square test of independence between the adaptation strategy and the climatic shock

Case processing summary

	Cases					
	Valid		Missing		Total	
	N	Percent	N	Percent	N	Percent
Type of climatic hazard *Type of adaptation strategy	1558	100.0%	0	0.0%	1558	100.0%

Chi-Square tests

	Value	df	Asymptotic significance (2-sided)	Exact sig. (2-sided)	Exact sig. (1-sided)
Pearson Chi-Square	.432[a]	1	.511		
Continuity correction[b]	.313	1	.576		
Likelihood ratio	.425	1	.514		
Fisher's exact test				.496	.285
N of valid cases	1558				

a. 0 cells (0.0%) have expected count less than 5. The minimum expected count is 39.64
b. Computed only for a 2 × 2 table
Source: Author

Table 3 Multiple logistic regression analysis on how socio-economic variables influence adaptation strategy

Variables in the equation

		B	S.E	Wald	df	Sig.	Exp(B)	95% C.I.for EXP(B)	
								Lower	Upper
Step 1[a]	Subjective well-being	.260	.234	1.235	1	.266	1.296	.820	2.049
	Basic education dummy	.458	.174	6.936	1	.008	1.581	1.124	2.222
	Advance education dummy	.329	.273	1.453	1	.228	1.389	.814	2.371
	Credit access	−.254	.141	3.258	1	.071	.775	.588	1.022
	Radio access	.318	.144	4.921	1	.027	1.375	1.038	1.822
	TV access	.696	.292	5.667	1	.017	2.006	1.131	3.559
	Constant	.613	.234	6.855	1	.009	1.846		

Hosmer and Lemeshow test

Step	Chi-square	df	Sig.
1	5.629	7	.584

a. Variable(s) entered on step 1: Subjective well-being, Basic Education (Dummy), Advance Education (Dummy), Credit Access, Radio Access, TV Access.
Source: Author

likely compared to women without any form of education to adopt autonomous adaptation strategy, hence this acts as the main initiators of climate change adaptation. In similar manner, women with advanced education have an OR of 1.389 making them more likely to adapt autonomously compared to women with no education. This result was controlled for all other factors included in the analysis. A p-value of 0.008 for basic education at 0.1% level of significance indicates that there are statistically significant associations between basic education and adaptation strategy. However, this relationship changes for women with advanced education level (p-value = 0.228) where a statistically significant relationship is not realized between advanced education and adaptation strategy. It can be concluded that higher education level among women leads to adapting autonomously to climatic shocks in Malawi.

Women who perceive their subjective well-being to be poor are more likely to adopt autonomous adaptation strategy (OR = 1.296) holding all other factors constant. Subjective well-being has a p-value of 0.266, hence there is no statistically significant association between subjective well-being and adaptation strategy at 0.1% level of significance. A possible explanation for rich women being less likely to adapt autonomously compared to poor women is that, rich women are likely to use their resource to insure themselves, hence they depend on external support in times of disasters. Moreover, power relations is such that rich people are able to build important networks with people of higher positions which may influence their access to support from them. It is also suggested that rich women have the required collateral to secure credits that aid in their adaptation, hence the external support. With limited literature evidence to juxtapose these results with, the points raised should be considered as propositions that require further empirical research in other countries to rightly justify.

Access to credit was analyzed to have an OR of 0.775, hence women who have access to credit are less likely to adapt to climatic shocks autonomously. The results are adjusted for all other variables in the analysis. With a p-value of 0.71, access to credit has no statistically significant relationship with adaptation strategy. This result reflects the deduction made from the work of Deressa et al. (2009) in this chapter. OR of radio and TV access were 1.375 and 2.006, respectively, hence women with access to radio and TV are more likely to adopt autonomous adaptation strategies. Both access to radio and TV have a statistically significant relationship with the type of adaptation strategy.

The Hosmer and Lemeshow Test produce a p-value of 0.584 which is greater than 0.05 hence statistically significant (at 0.1% level). Thus, the estimated model has adequate fit.

6 Conclusion

Progress on adaptation research has attained a commendable level in terms of theoretical and conceptual framing, thus it needs to be better utilized in empirical research. The complexity to understand the human–environment interaction in the context of a changing climate warrants the need for further conceptual development for which existing frameworks can only serve as a starting point. The development of new conceptual frameworks needs to be deliberately undertaken to avoid the challenges of being overly broad and plagued with unfounded assumptions, hence their restricted applicability. This requires the contextualization of frameworks to further specify how various components should be conceptualize with clear-cut examples.

Reviewing the existing climate change adaptation frameworks and concepts, the much-needed work of drawing a linkage between the dominant themes with the key element of adequately contextualizing them for applicability, have been addressed in this chapter. Accordingly, how characteristics of human system interact with the hazards in the natural system (climate) has not only been conceptualized, but also contextualized and supported with an empirical testing. Utilizing this to fill another research gap, the analysis was viewed from a gendered lens. Deliberate attempt was made to avoid the shortcoming of a generalized inequality evident in comparative assessments between men and women, hence the focus on women in Malawi. The conclusion drawn from this research is that, with the exception of access to credit, all other socio-economic variables assessed influenced women to adopt autonomous adaptation strategy. Hence, the need to adopt a bottom-up approach to adaptive capacity building of women is well expressed at the expense of other forms of external support.

References

Adeloye AJ, Mwale FD, Dulanya Z (2015) A metric-based assessment of flood risk and vulnerability of rural communities in the Lower Shire Valley, Malawi. Proc Int Assoc Hydrol Sci 370:139

Agrawal A (2008) The role of local institutions in adaptation to climate change. World Bank

Agrawal A (2010) Local institutions and adaptation to climate change. Soc Dimens Clim Change Equity Vulnerabil Warming World 2:173–178

Alhassan SI, Shaibu MT, Kuwornu KMJ, Osman TD (2018) Assessing smallholder women farmers' adaptive capacity to climate change and variability in the Northern Region of Ghana: a composite index approach. J Energy Nat Res Manag (JENRM) 1(1)

Althor G, Watson JE, Fuller RA (2016) Global mismatch between greenhouse gas emissions and the burden of climate change. Sci Rep 6:20281

Asfaw A, Admassie A (2004) The role of education on the adoption of chemical fertiliser under different socioeconomic environments in Ethiopia. Agric Econ 30(3):215–228

Asfaw S, McCarthy N, Lipper L, Arslan A, Cattaneo A, Kachulu M (2015) Climate variability, adaptation strategies and food security in Malawi. No. 1008–2016–80228

Asante FA, Boakye AA, Egyir IS, Jatoe JB (2012) Climate change and farmers' adaptive capacity to strategic innovations: The case of northern Ghana

AU, A. U. (2012) African gender, climate change and agriculture support program (GCCASP)

Babugura A, Mtshali N, Mtshali M (2010) Gender and climate change: South Africa case study. Heinrich Böll Stiftung Southern

Berman R, Quinn C, Paavola J (2012) The role of institutions in the transformation of coping capacity to sustainable adaptive capacity. Environ Dev 2:86–100

Boon HJ, Cottrell A, King D, Stevenson RB, Millar J (2012) Bronfenbrenner's bioecological theory for modelling community resilience to natural disasters. Nat Hazards 60(2):381–408

Bosello F, Carraro C, De Cian E (2010) Climate policy and the optimal balance between mitigation, adaptation and unavoided damage. Clim Change Econ 1(02):71–92
Bowen A, Cochrane S, Fankhauser S (2012) Climate change, adaptation and economic growth. Clim Change 113(2):95–106
Brooks N (2003) Vulnerability, risk and adaptation: A conceptual framework. Tyndall Centre for Climate Change Research Working Paper 38(38): 1–16
Brooks N, Adger WN (2005) Assessing and enhancing adaptive capacity. B Adaptation policy frameworks for climate change: Developing strategies, policies and measures, 165–181
Brooks N, Adger WN, Kelly PM (2005) The determinants of vulnerability and adaptive capacity at the national level and the implications for adaptation. Glob Environ Chang 15(2):151–163
Brown HCP, Nkem JN, Sonwa DJ, Bele Y (2010) Institutional adaptive capacity and climate change response in the Congo Basin forests of Cameroon. Mitig Adapt Strat Glob Change 15(3):263–282
Burton I (1994) Costs of atmospheric hazards. In: Proceedings of a Workshop on Improving Responses to Atmospheric Extremes: The Role of Insurance and Compensation, Ottawa, Canada, Environment Canada, pp 2–11
Burton I, Kates RW, White GF (1993) The Environment as Hazard Guilford Press. New York
Change GE (2016) Place, well-being, and fairness shape priorities for adaptation to climate change. Glob Environ Chang 38:A1–A3
Chen RS, Kates RW (1994) Climate change and world food security
Chigwada J (2004) Adverse Impacts of Climate Change and Development Challenges: Integrating Adaptation in Policy and Development in Malawi. IIED
Cohen PJ, Lawless S, Dyer M, Morgan M, Saeni E, Teioli H, Kantor P (2016) Understanding adaptive capacity and capacity to innovate in social–ecological systems: applying a gender lens. Ambio 45(3):309–321
Coulibaly JY, Gbetibouo GA, Kundhlande G, Sileshi GW, Beedy TL (2015) Responding to crop failure: understanding farmers' coping strategies in Southern Malawi. Sustainability 7(2):1620–1636
Demetriades J, Esplen E (2010) The gender dimensions of poverty and climate change adaptation. In: Social dimensions of climate change: Equity and vulnerability in a warming world, pp 133–144
Denton F (2002) Climate change vulnerability, impacts, and adaptation: why does gender matter? Gend Dev 10(2):10–20
De Nys E, Engle N, Magalhães AR (eds) (2016) Drought in Brazil: proactive management and policy. CRC Press, Boca Raton
Deressa TT, Hassan RM, Ringler C, Alemu T, Yesuf M (2009) Determinants of farmers' choice of adaptation methods to climate change in the Nile Basin of Ethiopia. Glob Environ Chang 19(2):248–255
Dilley M, Chen RS, Deichmann U, Lerner-Lam AL, Arnold M (2005) Natural disaster hotspots: a global risk analysis. The World Bank
Few R (2007) Health and climatic hazards: framing social research on vulnerability, response and adaptation. Glob Environ Chang 17(2):281–295
Fisher M, Chaudhury M, McCusker B (2010) Do forests help rural households adapt to climate variability? evidence from Southern Malawi. World Dev 38(9):1241–1250
Garrity DP, Akinnifesi FK, Ajayi OC, Weldesemayat SG, Mowo JG, Kalinganire A, Bayala J (2010) Evergreen Agriculture: a robust approach to sustainable food security in Africa. Food Secur 2(3):197–214
Gebrehiwot T, Van Der Veen A (2013) Farm level adaptation to climate change: the case of farmer's in the Ethiopian Highlands. Environ Manage 52(1):29–44
Glad PA (2010) Meteorological and hydrological conditions leading to severe regional drought in Malawi (Master's thesis)
Glade T, Alexander D (2013) Classification of natural disasters. In: Bobrowsky PT (ed) Encyclopedia of natural hazards. Springer, Dordrecht, pp 78–82
Glantz MH (1992) Global warming and environmental change in sub-Saharan Africa. Glob Environ Chang 2(3):183–204

Goh AH (2012) A literature review of the gender-differentiated impacts of climate change on women's and men's assets and well-being in developing countries. International Food Policy Research Institute, CAPRi Work
Grothmann T, Grecksch K, Winges M, Siebenhüner B (2013) Assessing institutional capacities to adapt to climate change: integrating psychological dimensions in the Adaptive Capacity Wheel. Nat Hazard 13(12):3369
Guloba M (2014) Adaptation to climate variability and change in Uganda: Are there gender differences across households?, No. 2014/107. WIDER Working Paper
Habtezion S (2013) Overview of linkages between gender and climate change. Policy Brief. United Nations Development Programme, New York, 20
Handmer J (2003) Adaptive capacity: what does it mean in the context of natural hazards? In: Climate change, adaptive capacity and development, pp 51–70
Hassan RM, Nhemachena C (2008) Determinants of African farmers' strategies for adapting to climate change: Multinomial choice analysis. Afr J Agric Res Econ 2(311-2016-5521):83–104
Holland T, Smit B (2014) Recent climate change in the Prince Edward County winegrowing region, Ontario, Canada: implications for adaptation in a fledgling wine industry. Reg Environ Change 14(3):1109–1121
Huynh PT, Resurreccion BP (2014) Women's differentiated vulnerability and adaptations to climate-related agricultural water scarcity in rural Central Vietnam. Climate Dev 6(3):226–237
Ivey JL, Smithers J, De Loë RC, Kreutzwiser RD (2004) Community capacity for adaptation to climate-induced water shortages: linking institutional complexity and local actors. Environ Manage 33(1):36–47
Jørstad H, Webersik C (2016) Vulnerability to climate change and adaptation strategies of local communities in Malawi: experiences of women fish-processing groups in the Lake Chilwa Basin. Earth Syst Dyn 7(4):977–989
Kelly PM, Adger WN (2000) Theory and practice in assessing vulnerability to climate change and facilitating adaptation. Clim Change 47(4):325–352
Klein RJ, Nicholls RJ, Ragoonaden S, Capobianco M, Aston J, Buckley EN. (2001) Technological options for adaptation to climate change in coastal zones. J Coast Res 531–543
Kotir JH (2011) Climate change and variability in Sub-Saharan Africa: a review of current and future trends and impacts on agriculture and food security. Environ Dev Sustain 13(3):587–605
Lowe R, Chirombo J, Tompkins AM (2013) Relative importance of climatic, geographic and socio-economic determinants of malaria in Malawi. Malar J 12(1):416
Malik A, Qin X, Smith SC (2010) Autonomous adaptation to climate change: A literature review. Institute for International Economic Policy Working Paper Series, pp 1–25
McCarthy JJ, Canziani OF, Leary NA, Dokken DJ, White KS (eds) (2001) Climate change 2001: impacts, adaptation, and vulnerability: contribution of Working Group II to the third assessment report of the Intergovernmental Panel on Climate Change, vol. 2. Cambridge University Press, Cambridge
Melching CS, Pilon PJ (2006) Comprehensive risk assessment for natural hazards. World Meteorological Organization, WMO/TD, 955
Monnereau I, Abraham S (2013) Limits to autonomous adaptation in response to coastal erosion in Kosrae, Micronesia. Int J Glob Warm 5(4):416–432
Murtinho F, Hayes TM (2008) Insights from the field for measuring and analyzing adaptation in common-pool resource management. In: Biennial Conference of the International Association for the Study of the Commons.
Murray U, Gebremedhin Z, Brychkova G, Spillane C (2016) Smallholder farmers and climate smart agriculture: technology and labor-productivity constraints amongst women smallholders in Malawi. Gend Technol Dev 20(2):117–148
Nangoma E (2007) National adaptation strategy to climate change impacts: A case study of Malawi. Human Development Report, 2008(8)
Nelson G (2015) Gender-Responsive National Communications Toolkit. UNDP, New York

Nelson R, Howden M, Smith MS (2008) Using adaptive governance to rethink the way science supports Australian drought policy. Environ Sci Policy 11(7):588–601

Obayelu OA, Adepoju AO, Idowu T (2014) Factors influencing farmers' choices of adaptation to climate change in Ekiti State, Nigeria. J Agric Environ Int Dev (JAEID) 108(1):3–16

Onta N, Resurreccion BP (2011) The role of gender and caste in climate adaptation strategies in Nepal. Mt Res Dev 31(4):351–357

Panganga P, Jumbe CB, Kanyanda S, Thangalimodzi L (2012) Policy implications of droughts and floods adaptation on household crop production and food security in southern Malawi. Brit J Environ Clim Change 2(3):245

Parry ML, Carter TR, Hulme M (1996) What is a dangerous climate change? Glob Environ Chang 6(1):1–6

Parry M, Parry ML, Canziani O, Palutikof J, Van der Linden P, Hanson C (eds) (2007) Climate change 2007-impacts, adaptation and vulnerability: working group II contribution to the fourth assessment report of the IPCC, vol. 4. Cambridge University Press, Cambridge

Pauw K, Thurlow J (2009) Economic losses and poverty effects of droughts and floods in Malawi, no. 2. International Food Policy Research Institute (IFPRI)

Pauw K, Thurlow J, van Seventer D (2010) Droughts and floods in Malawi. International Food Policy Research Institute, Washington

Phan LT, Jou SC, Lin JH (2019) Gender inequality and adaptive capacity: the role of social capital on the impacts of climate change in Vietnam. Sustainability 11(5):1257

Pielke RA Jr (1998) Rethinking the role of adaptation in climate policy. Glob Environ Chang 8(2):159–170

Polsky C, Easterling WE III (2001) Adaptation to climate variability and change in the US Great Plains: A multi-scale analysis of Ricardian climate sensitivities. Agr Ecosyst Environ 85(1–3):133–144

Piya L, Maharjan KL, Joshi NP (2013) Determinants of adaptation practices to climate change by Chepang households in the rural Mid-Hills of Nepal. Reg Environ Change 13(2):437–447

Pratiwi NAH, Rahmawati YD, Setiono I (2017) Gender equality in climate change adaptation: a case of Cirebon, Indonesia. Ind J Plann Dev 2(2):74–86

Ranger N, Millner A, Dietz S, Fankhauser S, Lopez A, Ruta G (2010) Adaptation in the UK: a decision-making process. Environ Agency 9

Ravankhah M, de Wit R, Argyriou AV, Chliaoutakis A, Revez MJ, Birkmann J, Giapitsoglou K (2019) Integrated assessment of natural hazards, including climate change's influences, for cultural heritage sites: the case of the historic centre of Rethymno in Greece. Int J Disast Risk Sci 10(3):343–361

Roehr U (2007) Gender, climate change and adaptation. Introduction to the gender dimensions. In: Background Paper prepared for the Both Ends Briefing Paper Adapting to climate change: How local experiences can shape the debate

Shikuku KM, Winowiecki L, Twyman J, Eitzinger A, Perez JG, Mwongera C, Läderach P (2017) Smallholder farmers' attitudes and determinants of adaptation to climate risks in East Africa. Clim Risk Manag 16:234–245

Smakhtin VU, Schipper ELF (2008) Droughts: the impact of semantics and perceptions. Water Policy 10(2):131–143

Smit B, Burton I, Klein RJ, Wandel J (2000) An anatomy of adaptation to climate change and variability. In: Societal adaptation to climate variability and change, pp 223–251. Springer, Dordrecht.

Smith K, Petley DN (2009) Environmental hazards: Assessing risk and reducing disaster. Routledge, London

Smit B, Pilifosova O (2003) From adaptation to adaptive capacity and vulnerability reduction. In: Climate change, adaptive capacity and development, pp 9–28

Smit B, Wandel J (2006) Adaptation, adaptive capacity and vulnerability. Glob Environ Chang 16(3):282–292

Smithers J, Smit B (1997) Human adaptation to climatic variability and change. Glob Environ Chang 7(2):129–146
Smith JB, Vogel JM, Cromwell JE III (2009) An architecture for government action on adaptation to climate change. Clim Change 95(1–2):53–61
Srex I (2012) Managing the risks of extreme events and disasters to advance climate change adaptation. A special report of working groups I and II of the intergovernmental panel on climate change. In: Field CB, Barros V, Stocker TF, Qin D, Dokken DJ, Ebi KL, Mastrandrea MD Mach KJ, Plattner GK, Allen SK, Tignor M, Midgley PM. Cambridge University Press, Cambridge
Stevens T, Madani K (2016) Future climate impacts on maize farming and food security in Malawi. Sci Rep 6:36241
Syroka J, Nucifora A (2010) National drought insurance for Malawi. The World Bank
Tadross M, Suarez P, Lotsch A, Hachigonta S, Mdoka M, Unganai L, Muchinda M (2007) Changes in growing-season rainfall characteristics and downscaled scenarios of change over southern Africa: implications for growing maize. In:IPCC regional Expert Meeting on Regional Impacts, Adaptation, Vulnerability, and Mitigation, Nadi, Fiji, pp 193–204
Tazeze A, Haji J, Ketema M (2012) Climate change adaptation strategies of smallholder farmers: the case of Babilie District, East Harerghe Zone of Oromia regional state of Ethiopia. J Econ Sustain Dev 3(14):1–12
Tenge AJ, De Graaff J, Hella JP (2004) Social and economic factors affecting the adoption of soil and water conservation in West Usambara highlands, Tanzania. Land Degrad Dev 15(2):99–114
Terry G (2009) Climate change and gender justice. Oxfam GB
Thornton TF, Puri RK, Bhagwat S, Howard P (2019) Human adaptation to biodiversity change: an adaptation process approach applied to a case study from southern India. Ambio 48(12):1431–1446
Tobey JA (1992) Economic issues in global climate change. Glob Environ Chang 2(3):215–228
Tompkins EL, Adger WN, Boyd E, Nicholson-Cole S, Weatherhead K, Arnell N (2010) Observed adaptation to climate change: UK evidence of transition to a well-adapting society. Glob Environ Chang 20(4):627–635
Tschakert P, Machado M (2012) Gender justice and rights in climate change adaptation: opportunities and pitfalls. Ethics Social Welfare 6(3):275–289
Uddin MN, Bokelmann W, Entsminger JS (2014) Factors affecting farmers' adaptation strategies to environmental degradation and climate change effects: a farm level study in Bangladesh. Climate 2(4):223–241
UNDP (2011) Africa adaptation programme experiences gender and climate change: advancing development through an integrated gender perspective. In: Discussion Paper Series, vol. 1. UNDP, New York
UNICEF (2020) Africa's first humanitarian drone testing corridor launched in Malawi by Government and UNICEF. Accessed 10th June 2020, https://www.unicef.org/press-releases/africas-first-humanitarian-drone-testing-corridor-launched-malawi-government
Urry J (2015) Climate change and society. In: Why the social sciences matter, pp. 45–59. Palgrave Macmillan, London
Van Aelst K (2016) Gender, households and climate change: adaptation decision-making in the Morogoro Region of Tanzania. Doctoral dissertation, University of Antwerp
Waggoner PE (1994) Preparing US agricultural for global climate change. Philadelphia Society for Promoting Agriculture, USA
Webb J (2015) Gender dynamics in a changing climate: how gender and adaptive capacity affect resilience
Wilhelmi OV, Morss RE (2013) Integrated analysis of societal vulnerability in an extreme precipitation event: A Fort Collins case study. Environ Sci Policy 26:49–62
Wilhite DA (2000) Drought as a natural hazard: concepts and definitions
Wise RM, Fazey I, Smith MS, Park SE, Eakin HC, Van Garderen EA, Campbell B (2014) Reconceptualising adaptation to climate change as part of pathways of change and response. Glob Environ Chang 28:325–336

Wood A, Mayer J (2006) Malawi Poverty and Vulnerability Assessment: Investing in our Future. World Bank, Washington

World Bank (2018) Systematic country diagnostic: breaking the cycle of low growth and slow poverty reduction. The World Bank, Washington

World Bank (2018b). The Fourth Integrated Household Survey of Malawi, 2016–2017. Retrieved from https://microdata.worldbank.org/index.php/catalog/2936/study-description, Accessed 27 Aug 2019

The Effects of Land Cover Change on Sustainability: Human Security and Environmental Change in Semi-arid Ecosystems

Taisser H. H. Deafalla, Elmar Csaplovics, Osman Elkhair, and Mustafa M. El Abbas

Abstract The links between Environmental Change (EC), seen as a destabilizing interference in the ecosystems' equilibrium, on the one hand, and the outbreak of violent conflicts, on the other, remain ambiguous and very complex. Till now, there is a gap in the researches that focus on the intersections between EC and human security based on Land Use/Land Cover (LU/LC) changes. Therefore, this study will examine the interplay between environmental factors, LU/LC, and conflict in South Kordofan of Sudan. Satellite data and high-resolution Digital Elevation Model (DEM) were used to address the LC changes in the study area over the past three decades (1984–2014). Furthermore, household survey, group discussions and key informants techniques were applied. The qualitative and quantitative techniques were used to analyze the socio-economic data. Geographic Object-Based Image Analysis (GEOBIA) based on a developed model of integrated multi-features was utilized. The study showed that the pressures imposed resulted in the complex of spatial and temporal interactions within topographical systems at South Kordofan, where it has led both to a new rapidity and depth in rural transformation and a significant impact on urban areas as well. More information exchange is needed to inform actors and decision makers regarding capacity gaps and knowledge to address the EC.

Keywords Environmental change · Conflict · Semi-arid region · Remote sensing,· control of resources

T. H. H. Deafalla (✉) · E. Csaplovics
Institute of Photogrammetry and Remote Sensing, University of Dresden, Dresden, Germany

E. Csaplovics
e-mail: elmar.csaplovics@tu-dresden.de

O. Elkhair
Faculty of Architecture, University of Khartoum, Khartoum, Sudan

M. M. El Abbas
Faculty of Environmental Sciences, TU Dresden, Dresden, Germany

Faculty of Forestry, University of Khartoum, Khartoum, Sudan

M. Behnassi et al. (eds.), *Social-Ecological Systems (SES)*,
https://doi.org/10.1007/978-3-030-76247-6_7

1 Introduction

Exploring the linkages between the environment, security and conflict is not a novelty. It is theoretically backed up and scientifically evidenced (Behnassi 2017) even if it has not yet reached the satisfaction and acceptable level. The Environmental Change (EC) is a complex dynamic system. It is defined as a change or disturbance of the environment caused by human influences or natural ecological processes (Khanna and Palepu 1997; Suarez and Oliva 2005). The characteristics and importance of each driving force may vary from one area to another. However, the long-range interactions of pressures may lead to the EC (Deafalla et al. 2017). In addition to that, the ability to create and transfer environmental pressures or stress onto the environment of other societies varies from one area to another. For example, developed countries are significantly contributing to global and transboundary environmental pressures than the less affluent societies which have minimal interaction with the environment in which they live (UNEP 2007a). In his study, Lein (2012) referred to the transformations characterizing EC as subtle and slow to emerge, or dramatic and quick to materialize. In either case, they reflect the consequence of disturbances that alter material and energy flows within the environmental system. Here, the concept of a "disturbance" becomes a convenient way to connect environmental stress to actions that will display both in temporal and spatial dimensions (White and Picket 1985).

Nowadays, EC is the most important one, of the security, climatic and developmental challenges facing humanity, particularly in developing countries (Deafalla et al. 2018). In effect, the 'threat multiplier' such as drought, desertification, water scarcity, food insecurity and land degradation, that makes existing concerns more complex and intractable. However, it is the non-climate factors, such as poverty, conflict management, governance, regional diplomacy etc., which will largely determine whether and how EC moves from being a development challenge to presenting a security threat (Brown et al. 2007; Brown and Crawford 2009). Inreality, LC patterns play an important role on EC detection studies, where they reflect the underlying natural and social processes, thus providing essential information for modelling and understanding many phenomena on the earth, including the complex interactions between global EC and human activities.

In recent times, many concerns have been raised about the relationship between EC and human activities. Within the concept of this relationship, it is found that they simultaneously reinforced each other and formed a vicious circle. Several studies, such as Foley et al. (2004); (Deafalla et al. 2014) and Deafalla (2019), confirmed that there is a strong and complex link between environmental variability and LU/LC. This relationship takes more important dimension when one starts thinking of building resilience against insecurity or problems related to sustainable management of ecosystem services. The changes caused by human-induced impact on atmospheric composition, climate, land, water, and biodiversity are occurring so rapidly that the natural systems are increasingly losing their adaptive/resilience capacity. The resulting degradation of the planet's resources and life-support systems may be irreversible at scales relevant to present human society. Moreover, the inertia in many natural processes means that the damage may take only years or decades later to become fully apparent (Melese 2016).

Numerous studies – for example: Collier (2003); Bannon and Collier (2003) and Deafalla (2019) – have found that, the Natural resources play a key role in increasing the likelihood of onset, the duration and the return to conflict; particularly in African developing countries. The links between environmental resources and conflicts are, however, highly complex, non-linear and influenced by a combination of factors, including political, social, economic, environmental, historical, and the different aspects of vulnerability (Notaras 2009). Actually, the interaction of these factors plays a role in either preventing or stimulating conflict. Nevertheless, large open access natural resources can undermine the quality of governance, aggravate corruption, weaken economic performance and, thereby, increase the vulnerability of countries to conflicts. Moreover, conflicts can also occur over the control and exploitation of resources (Deafalla 2012). On the other hand, the forced movement of people in those areas undermines, sometimes for decades, the economic development, sustainable livelihoods as well as the capacities of societies and nations. The shortage or degradation of natural resources contributes directly to lower levels of well-being and higher levels of vulnerability. However, in many developed societies, they clearly recognized the contribution of natural resources to poverty reduction efforts (UNEP 2007b). Environmental factors such as climate change are rarely the main driver of conflict. However, the related environmental stresses have a determining influence on peace and security. Furthermore, resource abundance or possession of specific high-value resources in socio-political contexts of weak institutions and poor governance are not only associated with low economic growth, but can contribute in increasing the likelihood and incidence of civil or armed conflicts, stimulates violence between rival groups as well as the stakeholders of the resource (UNEP 2007b; Ekbom 2009; Deafalla et al. 2014a). The changes in and depletion of natural resources linked to climate change have been considered as a causal factor in the current crisis in many regions of Sudan. Even through this brief mention, Sudan's Darfur region is a striking example of a vicious cycle where environmental sources are being fought over and at the same time being destroyed as a result of violence, one that highlights that peace is not possible unless underlying and closely linked environmental and livelihood issues are resolved (Deafalla et al. 2017).

As mentioned earlier, the links between EC, seen as a destabilizing interference in the ecosystems' equilibrium, on the one hand, and the outbreak of violent conflicts, on the other, remain ambiguous and very complex (Deafalla et al. 2017). Unfortunately, up to date, there is a gap in the researches that focus on the intersections between global EC and human security based on LU/LC changes (Behnassi 2017). In reality, LC patterns play an important role on EC detection studies, where they reflect the underlying natural and social processes, thus providing essential information for modeling and understanding many phenomena on the earth, including the complex interactions between global EC and human activities (Running 2008; Jia et al. 2014; Miller 2017). Timely and accurate LC data are, therefore, an essential factor for improving the performance of ecosystem (Running 2008; Gong et al. 2013) and have a permanent place in the international policy of sustainable development and ecosystem preservation.

Given the history of Nuba Mountains of Sudan, as a case study, the communities there live in fragile and unstable conditions, high poverty, high dependence upon natural resources as well as ethnic and religious fractionalization. The geography of the area, as well as the ethnic and political conflicts, makes them more vulnerable to the risk of violent conflict and the effects of environmental change. Unfortunately, the EC in the area based on the interaction of these factors could aggravate territorial and border disputes and complicate conflict resolution as well as mediation processes in future, in addition to hindering the process of development in the region. Some concerned scientists have warned of this prospect for several decades (e.g., Burr 1998; Faki et al. 2011; Totten et al. 2015), but the debate has been constrained by lack of carefully compiled evidence.

To address this shortfall of data, the study will examine the interplay between environmental factors, human security, and conflict on Nuba Mountains Region. As well as detecting the LU/LC in the area and their changes during the past decades. It also outlines some of the actions being taken to help a country adapting to the changing environment, and makes recommendations for how such actions could become more effective.

2 Study Site

The study site is located in Southern Kordofan State (Fig. 1). The State is bordered by Darfur in the west, Abyei and Republic of South Sudan in the South, White Nile and North Kordofan state in the North. The state is subdivided into 5 districts, Dilling, Kadugli, Rashad, Talodi and Abujubayha (Deafalla et al. 2017). The present research was carried out in Rashad district as shown in Fig. 1 below. It lies between latitudes 11° and 12° N, and longitudes 30° and 32° E, with a total area of 135,696 km^2. The study selected three different locations in the Eastern Nuba Mountains namely; Elabbassia, Rashad and Abu Karshola localities. Each locality is divided further into smaller administrative units. A commissioner is responsible for administering the locality, through legislative and executive bodies. At the village level, the government is represented by the tribal system. The region has a varying climate, ranging from semi-desert in the north to rich savanna in the south. Annual rainfall ranges from less than 50 mm on the northern border to more than 800 mm on the southern border. The rainy season varies from about five months or less, with rains occurring between May and October. The average daily temperature ranges from 10 to 35 °C with an annual variation of 15 °C. April to June is the hottest period and December to February is the coldest. Wind direction differs according to seasons: northeast in winter and southwest in summer (El Tahir et al. 2010). Total population of Nuba Mountains in 2008 was 1.3 million distributed into 120,986 households (CBS 2009). The livelihood activities found in the area are agro-pastoralism, nomadic pastoralism and rain-fed agriculture; both traditional farming for subsistence and mechanized farming for

Fig. 1 Location of the study area *Source* DIVA-GIS, developed by the authors

commercial operations. In addition to that, a third source of livelihood is related to natural forests in the form of woody and non-woody production derived from various tree species (UNDP 2006).

3 Research Methods

3.1 Remotely Sensed Data

Earth observation data provides an immense amount of information on LU/LC (Masek et al. 2001; Phiri and Morgenroth 2017), accordingly it is used as a series of temporal satellite data to study the LC changes in the study area for the past decades. Landsat (1984, 1994, 2002 and 2014) imagery, in addition to high resolution Digital Elevation Model (DEM) data from Advanced Spaceborne Thermal Emission and Reflection Radiometer (ASTER) and meteorological information, were used to gather the spatial characteristics of the region such as; vegetation cover (density and type), terrain features and surface temperature.

Moreover, to describe and verify the ground-types adequately, several ground-truthing methods were used. The first way involved the field points or ground-truth. 325 points were gathered during the field visit randomly from the secure areas in the study site representing different LC types, and to create a "test set" for the classification accuracy. The ground reference points were collected by using GARMIN eTrex Venture HC GPS device to record the coordinates of each visited point.

3.1.1 Data Pre-processing

The pre-processing is necessary for extracting and quantifying meaningful information from remotely sensed data (Iqbal and Khanb 2014; Naqvi et al. 2014; Butt et al. 2015). The typical pre-processing operations in the present study included: geometric correction, radiometric correction and topographic normalization and image enhancement by using Environmental Visualization (ENVI), Earth Resources Data Analysis System (ERDAS) Imagine and ArcGIS software.

3.1.2 Image Classification

a. Image Segmentationa. Image Segmentation

Multiresolution segmentation strategy was applied to the dataset in the present study as a preliminary step of GEOBIA approaches to create meaningful unclassified image objects (object primitives) and to generate different abstraction levels (Fig. 2). This segmentation algorithm characterized as a bottom-up region-merging technique, applying a threshold optimization procedure for image segments, which categorizes the image objects according to a maximum acceptable heterogeneity based on defined threshold to maximize their respective similarity (Baatz et al. 2002, 2004).

b. Geographic Object-Based Classification (GEOBIA)

In recent years GEOBIA classification for LC mapping purposes using Earth observation data has attracted significant attention (Lang 2008; Blaschke 2010; Li and Shao 2014; Ma et al. 2017), where the RS community has undertaken considerable efforts to promote the use of this technology for LU/LC mapping (Blaschke and Strobl 2001; Blaschke et al. 2004; Walker and Blaschke 2008; Blaschke 2010). In semi-arid regions, where areas are spatially heterogeneous and with a similar spectral response, LU/LC mapping with remotely sensed data encounters a complexity of problems when applying methods based on spectral information and ignoring spatial information (Chambers et al. 2007; De Sy et al. 2012; El-Abbas 2015). Therefore, for

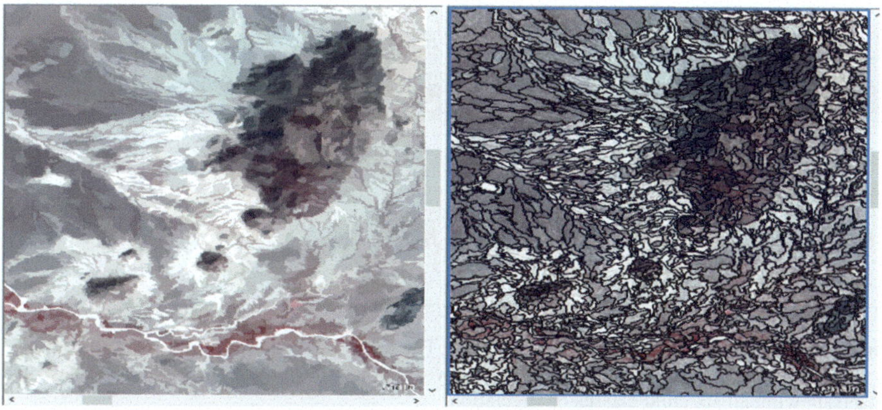

Fig. 2 The example segmentation result for the image year 1994 with scale parameter of 15, shape 0.1, smoothness 0.5 (left), Original image without segmentation, (right)

this study, the object-based approach has been proposed for discriminating different LC classes based on group pixels with analogous spectral and spatial response, based on predefined criteria to extract features of interest.

A GEOBIA hierarchical system was proposed in this study to attain LC classes. The rule-based classification of the study area was built on three types of knowledge: spectral domain knowledge, it was applied to construct the hierarchical structure of LC classes. Secondly, spectral classification rules based on training data, where the training data helps to generate thresholds to be used later as rules for discriminating and classifying LC categories more accurately. Finally, to support spectral knowledge in the classification of LC, spatial rules based on user experiences were used to increase the resultant accuracy. Trimble eCognitionTM Developer 8.7 software was used to analyze the images. The first level was created to discriminate between vegetation and non-vegetation parent classes. Subsequently, the second level of non-vegetation classes aimed to separate water from land classes, which allowed for the separation of one (Water bodies) and three land cover child classes (Bare land, Settlements and Rocky area). Meanwhile, vegetation categories were divided into five subcategories in a hierarchical format, which effectively reduced vegetation type-related confusions, these child classes comprised of; Agriculture lands, Dense forest, Grasslands, Horticulture land and Shrublands respectively in the second level. Furthermore, one child class (scattered forests) was generated from Dense forest class and it represented in the third level as shown in Fig. 3. Subsequently, after creating a nomenclature for LC, the research examined multi-temporal LU/LC changes in the study area.

c. Change Detection

After classifying the imagery of the selected dates, multi-temporal classified maps (1984, 1994, 2002 and 2014) were introduced to the Post Change Detection technique (PCD) to determine changes in LU/LC during the study period using mean-shift and

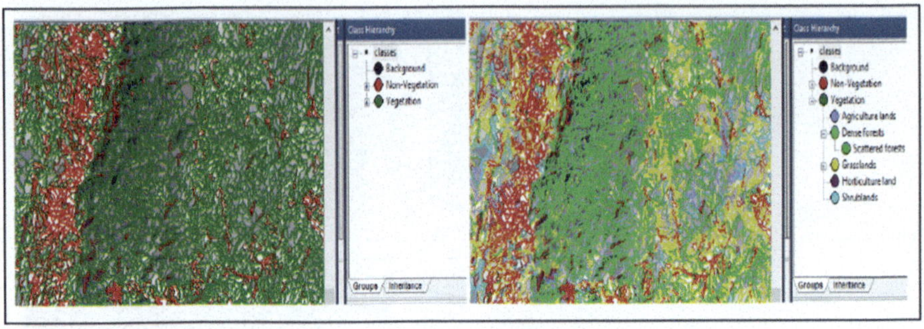

Fig. 3 The stratified classification diagram Note: The figure on the left shows the classified image to the top two parent classes – vegetation and non-vegetation. The figure on the right is the classified image showing all six child classes of vegetation.

outlier-distance metrics. In order to perform an appropriate multi-temporal analysis, PCD might reduce the possible effects of atmosphere, sun angle, seasonal variation of acquired date and multi-sensor variability (Singh 1989). However, rather than using multi-spectral imagery, the classified images were used. In addition, the post classification comparison provides class changes from-to and a change matrix of classes.

3.1.3 Accuracy Assessment

Classification accuracy assessment in this research was conducted for the four LU/LC images and for the three LU/LC change images based on the 60 reference TTA Mask samples utilizing the best classification result method (Zhan et al. 2005; Tiede et al. 2006). Additionally, for more accurate result of each class of LU/LC the overall accuracy, producer's accuracy, user's accuracy and Kappa statistics methods (Foody 2002; Schöpfer and Lang 2006; Grenier et al. 2008) were applied.

The result showed that Settlements class has higher accuracy (100% accuracy) which is attributed to the uniqueness of its spectral value as the remaining areas of the image represent very different spectral information compared to this class. The lowest accuracy of 81% has been observed in the water class for the image of 2014. This was confused with Bare lands and Rocky area classes. The confusion is attributed to units of pixels within the same class with heterogeneous values, due to the increased class variances of spectrum in high resolution imagery. Another confusion was observed between the Grassland class and Shrublands for the image of 2014 (86% accuracy). This is due to the fact that object information contained in these classes is not easily distinguishable. Even during the field visit for ground-truthing, it was difficult to differentiate between these classes.

The producer's accuracy and user's accuracy measures are useful to evaluate the classification accuracy for individual classes. Where, the accuracy assessment for all LU/LC categories were found, by using these methods, to be good, the only exception was for the class of Rocky areas (for image 1984) which achieved value of 0.71 and the class of Grasslands (for images 1984 and 2002) that showed precision values 0.83 and 0.86 respectively. The performance was poor to some extent. The current finding agreed with the result of best classification result method, regarding the class of Grassland, where it represented lower percentage as well. In general, the overall classification accuracy of both methods resulted in strong and perfect agreement. Meanwhile, these maps achieved an overall accuracy with values of 98, 99, 97 and 100% for years 1984, 1994, 2002 and 2014 respectively by using overall accuracy and kappa methods. Additionally, the kappa value of 0.97, 0.99, 0.97 and 1 for maximum likelihood classifier of the selected years respectively indicates better classification.

3.2 Socioeconomic Data

Semi-structured interviews of household heads, as well as Rapid Rural Appraisal (RRA), with focus on group discussions, informal meetings, free listing and key informant's techniques, and direct observations were applied, to address and detect the war's impact, demographic and LU/LC changes. The total sample size was 200 questionnaires, 200 for heads of households (covered displaced and non-displaced respondents) distributed among different units of Nuba Mountains according to the Principle of Population Proportional to Size (PPS) in the selected sites. Additionally, the study utilized observation questionnaire to identify the LU/LC, forest characteristics, topography, water and vegetation information. Furthermore, a formal literature review was undertaken to cover the areas which are directly related to difficulties of data collection due to security reasons.

Regarding data analyses, the Statistical Package of Social Sciences (SPSS) was used. A descriptive statistics method was applied to describe the basic features of the community as well as the social characters of respondents. Moreover, correlation analysis was applied to describe and test the relationship between these factors.

4 Results and Discussion

In the study site, the main impediment to carrying out the requisite field work was the risk of war. However, satellite sensor data have proven useful for mapping LC, estimating geophysical and biophysical characteristics of terrain features, as well as monitoring changes in LC, especially in conflict areas and post-war zones.

4.1 Changes of the First Period from 1984 to 1994

Unfortunately, the history of the study area during the last decade shows that it has been exposed to a series of recurring conflicts for years. The cataclysmic Sudanese civil war began in 1983 that pitted the North against the South. Since that date, till end of 1988, the Nuba generally sought to refrain from choosing any sides, despite that, the area has been affected and it was a silent casualty of war and armed conflict, where, the war was a major driving force in changing the rate and patterns of the local LU/LC dynamics.

In 1985, the situation changed when the Nuba people felt distrust and doubted that the government instituted policies aimed to marginalize or eliminate educated Nuba's leadership from participation in the Sudanese polity. Nevertheless, given government neglect and militia raiding, as well as the myriad and not so subtle forms of racial discrimination to which they were subjected, by 1989 many Nubas were allied with the Sudan People's Liberation Army (SPLA), and some had even joined its military (Burr 1998). The war in Sudan, since the end of 1989, has been characterized by an incremental fortunately in that time military activity and social policy directed

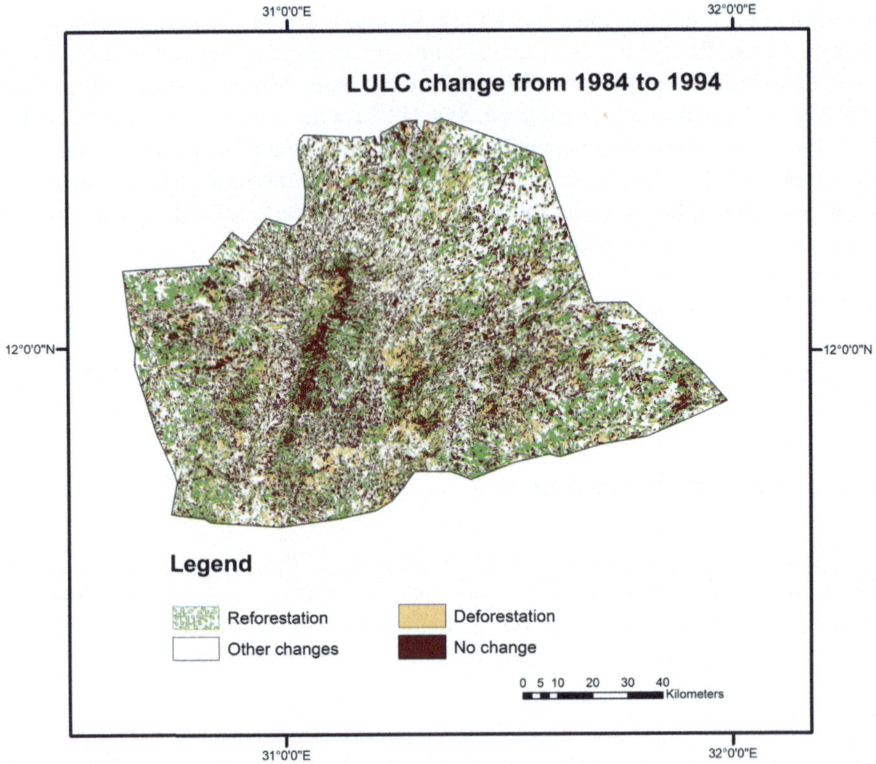

Fig. 4 LU/LC change from 1984 to 1994

against the Nuba peoples of South Kordofan have been nothing short of genocidal. The current results indicated that, the study area has undergone significant LU/LC alterations and transformations since late 1984s as illustrated in Fig. 4 above.

The initial clustering analysis of the period from 1984 to 1994 showed there is a considerable recovery in vegetation cover during the war period as shown in Fig. 4 above. Where, Dense forests areas coverage increased approximately to be 206,537.9 ha (16.4%) while Bare land decreased to be 100,532.7 ha (4.7%). Shrublands, Scattered forest and Grassland areas were declined slightly as well to 59,607.1 ha (4%), 225,005.4 ha (17.89%) and 356,203.3 ha (19.6%) respectively.

Relatively, cultivated lands increased from 15.7% to 16.5% during that period. It's important to note, there was a huge change in Settlements area, which increased from 1379.0 ha (0.2%) to be 35,929 ha (2.9%) in 1994. This is mainly due to the established villages "22 peace units" by Sudanese government in 1992 to house 70,000 returnees and resettlement of 500,000 Nuba internally displaced people (Burr 1998). Furthermore, the reluctance of Internally Displaced People (IDPs) to return to their rural homes after the war caused an increased demand for land driven by housing needs.

4.2 Changes in the Second Period from 1994 to 2002

Unfortunately, the analysis of the period 1994 to 2002, detected a pervasive environmental degradation in the area, which revealed an intensive and dynamic rate of deforestation, mainly related to admixture of dynamic interactions between social and ecological systems. Many empirical geospatial researches proved that, civil wars and other forms of conflict cause human displacement, especially when such displacement occurs within, rather than across, national boundaries. Civil wars are influential underlying drivers of human-induced environmental change in those environments (Geist and Lambin 2002; Tejaswi 2007; Gorsevski et al. 2012; Gbanie et al. 2018). Several authors (such as Dudley et al. 2002; Witmer 2008; Wilson 2014; Mansaray et al. 2016) have demonstrated that IDPs crises are contributing to increased LC change rates in areas where IDPs have relocated. These negative environmental consequences compromise the environment's potential to provide ecosystem services for both the IDPs as well as to non-displaced people (Gorsevski et al. 2012; Baumann et al. 2015). In reality, refugee villages that were created to take in those fleeing unstable areas, placed significant stress on the surrounding land and water resources in the study area. That led to uncontrolled exploitation of forest resources and a concomitant increase in LC change. As a result, natural vegetation, like forests, was removed, degraded and substituted mainly by crop fields which covered approximately 299,978.4 ha as clarified in Fig. 5 below.

The results indicate that severe land cover changes have occurred resulting from agricultural intensification, where Cultivated land and Bare land increased in the years 1994 and 2002 to 24 and 11.40% respectively. This is mainly due to urban and peri-urban agriculture became a major livelihood activity for IDPs to make the

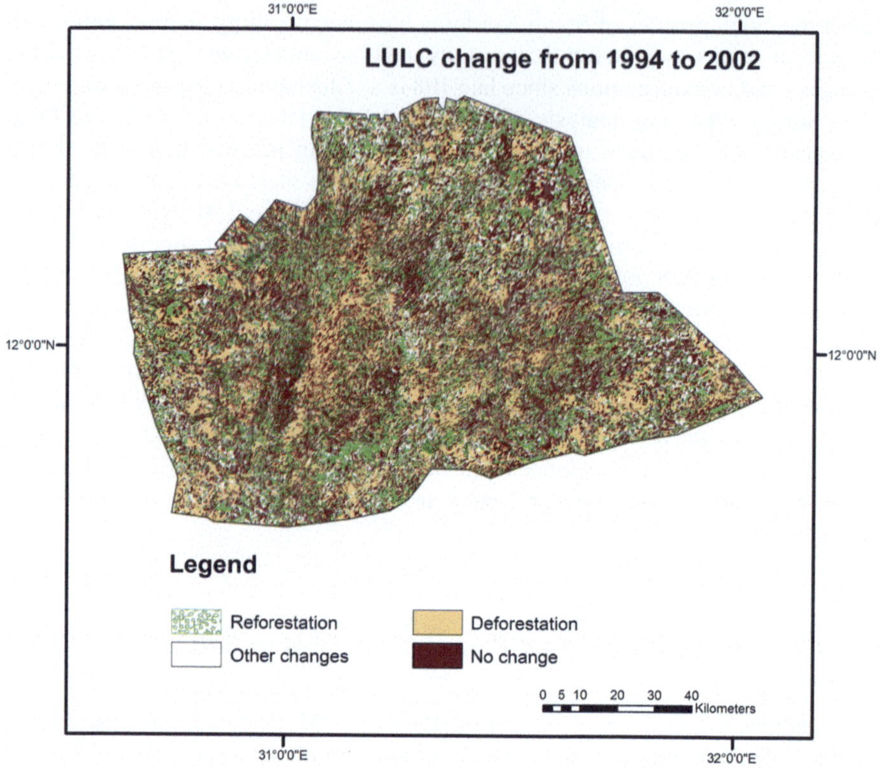

Fig. 5 LU/LC change from 1994 to 2002

region food self-sufficient, especially when the war destabilized food production activities. In contrast, Dense and Scattered forest lands were decreased from 16.4 and 17.89% to be 13.10 and 11.9% respectively. Indeed, that supports Ghazoul and Evans (2001) who referred to the fact that some types of small-scale agriculture undoubtedly cause deforestation. In fact, these changes directly contributed to the degradation and depletion of renewable resources of the area (Deafalla 2012; Deafalla et al. 2014b). The result of the case study revealed that, the area has been severely hit by sand encroachment, soil erosion especially in the northeast part of the area.

On the other hand, the analysis of the matrix of LU/LC changes demonstrated the rapid decline of Grasslands (10.65%). It was replaced mainly by Shrublands that became (20.49%). Actually, this was nearly five times the size of the areas in 1994. That was related to the ecological succession, which is the term used to describe the directional non-seasonal cumulative change in the types of plant species that occupy a given area over time. It involves the processes of colonization, establishment, and extinction which act on the participating plant species (Sahney and Benton 2008). Figure 6 below is an example for progress in ecological succession.

Fig. 6 An example for progress in ecological succession *Source* Pidwirny (2006)

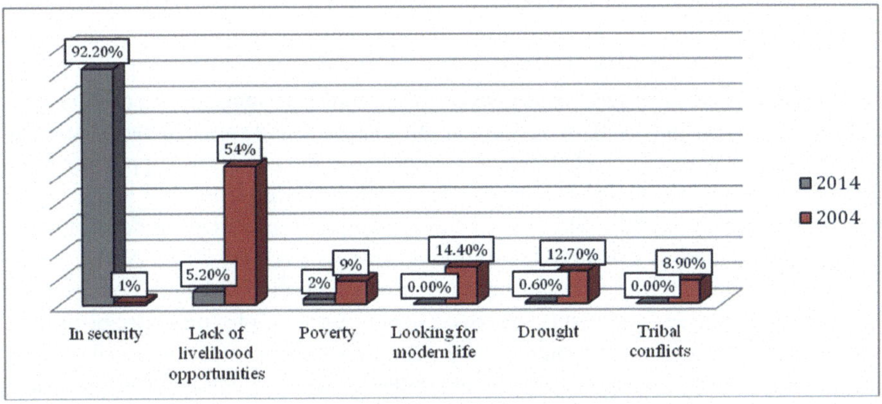

Fig. 7 Reasons of migration and mobility in Nuba Mountains

4.3 *Changes in the Third Period from 2002 to 2014*

In this period, the recent changes in the global political economy and environmental systems, as well as local dynamics of the state, have increasingly brought rapid changes in LC, social, institutional and livelihood transformation across broad areas of the study area.

4.3.1 The Demographic Change

The underlying causes of respondent's migration in 2004 were totally different than in 2014 (Fig. 7), where they were mainly attributed to the lack of livelihood opportunities (54%), followed by drought (12.7%), poverty (9%), tribal conflicts (8.9%)

and search for modern life (14.4%). The current study agreed with IOM (2005) and Abrha (2017) who noted that, the youth employment crisis pushes millions of people, especially youth, women and men, towards the decision to migrate with the purpose of seeking alternatives to improve their job prospects. Many of them migrate to urban areas and big cities within their countries or seek new opportunities in foreign countries. According to the statistics of UNDESA (2016), roughly 71 million unemployed youth leave their countries of birth to seek employment aboard as international migration.

The proportion of young migrants was higher in Nuba Mountains, estimated to be 80.4%, comparing with the proportion of girls and young women who migrated which represented only 19.6%. The average age of men migrants was between 15 to 25 years. Meanwhile, the average age of girls or women were 20 to 35 years.

Migration is often misperceived as a failure to adapt to the changing environment. Instead, migration can also be an adaptation strategy to climate and environmental change, if managed correctly, because it is an essential component of the socio-environmental interactions (IMO, 2017). Based on FAO (2016), (2018), migration has many challenges and opportunities. For example, at the rural level, losses in human capital and agricultural labour may have negative impact on crop production and food availability. Moreover, the labor migration experiences can end up representing either an opportunity or a risk to youthful people and can lead them to decent work, or it's very opposite. That depends on policies and measures supporting them e.g.; provision of better knowledge of the world of work, more and better social protection, education and training for employment, entrepreneurship development, social inclusion and an effective institutional framework (IOM 2005; Abrha 2017). On the other hand, migration fostering rural–urban economic linkages, could reduce pressure on resources, enhancing and diversifying rural employment opportunities, and helping the poor to better manage risks through social protection. Furthermore, the leveraging remittances for investments in the rural sector can be effective strategies for improving livelihoods and alleviating distress-induced migration. In addition, the migration can bring new opportunities for immigrants, through participating in higher education, finding a better and decent job, getting a chance to gain professional experience and an opportunity to develop their personal competencies (UNDESA 2016). The results of the case of Nuba community in Khartoum state were the best example supporting the above studies. The equality (to some extent) of opportunity and treatment of migrants, allowed them to contribute as productive members of the new community at different levels of income, as workers, entrepreneurs, students, and consumers. Moreover, the migration became a productive and empowering experience and opened up new opportunities for them, where migrants, both women and men, gained more skills through education, and/or work experience, and earned higher wages, allowing them to support their families and to contribute to the development of their communities of origin, as well as the societies in which they live and work. For young women, migration allowed them socially empowering benefits, as the recipients of remittances or as breadwinners or students in a new state. In the future, young women may gain decision-making power and experience greater personal autonomy.

4.3.2 LU/LC

Unlike the analysis during the study period, 1994 to 2002, the investigation of the third period (Fig. 8) showed a considerable recovery of the natural resources, due to the abandonment of agricultural lands during the war period. That agreed with Stevens et al. (2011); Ordway (2015); Gbanie et al. (2018), who have referred to that conflicts sometimes help protect forest environments and their biodiversity, depending upon their nature and location. For instance, the demilitarized zone between North and South Korea has provided a safe haven for migratory birds (Brady 2008). Similarly, forests along the Thailand-Malaysia border, which remained untouched during periods of insurgency (1960–1970), have supported abundant wildlife and are being converted into a National Park by the Malaysian Government (McNeely 2003). The results of changing in LU/LC from 2002 to 2014 reveals that, the natural vegetation has been reforested and modified during the last years, where Forest lands (Dense and Scattered) increased from 316,048.7 ha in 2002 to 424,002 ha in 2014 (Fig. 8 below). The observed increase was due to the transformation of 121,097 ha of Cultivated land, 24,989.3 ha of Bare land, 5009.75 ha of Horticulture land and 36,865.56 ha of Rocky land.

Nuba Mountains is rich with various naturally growing tree species, where the forests cover 33.76% of the study area. The region has diverse and rich vegetation, resulting from the variability in climate, soils and rainfall (Harrison and Jackson 1958; El Tahir et al. 2010; Deafalla et al. 2018), where the area covered with a mixture of different tree families such as: Acacia senegal savanna, broad-leaved savanna woodland Acacia seyal and Balanites aegyptiaca. Other thorny and broad-leaf non-thorny woody species are also present. The study indicated three types of forest tenure systems in the area; reserved forests, community forests and natural woodlands. The forest reserves in the study area constitute a basic source of a variety of environmental goods and services most needed by villagers in addition to providing

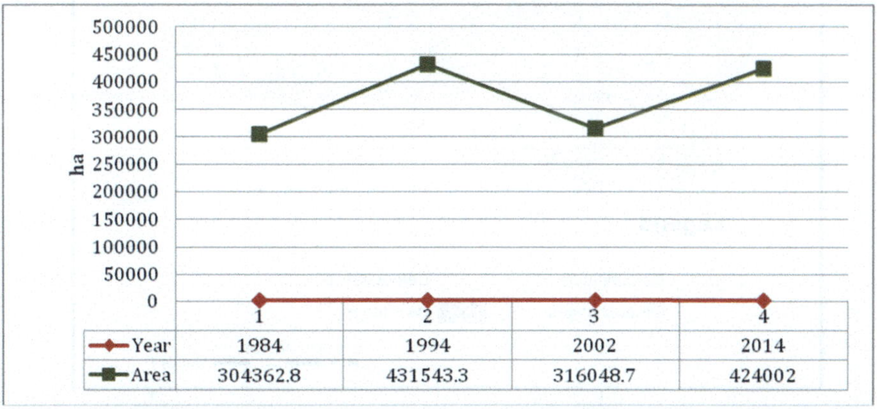

Fig. 8 Forest lands in study area

them with critical subsistence, income generation and job opportunity. Furthermore, the forestry and agricultural residues of these communities typify the fundamental source of energy for 96% of citizens around the forest.

By contrast, as illustrated in Fig. 8 above, Bare land in the study site has decreased from 11.40% to 9.7%. A huge decrease of Cultivated land and Horticulture land was observed during the study period, where they represented only 13.16 and 0.18% respectively. Generally, all these areas became reforested. The analysis, as well, indicated a slight decline in Shrublands (20.06%), with increase in Grasslands that became (20.54%). Actually, this was almost double their area in 2002. Figure 9 below shows the LU/LC change dynamics in the study area, 2002 to 2014.

Indeed, economic conditions have a major influence on the situation of forests in terms of exploitation, conservation possibilities and regulations, where poor communities depend heavily on natural resources, as well, it is to be noted that economic development increases the pressure on the environment (FAO 1994). The results showed a rate of deforestation in some locations, such as the findings in Fig. 9 below.

Sudan, like other developing countries, has weak strategies, policies and legal frameworks that support the sustainability of natural resources (Atta El Moula 1985;

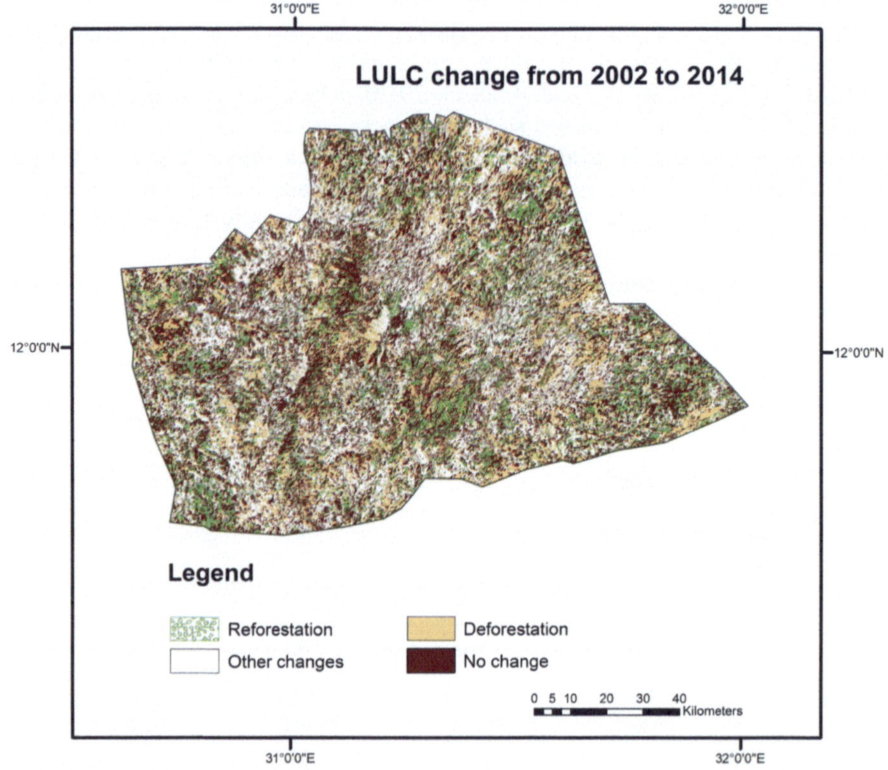

Fig. 9 LU/LC change from 2002 to 2014

Tolentino 1991; Deafalla et al. 2015). The criteria of forest resource management were largely carried out informally through local community leaders prior to the colonial era. Since 1986, governments intervened and withdrew the managing control from villagers and followed new approaches to manage this resource. The study found that the largest parts of forest benefits were obtained through illegal access, leading to negative impact on forest resources. Moreover, 82.9% of respondents had no relationship with Sudan's Forests National Corporation, while only 17.1% had this relationship, implying: access to extension services, licenses, establishing relationships, training on awareness, as well as rights and properties. On the other hand, and despite the small area and limited contribution of community forests, farmers possessed a huge experience in the establishment of plantations and their management. Therefore, opposite to the expectations of the government, multiple-use management policy proved to be ineffective in the absence of coordination between all stakeholders.

Although the study area has heavy rain, there are no permanent watercourses in the area, where the water represents only 0.07% of the land cover. However, it forms seasonal streams, the most prominent among which are Khor Aldelib, Khor Kadada, Khor Tagmala and Khor Umbrumbita, which carry substantial run-off during July through October. The degradation of farmland hit the poor first, the wealthy monopolized the fertile growing areas known as "Wadis" in the study area, which are areas of land with subsurface water, and therefore more capable of producing crops. To overcome these problems of irrigation there, farmers have worked through the rehabilitation and expansion of traditional water harvesting techniques which led to the spread of water harvesting techniques called 'Hafir'. The Hafir is the local name in Sudan for water reservoir. It is a hollow dug in the ground designed to store water runoff after a rainy season, it can be natural or manmade. The Hafir is largely used in semi-arid regions, where rainfall is annual but over short periods, and storage is required for the rest of the year and preferred in clay soils, so filtration is reduced, allowing maximum storage and less labour. The water is used by the entire community, farmers, nomads, and livestock for various purposes, including domestic drinking water. The number of Hafirs in the area increased during the last years from 0 in 1984 to 22 in 1994, the huge spread was in 2002, when the number was estimated to be 90 Hafirs, but unfortunately this number retreated to 79 in 2014, due to devastation by civil war in some sites. The other water sources in the area are shallow wells and deep boreholes. Water resources of Nuba Mountains were affected during the last years, where groundwater recharge was decreased by soil erosion, precipitation and/or increased temperatures, evaporation and war. Interestingly, the time consumed in a trip of water collection for household uses ranged from 2 to 3 h in 2003 and increased to 3 to 5 h in 2008 (Deafalla et al. 2018). Water-related problems are likely to worsen as a result of environmental change during upcoming years.

5 Conclusions

The spatial heterogeneity and rapid changes observed in the Nuba Mountains region of Sudan motivate the inspection for more efficient, reliable and accurate methods to update the required information for sustainable development. At present, planners and decision makers are not provided with automated data sources of the desired information levels. Over large areas, field surveys are often tedious and time consuming and other innovative methods of gathering the information need to be developed. In this sense, the study reveals that the imagery from mid/higher resolution sensors (i.e., Landsat imagery) linked with GIS data and a socio-economic survey, has great potential in identifying the key of categorical LU/LC patterns. This method has proven its capability, as an efficient and accurate methodological framework for gaining knowledge about land features, and for identifying areas which are most vulnerable to EC in a way that would be relevant to policy makers and other stakeholders. More information exchange is needed to inform actors and decision makers regarding specific experiences, capacity gaps and knowledge to address the environmental change, in addition to developing viable policy responses. Furthermore, dynamic innovative research, strategies, management and policies that focus on local communities are required.

References

Abrha FW (2017) Determination of youth migration the case of tsegedie wereda. J Bus Fin Aff 5: 234. an open access journal. vol. 6 (1), p 1000234

Atta El ME (1985) On the problem of resource management in the Sudan. Institute of Environmental Studies, University of Khartoum, Environmental Monograph No, p 4

Baatz M, Benz U, Dehghani S, Heynen M, Höltje A, Hofmann P, Lingenfelder I, Mimler M, Babu TP, Singh PK, Nigam RK (2002) Mapping of forest cover in Rohtak district, Haryana using remote sensing techniques. In: Proceedings of Geomatics 2002 conference on IT enabled spatial data services, Tiruchirapalli, India, pp 115–120

Baatz M, Benz U, Dehghani S, Heynen M, oltje A H, Hofmann P, Lingenfelder I (2004) eCognition professional user guide, version 4.0. Definiens Imaging GmbH. Munchen, Germany

Bannon I, Collier P (2003) Natural resources and conflict: What we can do. In Natural Resources and Violent Conflict: Options and Actions. In: Bannon I, Collier P (eds) Washington, D.C.: World Bank. pp 1–16

Baumann M, Radeloff VC, Avedian V, Kuemmerle T (2015) Land-use change in the Caucasus during and after the Nagorno-Karabakh conflict. Regional Environ Change 15:1703–1716

Behnassi M (2017) Environmental Change and Human Security in Africa and the Middle East. Springer, pp 3–14

Blaschke T (2010) Object based image analysis for remote sensing. ISPRS J Photogramm Remote Sens 65(1):2–16

Blaschke T, Strobl J (2001) What's wrong with pixels? some recent developments interfacing remote sensing and GIS. GIS–Zeitschrift für Geoinformationssysteme 14(6):12–17

Blaschke T, Burnett C, Pekkarinen A (2004) New contextual approaches using image segmentation for object-based classification. In: De Meer F, de Jong S. (eds.) Remote Sensing Image Analysis: Including the spatial domain. Kluver Academic Publishers, Dordrecht, pp 211–236

Brady LM (2008) Life in the DMZ: Turning a diplomatic failure into an environmental success. Dipl Hist 32:585–611

Brown O, Hammill A, McLeman R (2007) Climate change as the 'new' security threat: implications for Africa. Int Affairs J 83(6):1141–1154

Brown O, Crawford A (2009) Climate Change and Security in Africa. Published by the International Institute for Sustainable Development, Manitoba, Canada, A Study for the Nordic-African Foreign Ministers Meeting, pp 1–20

Burr J M (1998) Quantifying Genocide in Southern Sudan and the Nuba Mountains, 1983–1998. In: Conflict in the Nuba Mountains. From Genocide-by-Attrition to the Contemporary Crisis in Sudan. Chapter 5. p 23

Butt A, Shabbir R, Ahmad SS, Aziz N (2015) Land use change mapping and analysis using Remote Sensing and GIS: a case study of Simly watershed, Islamabad, Pakistan. Egyptian J Remote Sensing Space Sci (EJRS) 18:251–259

Central Bureau of Statistics (CBS) (2009) Fifth Sudan National Population and Housing Census. Census tables. www.cbs.gov.sd. Accessed 1 June 2014

Chambers JQ, Asner GP, Morton DC, Anderson LO, Saatchi SS, Espírito FD, Palace M, Souza C (2007) Regional ecosystem structure and function: ecological insights from remote sensing of tropical forests. Trends Ecol Evol. 22(8):414–423

Collier P (2003) Natural Resources, Development and Conflict: Channels of Causation and Policy Interventions, World Bank, Washington D.C., pp 1–13

De Sy V, Herold M, Achard F, Asner GP, Held A, Kellndorfer J, Verbesselt J (2012) Synergies of multiple remote sensing data sources for REDD+ monitoring. Curr Opin Environ Sustain 4:696–706

Deafalla THH (2012) Non-Wood Forest Products and Poverty Alleviation in Semi-Arid Region. By LAP LAMBART Academic Publishing GmbH & Co. KG, Germany, pp 15–35

Deafalla THH (2019) Integrative Assessment and Modelling of the Non-Timber Forest Products Potential in Nuba Mountains of Sudan by Field Methods, Remote Sensing and GIS (PhD. Thesis). Submitted to Faculty of Environmental Sciences, University of Dresden, June 2019, Germany, 319 p

Deafalla THH, Csaplovics E, El-Abbas MM (2014a) Analysis of Environmental Change Dynamics in Arid and Semi-Arid climatic zones. Int J Performability Eng Sustainability Balancing Technol Environ Life Styles **10**(4):401–410

Deafalla THH, Csaplovics E, El-Abbas MM (2014b) The application of remote sensing for climate change adaptation in Sahel region. In: Proceedings of SPIE International Conference, Earth Resources and Environmental Remote Sensing/GIS Applications V, 92451R, Amsterdam, Netherlands

Deafalla THH, Csaplovics E, El-Abbas MM, Baba CAK (2015) Constraints of sustainable forest management: a case of nuba mountains region. In: Tropentag 2015 international conference, management of land use systems for enhanced food security – conflicts, controversies and resolutions; 16–18 September, Berlin, Germany. Edited by Tielkes, E. ID. 1041. p 341

Deafalla THH, Csaplovics E, El-Abbas MM (2017) Impact Assessment of the environmental change in nuba mountains of sudan. In: Behnassi M, McGlade K. (eds) Environmental Change and Human Security in Africa and the Middle East, Chapter 5, pp 95–107

Deafalla THH, Csaplovics E, El-Abbas MM (2018) Environmental change impacts and human security in semi-arid region, the case of nuba mountains of sudan. In: Behnassi M, Olf P (eds) Human and environmental security in the era of global risks, pp 187–203

DIVA-GIS. http://www.diva-gis.org/gdata

Dudley JP, Ginsberg JR, Plumptre AJ, Hart JA, Campos LC (2002) Effects of war and civil strife on wildlife and wildlife habitats. Conservation Biol J 16:319–329

Ekbom A (2009) Africa Environment and Climate Change Policy Brief, Sida's regional Climate Change and Development-training workshop, 5–7 May 2009, Nairobi Kenya, pp. 1–27

El-Abbas MM (2015) Assessing, monitoringand mapping forest resourcesin the blue nile regionof sudan usingan object-based image analysis approach. Doctoral thesis, Faculty of Environmental Sciences, University of Dresden. Germany. 262 p

El Tahir BA, Fadl KM, Fadlalmula AGD (2010) Forest biodiversity in Kordofan Region, Sudan: effects of climate change, pests, disease and human activities. Biodiversity Conservancy 11(3–4):34–44

Faki H, Nur ME, Abdelfattah A, Hassan AAA (2011) Poverty Assessment. Northern Sudan. ICARDA, Aleppo, Syria. pp vi + 62

FAO (1994) The state of food and agriculture 1994. FAO Agriculture Series, no (27). http://www.fao.org/docrep/t4450e/T4450E00.htm. Accessed 9 May 2013

FAO (2016) Migration, Agriculture and Rural Development, Addressing the root causes of migration and harnessing its potential for development. 20 p. http://www.fao.org/3/a-i6064e.pdf. Accessed 2 Jan 2018

FAO (2018) Decent Rural Employment, Migration. http://www.fao.org/rural-employment/work-areas/migration/en/. Accessed 2 Jan 2018

Foley JA, Kucharik CJ, Twine TE, Coe MT, Donner SD (2004) Land Use, Land Cover, and Climate Change Across the Mississippi Basin: Impacts on Selected Land and Water Resources, Ecosystems and Land Use Change, Geophysical Monograph Series vol. 153

Foody GM (2002) Status of land cover classification accuracy assessment. Remote Sens Environ 80:185–201

Gbanie SP, Griffin AL, Thornton A (2018) Impacts on the urban environment: land cover change trajectories and landscape fragmentation in post-war western area. Sierra Leone Remote Sens 10(129):1–25

Geist HJ, Lambin EF (2002) Proximate causes and underlying driving forces of tropical deforestation. Bioscience 52:143–150

Ghazoul J, Evans J (2001) Deforestation and land clearing. In: Levin SA (ed) Encyclopedia of Biodiversity, vol. 2

Gorsevski V, Kasischke E, Dempewolf J, Loboda T, Grossmann F (2012) Analysis of the Impacts of armed conflict on the Eastern Afromontane forest region on the South Sudan—Uganda border using multitemporal Landsat imagery. Remote Sens Environ 118:10–20

Grenier M, Labrecque S, Benoit M, Allard M (2008) Accuracy assessment method for wetland object-based classification. In: Proceedings of GEOBIA 2008, Calgary, AB, Canada, 6–7 August, pp 285–289

International Organization for Migration (IOM) (2005) Results of the survey engaging Diasporas as agent for development, pp 193–250.

IOM (2017) The World Migration Report 2018. ISBN 978-92-9068-742-9, 344 p. http://publications.iom.int/system/files/pdf/wmr_2018_en.pdf.

Iqbal MF, Khanb AA (2014) Spatiotemporal Land Use Land Cover change analysis and erosion risk mapping of Azad Jammu and Kashmir, Pakistan. Egyptian J. Remote Sens. Space Sci. 17(2):209–229

Jia K, Liang S, Zhang N, Wei X, Gu X, Zhao X, Ya Y, Xie X (2014) Land cover classification of finer resolution remote sensing data integrating temporal features from time series coarser resolution data. ISPRS J Photogramm Remote Sens 93:49–55

Khanna T, Palepu K (1997) Why focused strategies may be wrong for emerging markets'. Harv Bus Rev 74(4):41–51

Lang S (2008) Object-based image analysis for remote sensing applications: modeling reality – dealing with complexity. In: Blaschke T, Hay G, Lang S (eds) Object-based image analysis - spatial concepts for knowledge-driven remote sensing applications. Lecture Notes in Geoinformation & Cartography (LNG&C), Springer, Berlin. pp 3–28

Lein JK (2012) Analytical Techniques for Earth Observation. Springer, In Environmental Sensing, pp 23–49

Li X, Shao G (2014) Object-based land-cover mapping with high resolution aerial photography at a county scale in midwestern USA. Remote Sensing 6:11372–11390

Ma L, Manchun LI, Ma X, Cheng L, Du P, Liu Y (2017) A review of supervised object-based land-cover image classification. ISPRS J Photogramm Remote Sens 130:277–293

Mansaray LR, Huang J, Kamara AA (2016) Mapping deforestation and urban expansion in Freetown, Sierra Leone, from pre- to post-war economic recovery. Environ Monit **188**(8): 470

Masek JG, Honzak M, Goward SN, Liu P, Pak E (2001) Landsat-7 ETM+ as an observatory for land cover: Initial radiometric and geometric comparisons with Landsat-5 Thematic Mapper. Remote Sens Environ 78:118–130

Melese SM (2016) Effect of land use land cover changes on the forest resources of ethiopia. Int J Natural Resource Ecol Manage 1(2):51–57

McNeely JA (2003) (2003) Conserving forest biodiversity in times of violent conflict. Oryx 37:142–152

Miller PV (2017) Is there a future for survey? Public Opinion Quart 81(Special Issue):205–212

Naqvi HR, Siddiqui L, Devi LM, Siddiqui MA (2014) Landscape transformation analysis employing compound interest formula in the Nun Nadi Watershed, India. Egyptian J Remote Sens Space Sci 17(2):149–157

Notaras M (2009) Does Climate Change Cause Conflict? Our World, United Nation University. https://ourworld.unu.edu/en/does-climate-change-cause-conflict

Ordway EM (2015) Political shifts and changing forests: effects of armed conflict on forest conservation in Rwanda. Glob Ecol Conserv 3:448–460

Phiri D, Morgenroth J (2017) Developments in landsat land cover classification methods: a review. Remote Sens 9:967

Pidwirny M (2006) Plant Succession. Fundamentals of Physical Geography, 2nd edn. http://www.physicalgeography.net/fundamentals/9i.html

Running SW (2008) Ecosystem disturbance, carbon, and climate. Science 321:652–653

Sahney S, Benton MJ (2008) Recovery from the most profound mass extinction of all time (PDF). Proc Roy Soc Biolo 275(1636):759–765

Schöpfer E, Lang S (2006) Object fate analysis- a virtual overlay method for the categorisation of object transition and object-based accuracy assessment. In: Proceedings of 1st International Conference on Object-Based Image Analysis (OBIA 2006), Salzburg, Austria

Singh A (1989) Digital change detection techniques using remotely-sensed data. Int J Remote Sens 10(6):989–1003

Stevens K, Campbell L, Urquhart G, Kramer D, Qi J (2011) Examining complexities of forest cover change during armed conflict on Nicaragua's Atlantic Coast. Biodivers Conserv 20:2597–2613

Suarez FF, Oliva R (2005) Environmental change and organizational transformation. Ind Corp Chang 14(6):1017–1041

Tiede D, Lang S, Hoffmann C (2006) Supervised and forest type-specific multi-scale segmentation for a one-level-representation of single trees. In: 1st international conference on object-based image analysis (OBIA 2006). 5 p. http://citeseerx.ist.psu.edu/viewdoc/download?doi=10.1.1.374.3670&rep=rep1&type=pdf

Tejaswi G (2007) Manual on Deforestation, Degradation and Fragmentation Using Remote Sensing and GIS; MAR-SFM, Working Paper 5; Food and Agriculture Organization of the United Nations: Rome, Italy

Tolentino B (1991) Promoting forestry as a land use under the Philippines Social Forestry Programme. In: . Mathoo M, Chipeta M, (ed) Trees and forests in rural land use. Forestry Department, Food and Agriculture Organization of the United Nations, Rome, Italy

Totten S, Grzyb A (eds). (2015) Conflict in the nuba mountains. New York: Routledge. 1st edn, 314 p

United Nations Development Programme (UNDP) (2006) Pastoral production systems in south Kordofan, Sudan (study 2), pp 8–13

United Nations Environment Program (UNEP) (2007a) The fourth Global Environment Outlook–environment for development (GEO-4). UNEP paperback, Vulnerability of people and the environment: Challenges and opportunities, pp 33–342.

UNEP (2007b) Global Environment Outlook GEO environment for development4. 572 p. The fourth Global Environment Outlook - Environment for development - GEO 4
United Nations Department of Economic and Social Affairs (UNDESA) (2016) Youth Issue Briefs 2016 Migration and Youth. UN Department of Economic and Social Affairs. pp 2–5
White P, Picket S (1985) Natural disturbance and patch dynamics: an introduction. In: Picket S, White P (eds) The ecology of natural disturbance and patch dynamics. Academic Press, New York, pp 3–13
Walker JS, Blaschke T (2008) Object-based landcover classification for the Phoenix metropolitan area: Optimization vs. transportability. Int J Remote Sens **29**(7):2021–2040
Wilson C (2014) Spectral analysis of civil conflict-induced forced migration on land-use/land-cover change: the case of a primate and lower-ranked cities in Sierra Leone. Int J Remote Sens 35:1094–1125
Witmer FDW (2008) Detecting war-induced abandoned agricultural land in northeast Bosnia using multispectral, multitemporal Landsat TM imagery. Int J Remote Sens 29:3805–3831
Zhan Q, Molenaar M, Tempfli K, Shi W (2005) Quality assessment for geo-spatial objects derived from remotely sensed data. Int J Remote Sens 26:2953–2974

Resilience building Initiatives to Counter Shocks and Stressors Affecting Rural Communities in Chiredzi District, Zimbabwe

Rameck Defe and Mark Matsa

Abstract Zimbabwe has experienced innumerable shocks and stressors which are bedeviling economic and social development. Shocks in the form of natural hazards (droughts, floods), epidemics, environmental degradation, and cash crisis severity, scope and impact have necessitated coalition of stakeholders and development of resilience building initiatives to counter shocks and stressors in a manner that protects livelihoods and recovery gains. This study sought to assess the effectiveness of implemented resilience building initiatives to counter shocks affecting rural communities of Chiredzi District, Zimbabwe. Initiatives were implemented through the Enhanced Community Resilience and Sustainability (ECRAS) projects led by Care International and Plan International targeting to increase the capacities of Chiredzi District communities to withstand shocks and stressors. Questionnaires, interviews, and direct field observations were used to gather field data. Questionnaires were electronically administered using kobo collect to respondents randomly selected from conveniently sampled wards involved in resilience building initiatives in Chiredzi District. Interviews were conducted with purposively sampled key informants who included District Agritex officer, District Administrator, District Social Services, Care International Monitoring and Evaluation Officer, Plan Project facilitator and Ward Councilors. Findings revealed three main categories of resilience building initiatives in the rural communities of Chiredzi District, namely: crop production related initiatives; livestock production related initiatives; and social and well-being related initiatives. Resilience building initiatives are effective tools to deal with recurrent shocks and stressors in rural communities of Chiredzi District which are being projected to continue for the next decades.

Keywords Building resilience initiatives · Shocks and stressors · Rural communities · Chiredzi district

R. Defe (✉) · M. Matsa
Department of Geography and Environmental Studies, Midlands State University, Gweru, Zimbabwe

M. Matsa
e-mail: matsam@staff.msu.ac.zw

M. Behnassi et al. (eds.), *Social-Ecological Systems (SES)*,
https://doi.org/10.1007/978-3-030-76247-6_8

1 Background of the Study

'Building resilience' is defined as the ability of at-risk individuals, households, communities and systems to anticipate, cushion, adapt, bounce back better and move on from the effects of shocks and hazards in a manner that protects livelihoods and recovery gains, and supports sustainable transformation (Mitchelle 2013; UNDP 2015). The concept of 'resilience' has emerged as a plausible framework among development actors and governments' long term and more cost-effective strategy for substantially improving regional or local capacity to withstand shocks and stressors (Choularton et al. 2015; FAO 2015; Kuria and Nyaggah 2016).

Shocks and stressors may be interrelated or may occur simultaneously; one type of shock may contribute to another: high food prices can lead to social unrest and political instability (Lagi et al. 2011; Breisinger et al. 2014). Major shocks and stressors being felt at the global level are hydro-meteorological hazards (floods, droughts, storms and wildfires) and geological hazards, for example, earthquakes and landslides; these are threatening to reverse major developmental gains (World Bank 2003; FAO 2014; Obefreng 2016). The Global status quo of heightened risks and fragility is being aggravated by climate change which is projected to continue to change over this century and beyond (IPCC 2007; Brown et al. 2012; MacDemortt 2015). Climate change, environmental degradation, water scarcity, diseases, and natural disasters in the form of floods, hurricanes and droughts have called need for building resilience initiatives and this has drained international donors and national governments millions of dollars.

Many shocks are increasing in their severity, scope, and impact (Oxley 2013; Zseleczky and Yosef 2014; FAO 2015). Shocks frequency, severity, and duration can vary widely; they range from low intensity shocks with gradual onsets (droughts) to more intense and sudden onset shocks (earthquakes). They can affect large geographic areas or populations *(covariate)* or single households (*idiosyncratic*) (Carletto et al. 2015). In the face of these challenges, the World Food Programme (WFP), governments and other donors are collaborating, developing innovative tools and strategies to reduce and mitigate risks in order to overcome hunger, achieve food security, and enhance resilience.

In developing countries, more than 2.3 billion people live in rural areas and most of them live on less than $2 a day (World Bank 2014; WFP 2017). They depend on agriculture for their livelihoods (Burchi and De Muro 2007; WHO 2017). They are living in areas where natural resources quality is continuing to depreciate. The WFP and Oxfam America launched the R4 Rural Resilience Initiative in 2011 to enable vulnerable families to increase their food and income security by managing climate related risks (WFP 2017). The R4 is currently active in Ethiopia, Senegal, Malawi and Zambia; the initiative is also being piloted in Kenya and Zimbabwe (FSIN 2014). It has a combination of four risk management strategies (4 Rs): improved resource management through asset creation (risk reduction); insurance (risk transfer); livelihoods diversification and microcredit (prudent risk taking); and savings (risk reserves), (WFP 2017). Between 2015 and 2016, (as a consequence of the El Nino phenomenon) about USD$450 000 pay outs were distributed through the initiative in Ethiopia, Senegal, and Malawi (Mechler 2012; USAID 2015).

Over the past decade, the amount of rainfall received in Zimbabwe has deviated from the multi-decadal mean on a regular basis (UNEP 2010). Approximately 80% of Zimbabweans depend on agriculture which is mostly rain-fed (Madzwamuse 2010; FAO 2015). Food insecurity, epidemics like HIV/AIDS, poor soils, deforestation, water scarcity, crop pests (fall army worm and quelea birds), and environmental degradation are serious challenges in Zimbabwe, particularly in rural areas; and the situation has been exacerbated by climate change which is believed to have trebled community vulnerability (Kurtz and Scarborough 2012; WFP 2014; USACE 2016).

Development actors, government, communities, households and individuals developed resilience building initiatives as strategies for substantially improving their capacities to withstand shocks and stressors (Pelling et al. 2014; Brown 2016). In Zimbabwe, efforts are being made through the Zimbabwe Resilience building Fund (ZRBF); for example, transforming mode of farming, in Shurugwi an initiative known as the Shurugwi Partners Community Garden Initiative (SPCGI) (Matsa and Dzawanda 2014). It was initiated targeting climate change risk of food insecurity. Gardens were established and were under irrigation to supplement rainfall. Results indicated that it was a notable success.

Successive droughts in Chiredzi District have resulted in increased poverty, food insecurity, malnutrition, and environmental degradation (Benson 2012). The shocks have manifested in poor yields resulting in low income (Chitonga 2013). Shocks have attracted collaboration and partnership of the government departments and NGOs to design and implement resilient initiatives (WFP 2017). Some of the NGO's involved in resilience building in Chiredzi District include CARE International, PLAN International, WFP, United States Agency for International Development (USAID), and International Rescue Committee (IRC). Resilience building initiatives that have been implemented mainly targeted to alleviate poverty, food insecurity, income generation, environmental conservation, and sustainable development at large. Initiatives include Food Assistance for Asserts in 2016 facilitated by WFP and Plan, Food for work by CARE and District Social Welfare, Financial assistance by IRC, Water, and Sanitation programs and Community-Based Natural Resources Management (CBNRM) projects by PLAN. It is against this background that this study seeks to assess the effectiveness of the resilience building initiatives to counter shocks and stressors affecting rural livelihoods in Chiredzi District.

1.1 Physical Description of Study Area

Chiredzi District is in the Southern dry zone of Zimbabwe in Masvingo Province and borders with South Africa and Mozambique on the Southern side. It is in the agro-ecological region 5 which receives mean annual rainfall of 450–600 mm and mean lowest temperatures of 16 °C and highest temperatures averaging 36 °C. Chiredzi

District is more vulnerable to hydro-meteorological hazards namely floods, droughts, and heat waves.

Four major rivers—namely Save, Runde, Chiredzi and Mutirikwi cross-cut the District. They contain water for a considerable length of time though they do not flow continuously throughout the year, hence causing water stresses in dry seasons for domestic and livestock uses. These rivers, all of which are heavily silted, are responsible for flooding which sometimes occur in the District to areas situated by the river banks and also Chiredzi being at the lower course of the rivers where most flooding occurs.

Large areas of Chiredzi are covered by vertisols which are well suited for irrigation. Mopane and Acacia are the dominant trees in the District. The tree species have an overall density of 0.03 stands/m (Oxfam UNDP/GEF 2015). Mopane tree species are the more dominant with a density of 0.02 square metres. Approximately, 90% of the land cover is sparse vegetation, 5% dense vegetation and 5% bare ground (Unganai 2011). Wild animals like baboons, monkeys, hyenas, elephants, and buffaloes are increasingly raiding crops and livestock (UNDP 2015).

1.2 Socio-economic Description of the Study Area

Chiredzi Rural District is partitioned into 32 wards with a total of 276 842 people: 133 385 males, 143 457 females, and 64 865 households with an average size of 4.3 individuals per household (ZimStat 2012). Out of 32 wards, 17 are involved in resilience building. There are many schools, hospitals, and clinics in the District, but in some areas villagers move considerably long distances of more than 5 km to reach a service centers like schools and hospitals.

Frequent natural disasters such as droughts and floods have further exacerbated poverty levels in the District. The livelihoods of the poor majority, especially woman, are highly dependent on climate-sensitive sectors such as agriculture, forest resources for household energy, food security, and water supply, beer brewing, and traditional wine making from wild ilala *(hyphaenepetersiana).* Vastly altered weather patterns and climate extremes are undermining socio-economic development in the District.

Cereal crop production is another livelihood activity; however, crop yields are generally low averaging 0.5 tonnes/ha. Reasons behind such poor harvest being shocks and stressors. Major crops grown include maize, rapoko, millet, and sorghum (red and white). Agriculture in Chiredzi is seasonal and, as a result, people diversify livelihood means, some engage in poaching wildlife in Gonarezhou National Park and others migrate to neighbouring South Africa and Mozambique as economic refugees. Livestock production is also one amongst other livelihood means, but it is challenged by diseases like foot and mouth disease (FMD), anthrax, and even foot rot (Fig. 1).

1.3 Study Area Map

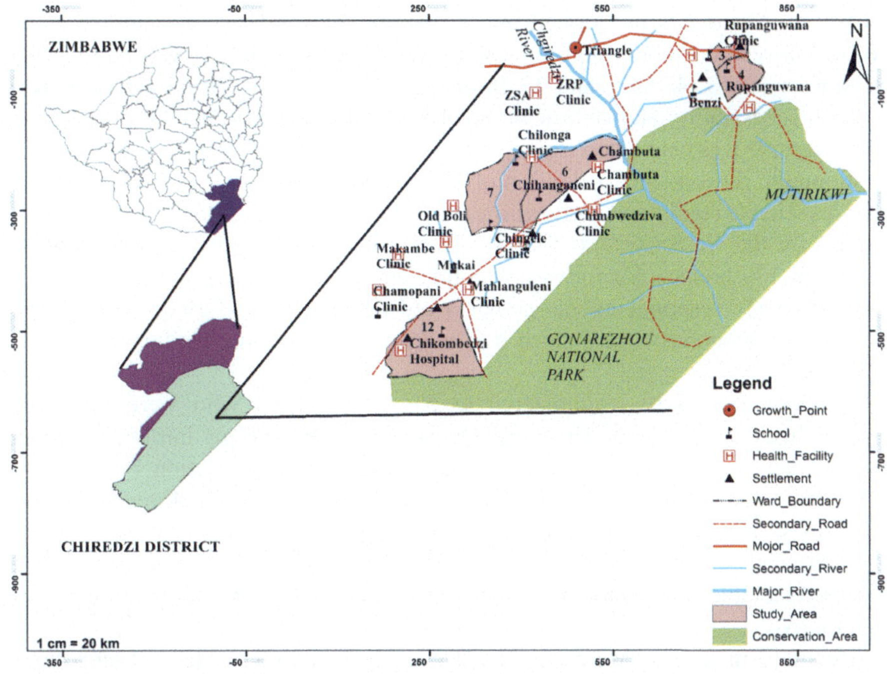

NB. Numbers 3, 4, 6, 7, and 12 represent selected wards

Fig. 1 Physical description of the study area

1.4 Methodology

This research was formulated and undertaken following the Resilience Building Analytical Framework. The Framework focuses on deepening the understanding of the distribution and characteristics of shocks and stresses identifying potential target areas that are affected by frequent and multiple shocks and stresses, and sets the stage for the process of identifying and selecting potential investments areas/activities that will confer resilience and sustain improved well-being outcomes. In this research, objectives are chronologically arranged according to the framework bases of implementation and procedures. Documentary analysis was done to identify wards where resilience building initiatives have been implemented.

The target population of this study was all rural households in Chiredzi District. The District was selected because it is located in the marginal areas of the semi-arid agro-ecological regions of Zimbabwe (region 5) where shocks and stressors are more magnified compared to other districts. The research also targeted all development actors in the wards working to initiate or promote humanitarian developmental gains among rural communities; these included District Social Welfare (DSW), Agriculture Technical and Extension Services (AGRITEX), Chiredzi Rural District Council (CRDC), CARE, PLAN, and traditional leaders.

Chiredzi District is partitioned into 32 wards; 17 wards were identified to be involved in resilience building initiatives. Convenient sampling was used to select five wards, viz. 3, 4, 6, 7, and 12 from the 17 total wards. Ward selection was based on accessibility and significant involvement in resilience building initiatives. A 10% sample was calculated from the total entire population of each ward to select questionnaire respondents. Adoption of 10% sample was based on the view that it provides reliable results and it saves time and resources for data collection. Calculated samples for the wards were as follows: 72 households in ward 7; 155 households in ward 6; 130 households in ward 4; 8 households in ward 12, and it constituted a sample of 510 respondents' altogether. Ward 12 has a total of 79 households and was selected as control ward since it is not involved into any resilience building initiatives.

Questionnaires were administered to 510 respondents who were randomly selected from the wards under study. Questions on the various initiatives implemented, successes yielded and even shocks and hazard manifestation were administered from a tablet pre-loaded with Kobo collect software. Both closed and open-ended questions were included in this research. Quantitative data was subjected to statistical analysis using SPSS version 20.0 and Microsoft excel, and results were presented on tables and charts. Chi square test was used to check for significance of the initiatives.

Structured interviews were conducted to gather data on the hazard and shock manifestation, implemented initiatives, successes being yielded and questions were directed by an interview guide. Purposive sampling technique was adopted to select interview respondents and identified relevant key stakeholders from the following organizations CARE, PLAN, District Social Welfare, AGRITEX and CRDC. CARE Monitoring and Evaluation Officer was interviewed to identify all initiatives implemented through CARE and their levels of success. PLAN Project Facilitator gave information on all projects led by PLAN and successes yielded. Chiredzi District Rural District Ward Councilors gave information on what transpired in their respective wards and to identify changes brought about by Resilience Building. District Social Welfare head of Department was interviewed to identify all forms of social assistance offered and how it improved local households' capacities to withstand shocks and stressors. AGRITEX ward extension officers gave information on all agro-related initiatives and assessment results of household capacities to withstand shocks and stressors. Qualitative data was analyzed using NVIVO software, thus stories presented by people from different organizations were formulated and presented into a more meaningful manner. The data acquired was emphasized by data obtained from observations through a checklist. Images were also captured to provide concrete evidences on the actual phenomena.

2 Results and Discussions

2.1 *Major Shocks and Stressors Affecting Rural Livelihoods of Chiredzi District*

Research findings have revealed that innumerable hazards have prevailed, and this has caused community economic instability across the district. Categories of identified shocks and stressors include hydrological hazards, socio-economic hazards, and livestock related hazards. Measures were designed in an attempt to minimize the impact of the above-mentioned hazards in the study area.

2.1.1 Hydrological Hazards

In some parts of Chiredzi District, flooding is a major hazard to community livelihoods. The 2016–17 Cyclone Dineo devastated critical infrastructures of the community and reversed development gains. Excessive rains destroyed local bridges, roads, houses and granaries. For example, in ward 7, 45 houses were destroyed and ten granaries collapsed. The District Agriculture Extension Officer reported flooding, hailstorms, and lightning as major hydrological hazards felt in Chiredzi District during rainy season. In most cases, Chiredzi District experiences drought due to shortage of rainfall. Meteorological services department reported that at times rainfall delays are experienced in the District, and most times it is insufficient for crop growth to maturity. The two moisture extremes of flooding and drought in Chiredzi District are major reasons for low yields and household vulnerability.

2.1.2 Livestock Diseases

There are many livestock diseases in Chiredzi District, and these reduce the quality and quantity of livestock production. The diseases affect different classes of livestock namely: goats, cattle, and chickens as illustrated in Table 1.

The Majority of the respondents in the studied communities of Chiredzi District expressed disappointment about the livestock loss due to diseases. Cattle in wards of Chiredzi District studied were affected with foot and mouth, theileriosis, and anthrax. Veterinary officers highlighted that livestock in close proximity to game parks are seriously affected by theileriosis or corridor disease, for example, in ward 4.

In ward 3, a total of 39.1% of the households lost their livestock due to the identified diseases. One hundred and forty (140) birds were lost due to a disease called '*chibhubhubhu*' in ward 3. Goats were lost mainly as a result of waterborne diseases like foot rot attributed to substandard housing structures that promote water logging conditions. A farmer in Chikondo village lost 15 goats due to foot rot. In ward 4,66% of the interviewed households lost their livestock. Cattle in this ward were seriously affected by theileriosis/corridor diseases,–which affected mostly livestock in areas

Table 1 Livestock diseases and classes of livestock affected livestock type

	Diseases
Goats	*Chindee*
	Heart water
	Chihudha
Cattle	Foot and mouth
	Black leg
	Anthrax
	Thereliosis
Poultry	New castle
	Infectious coryza
	Coccidiosis
	Chibhubhubhu

Source Survey results

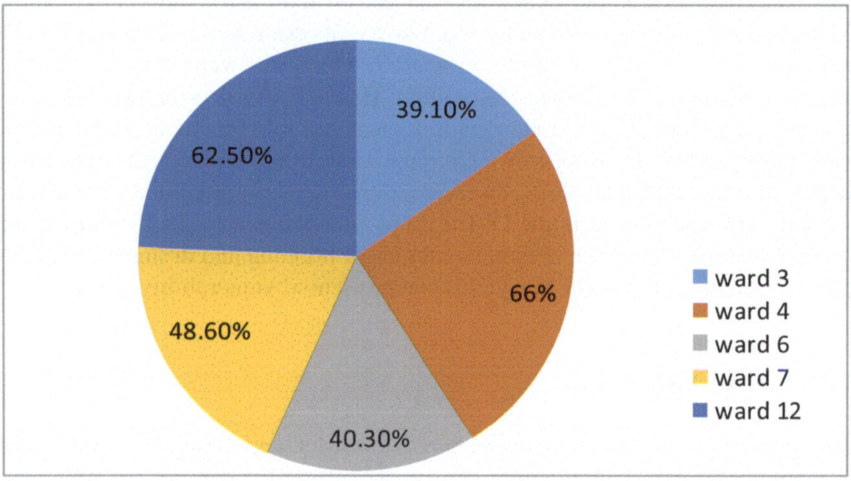

Fig. 2 Percentage of households per ward who experienced livestock losses *Source* The author

close to Gonarezhou National. Diseases like foot and mouth caused extensive losses of livestock. In ward 7, 48.6% of the interviewed households lost their livestock in the 2016/17 season. Losses experienced were attributed to the effects of flash floods which caused foot rot in goats and cattle, new castle in chickens. A greater proportion of the households in ward 12 (62.5%) experienced livestock losses including cattle to anthrax and foot and mouth and chickens to chibhubhubhu.

2.1.3 Crop Pests and Diseases

Survey results revealed that crop pests and diseases affect crops at different magnitudes in the District. Crop specialists and farmers in wards also identified many diseases affecting different crops as indicated in Table 2.

AGRITEX officers and some farmers interviewed indicated that crop pests and diseases have negatively affected agriculture production by 50%. Table 3 presents average yields obtained per ward on the selected main cereal crops. Expected yields were reduced by the impacts of cyclone Dineo, and also some identified pests and diseases. This has reduced food security status of many households, and it further aggravated poverty infestation levels in the communities of the study area. Musaverema irrigation scheme in ward 3 was completely destroyed by fall arm worm.

Agriculture produce for the season 2016/17 was generally low. Expected harvests were affected by the effects of cyclone Dineo, fall armyworm and armoured crickets. However, in wards 3, 4, 6, and 7 attempts were made to control the effects of crop pests and diseases and minimize losses through initiating disaster control initiatives. This helped to minimize the losses that could have been incurred by farmers. In ward 12, no attempts have been made to effectively control the disasters and this affected expected harvest.

Table 2 Crop pests and diseases experienced in studied wards

Crop pest, weed or disease	Crops affected
Bise (witch weed)	– white and red sorghum – maize
Fall army worm	– maize – sorghum – cotton
Armoured cricket (*mazekenene*)	– groundnuts – bambaranuts
Qualia birds	– sorghum – rapoko – pearl millet

Source Survey results

Table 3 Average crop yield harvested in the 2016\17 season per ward

Ward	Average crop yield obtained per crop in kilograms per ward			
	Maize	Sorghum	Pearl millet	Rapoko
Ward 3	1638	182.6	201.1	48.5
Ward 4	1521.6	231.9	149.9	116.1
Ward 6	1543.3	303.8	115.2	96.6
Ward 7	1432.3	329.4	101.7	96.8
Ward 12	767.4	148.7	53.4	15

Source Agritex (2017)

2.1.4 Socio-economic Hazards

Information from interviews with stakeholders operating in the District has shown inference that HIV/AIDS prevalence rate in Chiredzi District is high as compared to other districts in Masvingo Province. HIV prevalence rate stands at 15.5% which is comparatively higher than other districts in Masvingo province. Malaria cases were identified in some parts of the study area and at instances caused deaths of people.

During the 2016–17 season, there was loss of life and serious injuries due to drowning of people in flooded rivers Runde and Save. In ward 7, 10 people lost lives to water and a total of 65 households in the Chilonga area, which comprises wards 6 and 7, were seriously affected by flash floods. Extensive flooding was attributed to cyclone Dineo.

The salient economic hazard experienced in the District was the cash crisis which was also a nationwide phenomenon. Farmers adversely suffered cash crisis with highly rated eco-cash percentages of 25–30% from the black market.

2.2 Contribution of NGOs and Government Departments Operating in Chiredzi District

All government departments and NGOs operating in Chiredzi District targeted to overcome poverty and build resilience under the resilience board that is Zimbabwe Resilience Building Fund (ZRBF). They get back up from international agencies like DFID and UNDP. They offer services like risk management and mitigation across the District.

AGRITEX plays a pivotal role in attempting to enhance resilience building in Chiredzi District communal areas. It does all farmer trainings, field assessments, evaluations, monitoring, and compile progress reports. AGRITEX officers set demo plots in every ward and these are learning points for most farmers. The majority of farmers above 50% indicated AGRITEX to be the best in service delivery, farmer assistance, and information dissemination to successfully build resilience. Farmers indicated that they implement, check, and monitor progress of agro-based activities to cope with prevailing conditions. They do extensive visits to farmers disseminating knowledge and skills to enhance food security, eliminate poverty, and build resilience. However, AGRITEX staff have challenges to successfully deliver their services.

Veterinary Department in Chiredzi District was distinguished by farmers as one department which carries out its mandate of ensuring animal health care and production meticulously. It offers extensive health care on livestock across the District. They assist farmers through vaccination and diseases control on livestock. The Department is also given credit on the effort to minimize foot rot on cattle and goats and airborne diseases on chickens through the improved structures initiative.

Other departments involved include the Ministry of Woman Affairs, Gender and Community Developments, RDC, and CPU. The Ministry of Woman Affairs has mobilized women and youths to participate in projects and encourage formation of village savings and landings groups.

Care International is one of the NGOs active in resilience building initiatives in the communities of Chiredzi District through the Enhanced Community Resilience and Sustainability Projects (ECRAS). It offers services to reduce risks and ensure people in vulnerable communities are empowered enough to withstand shocks and stressors. The Monitoring and Evaluation Officer for Care International mentioned that, Care overlooks all resilience related activities and ensures smooth delivery of services across the District. The organization sources funds from international donors and support teams on the ground through funding. It is again responsible for coordinating with government stakeholders and give direction in terms of targeted goals to build resilience in vulnerable communities to mitigate identified risks that is food insecurity and environmental degradation (Benson 2012).

Plan International in Chiredzi District is implementing initiatives mainly targeting to alleviate poverty, food insecurity, income generation, environmental conservation, and sustainable development in general. The majority of the farmers (60%) expressed satisfaction with Plan's interventions in designing measures to counter shocks and stressors affecting rural livelihoods in Chiredzi District. Plan's specific roles encompass mobilization, training communities on resilience building projects, and linking producer groups with markets. It was Plan that facilitated the uptake of climate-smart techniques which improved food availability. The majority vote indicated that Plan ensured inclusion of gender dimensions in all activities. Plan facilitated asset creation in the District, that is, construction of dip tanks, community gardens, rehabilitation of boreholes, installation of solar powered boreholes, and designing of lower cost water harvesting techniques. In ward 4, three (3) solar-driven water pumps were set and this helped reduce water problems during dry seasons. A community garden was rehabilitated in ward 6 and was set to function. Plan also facilitated the creation of a dip tank in ward 7. In ward 3, a solar-powered borehole system (Tsamwisi borehole) was established; however, in ward 12, none of the above initiatives was implemented and households are struggling to access water for domestic use and even for livestock consumption.

2.3 *Effectiveness of Resilience Building Initiatives*

Research findings indicate that resilience building initiatives have increased or enhanced the capacities of Chiredzi district communities to sustain developmental gains and achieve improved well-being outcomes in the face of shocks and stressors. Conclusions were based on initiatives implemented and these included village savings and landings, improved livestock structures, indigenous chicken program, fodder production and preservation, farmer training, producer supplier engagements and community gardens initiative.

Table 4 Wards and numbers of VSL groups formed

Ward	Number of VSL groups formed
Ward 3	6
Ward 4	10
Ward 6	4
Ward 7	3
Ward 12	0

2.3.1 Village Savings and Lendings (VSL)

VSL in wards where it was implemented happened to be the most successful initiative and has contributed to resilience building. It happened to be the source of income and capital to purchase basic commodities at household level. According to the Ministry of Gender and Women Affairs, VSL addressed gender inequalities which existed prior to the ECRAS projects through inclusion of women in all programs. ECRAS was a project led by CARE and PLAN with many initiatives targeting to increase the capacities of communities to sustain development gains and achieve improved well-being outcomes in the face of shocks and stressors and strengthening beneficiary capacities to withstand climate change induced shocks and other adversities. Admission into the club was voluntary, however membership in all wards implemented was predominantly woman. In ward 4, ten VSL groups were formed each group constituting 20 to 25 females. A greater proportion of female respondents (70%) indicated that with VSL they managed to purchase their own properties, sustain families and even buy some basic commodities for the family and pay school fees and other requirements. Many VSL groups were formed in studied wards (Table 4).

VSL intervention has increased livestock ownership in the community. Proceeds obtained from VSL in some communities were used to purchase livestock, for example, goats, cattle and chickens. Such achievements were observed only in wards where ECRAS was implemented and absent in areas where it was not implemented, for example ward 12. Resilience capacity of the farmers in ward 12 was found to be minimal.

2.3.2 VSL and Indigenous Chicken Program

VSL was mentioned as the main source of revenue used to engage in the indigenous chicken program. A greater proportion of the farmers in the indigenous chicken program indicated that they got capital from the VSL clubs. Villagers view VSL as an enabler and funder of other initiatives. Proceeds generated from VSL were used to purchase basics for chicken production, for example, chicken feed, vaccines and construction of chicken structures. Inferences made from the VSL present it as a sustainable and reliable source of resilience building that sustains other projects as presented in Fig. 3. Chi square tests were done to test for the relationship between VSL membership and joining the indigenous chicken production scheme amongst

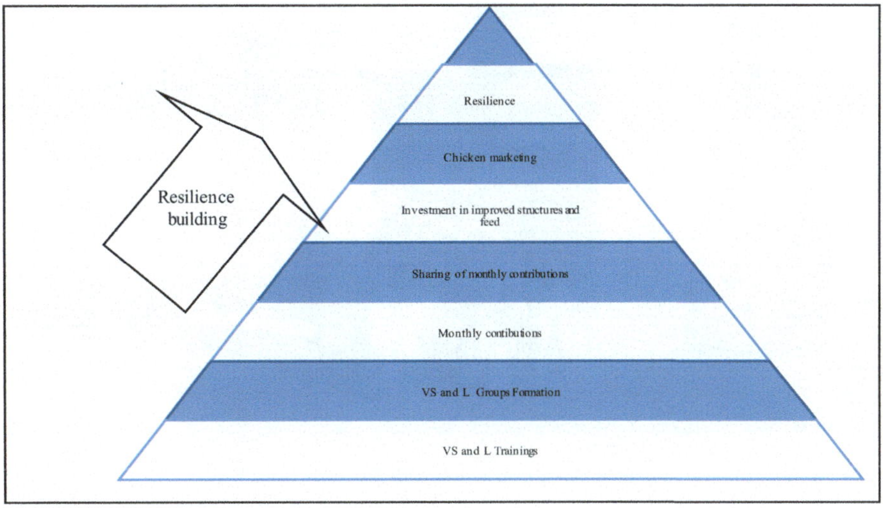

Fig. 3 The flow of VSL proceeds in building resilience

households in wards 3, 4, 6, and 7. Chi square results indicate that there is a strong relationship (P value = 0.00) between VSL membership and undertaking the indigenous chicken scheme. This helps explain that VSL is the benchmark of other resilience initiatives. However, in non-resilience building wards poverty infestation and challenges to successfully sustain livelihoods have been identified. Basically, this was due to failure to be capacitated sufficiently to engage in resilience building initiatives.

2.3.3 VSL and People's Welfare

VSL improved the well-being and living conditions of most households in Chiredzi District. It helped communities use modern building materials, codes and standards. Respondents (50%) indicated that they are now using bricks and iron/asbestos sheets for building better and modern structures instead of the traditional pole and dagga. Participants of VSL indicated that proceeds from this intervention enabled them to go for modern structures. Figure 4 presents a kitchen which is under construction using modern materials which are attributed to have been bought with proceeds generated from VSL. The owner of the kitchen was a female in ward 4 and was under Takunda VSL group. VSL has helped many households construct toilets at their homesteads and this has reduced outbreaks of epidemics caused by poor hygiene and sanitation, for example diarrhea, typhoid and cholera. Overall, across the study wards, 69.2% of interviewed households have toilets at their homesteads. Group representatives of VSL scheme mentioned that VSL is a benchmark of all developmental initiatives individuals are engaging in order to build resilience. However, it was observed that in ward 12, people are still constructing houses using traditional codes and standards

Fig. 4 A modern rural kitchen under construction with proceeds from VSL

and this is attributed to poor income and lack of awareness about resilience building initiatives. Some homesteads in ward 12 still do not have toilets or latrines, and only 0.5% households indicated that they have constructed toilets at their homesteads.

In some parts of ward 12, the rate of modernization is low. During the study, pole and dagga structures were identified as indicated in Fig. 5. Based on some identified structures, VSL (resilience building) is an important force to improve local livelihood and welfare.

Fig. 5 One of the structures in ward 12—A non-resilience building ward

2.3.4 Improved Livestock Structures

Resilience building initiatives implemented in Chiredzi District included aspects of improving livestock structures. Such innovations helped reduce livestock diseases due to sub-standard housing structures. The researcher observed improved structure for both goats, chicken, and cattle as presented by Fig. 6. On cattle and goats, the improved structures helped reduce foot rot and chicken aeration diseases. Many respondents indicated that the initiative reduced livestock loss, especially from diseases associated with poor housing structures. However, such positive achievements were only in wards where resilience building was introduced. Research findings reveal that in specific wards not involved in any initiative, livestock death and diseases infestations are high.

2.4 Resilience Building Through Trainings

Resilience building encompasses empowering farmers at risk through trainings. Trainings that have been conducted in Chiredzi include the following areas: disaster risk management; fodder production and preservation; and post-harvest management. These trainings are crucial in people's everyday life and, therefore, they largely influence their resilience capacity.

Fig. 6 An improved goat structure

2.4.1 Disaster Risk Management (DRM) Training

In Chiredzi District, 67% of the respondents received DRM training. The District is prone to disasters and risks like floods, droughts, livestock diseases, crop pests, and diseases. DRM trainings are crucial to ensure resilience building in rural communities. Information dissemination on this particular initiation has increased knowledge banks of at-risk households on how to successfully manage risks. Structured interviews with households in non-resilience building wards (ward 12) indicated that non-trained individuals lack knowhow in terms of disaster management.

2.4.2 Post-harvest Management Training

Survey results indicated that approximately 75% of the sampled households received training on post-harvest management. According to the AGRITEX Officer, some cases of food shortages being experienced in the District by farmers are not necessarily food crises, but rather poor management of agricultural produce after harvesting. Management and storage facilities used by farmers were seen to be inappropriate, such as structures made of pole and dagga. Post-harvest management trainings have led to the construction of grain banks and metal silos as observed in wards 3 and 7. However in non-trained wards people are still using primitive pole and dagga granaries.

2.4.3 Fodder Production and Preservation Training

Some cases of livestock loss experienced before resilience building by farmers were due to pasture shortage during dry seasons. During summer and late winter, vegetation dries up causing grazing challenges on both cattle and goats. However, resilience building initiatives brought the idea of fodder production and preservation. Farmers were taught to prepare and preserve low cost livestock food using locally available resources. This reduced food shortages which once prevailed in the District. It was observed that the nutritional value of the preserved fodder was higher than the general ordinary grazing. Fodder created by farmers increased beef quality and also made farmers to get better prices on the market. Chi square tests were done on the significance of fodder production and increase in beef quality. Results indicated that (P value = 0.001) there is a strong relationship between high quality beef production and fodder feeding.

3 Conclusion

Resilience building has been critical in improving lives of rural communities in Chiredzi District. It is a critical measure to address the shocks and stressors affecting rural livelihoods of Chiredzi District. The majority of farmers indicated to have attained considerable developmental gains once lost to prevailing disasters in the District.

It has been observed that resilience building initiatives are interrelated; successful implementation of an initiative will positively influence another initiative. Such findings are clearly illustrated by the VSL initiative whereby proceeds from VSL are invested to support another initiative.

Projects implementers have introduced some initiatives which addressed gender biases that existed in the communities. Initiatives introduced were open for every member, admission was regardless of being male or female. This gave females higher opportunities to be self-dependent rather than the traditional culture of women being regularly submissive to men. It again improved family life quality since many sources of income were made successful.

Community hygiene and sanitation was greatly improved. People began to use water from protected sources like rehabilitated boreholes and protected wells. Some farmers used proceeds from various initiatives to drill their own sources of water. People descended from using bush toilets and adopted the use of blair toilets. This greatly reduced outbreak of hygiene related epidemics like malaria, cholera, and typhoid.

4 Recommendations

Resilience initiatives implementers should try to cover the whole District. This is because there is a significant difference in the style and quality of life in wards under the projects. Significant suffering has been observed in non-resilience wards whilst there is a great change in wards covered by the projects.

Farmers need more trainings on resilience building. Education and training will help improve farmers' capacities to withstand shocks and stressors. The projected continued climatic change requires farmers to be stronger enough to withstand shocks.

More farmer field days should be held at successful farmers' homesteads. This would help create competition among farmers themselves and eventually improve resilience capacity of many farmers.

References

Baker LM (2006) Observations: a complex research method. Res Methods Library Trends 55(1):171–189. University of Illinois

Béné C, Frankenberger T, Nelson S (2015) Design, monitoring and evaluation of resilience interventions: conceptual and empirical considerations. IDS Working Paper Volume 2015: 495. Institute of Development Studies, Brighton, UK

Benson T (2012) An investigation of the spatial determinants of the local prevalence of poverty in rural Malawi. Food Policy 30:532–550
Black K (2010) Business statistics: contemporary decision making, 6th edn. Wiley , Hoboken
Breisinger C, Ecker O, Maystadt JF, Trinh Tan JF, Al-Riffai P, Bouzar K, Sma A, Abdelgadir M (2014) How to build resilience to conflict: The role of food security. International Food Policy Research Institute (IFPRI)
Brown K (2016) Resilience development and global change. Routledge , London
Brown C, Wilby RL (2012) An alternate approach to assessing climate risks. Eos Trans Am Geophys Union 93(41):401–402
Burchi F, De Muro P (2007) Education for rural people: neglected key to food security. Working Paper No. 78, 2007. Dipartimento di Economia, Università, Roma, Italy, http://host.uniroma3.it/dipartimenti/economia/pdf/wp78.pdf
Carletto G, Jolliffe D, Banerjee R (2015) From tragedy to renaissance: improving agricultural data for better policies. The World Bank
Chitonga L (2013) Towards comprehensive disaster risk management in zimbabwe: evaluating masvingo rural district's community drought management program. Int J Econ Manag Soc Sci 644–653
Choularton R, Frankenberger T, Kurtz J, Nelson S (2015) Measuring shocks and stressors as part of resilience measurement. Resilience measurement technical working group. Obtenido de technical series (5)
Driscoll DL (2011) Introduction to Primary Research: Observation, Survey and Interviews, Writing spaces: Readings and writing Volume 2, Califonia, USA
FAO (1997) Marketing research and information systems. www.fao.org. Accessed 06 July 2017
FAO (2015) The Impact of Disasters on Agriculture and Food Security
FAO (2014) Strengthening the links between resilience and nutrition in food and agriculture. Discussion paper, Rome
Feola G (2015) Societal Transformation in response to global and environmental a change: a review of merging concepts, Ambio 44(5):387
Fowler F (2009) Survey Research Methods. Sage Publications, Thousand Oaks
Food Security Information Network (FSIN) (2014) Resilience measurement principles. Toward an agenda for measurement design. Technical Series No. 1. Rome, Italy
IPCC (2007) Climate change the physical science basis. AGUFM, 2007, U43D-01
IPCC (2014). Annex II: Glossary (Mach, K.J., Planton, S. & von Stechow, C. (eds.). In: Climate Change 2014: Synthesis Report. Contribution of Working Groups I, II and III to the Fifth Assessment Report of the Intergovernmental Panel on Climate Change [Core Writing Team, Pachauri, R.K. & Meyer, L.A. (eds.)]. IPCC
Jerry W, Marcin S (2007) Guides for the erplexed: Research methods for the social sciences. Continuum International Publishing Group, London
Kohlbacher F (2006) The Use of Qualitative Content Analysis in Case Study Research, Journal on Forum Qualitative Sozialforschung / Forum: Qualitative Social Research 7. www.qualitative-research.net, Accessed on 17 February 2018
Kuria J, Nyaggah M (2016) Assessing Resilience Building Initiatives in Africa. Nairobi: Local Development Research Institute
Kurtz J, Scarborough, G (2012). From Conflict to Coping: Evidence from Southern Ethiopia on the Contributions of Peacebuilding to Drought Resilience among Pastoralist Groups. www.mercycorps.org/sites/default/files/from_conflict_to_coping
Lagi M, Bertrand KZ, Bar-Yam Y (2011) The food crises and political instability in North Africa and the Middle East. New England Complex Systems Institute, Cambridge, USA
Length RV (2009) Some Guidelines for effective sample size determination. Am Stat J 187–195
Local Development Research Institute (2016) Assessing Resilience Building Initiatives in Africa. Local Development Research Institute, Nairobi
MacDermott TKJ (2015) Investing in disaster risk management in an uncertain climate. World Bank Policy Research Papers. World Bank, Washington DC

Madzwamuse M (2010) Climate Governments in Africa: adaptation strategies and institutions. A synthesis report prepared for the Heinrich Boll Foundation
Matsa M, Mavugara R, Dzawanda B (2019) Urban domestic water supply crisis in the city of Gweru, Zimbabwe. GeoJournal 1–20
Mechler R (2012). Reviewing the economic efficiency of disaster risk management. review commissioned by the foresight project: reducing risks of future disasters. UK Government Office for Science, London
Mitchell N (2013) Risk and Resilience: From Good Idea to Good Practice. OECD Development Assistance Committee Working Paper 13/2013, OECD Publishing
Obefreng BP (2016) The property flood resilience action plan; Department for Environment Food and Rural Affairs
Oxfam UNDP/GEF (2015) Scaling up Adaptation in Zimbabwe, with a focus on rural livelihoods Annual report: UNDP, Zimbabwe
Oxley M (2013) A "People-centered Principles-based" post-hyogo framework to strengthen the resilience of nations and communities. Int J Disaster Risk Reduct 4:1–9
Pandey P, Pandey MM (2015) Research Methodology. Tools and Technique Bridge centre, Romania
Pelling K, Obrien D (2014) Adaptation and transformation. Climate Change 133(1): 113–127
Schipper ELF, Langston L (2015) A comparative overview of resilience measurement frameworks: analysing indicators and approaches. ODI Working Paper 422.ODI, London
Statistics Canada (2014) Data analysis (and presentation). www.statcan.gc.ca. Accessed 17 Feb 2018
UNEP (2010) Assessing the Environmental Impacts of Consumption and Production: Priority Products and Materials. A Report of the Working Group on the Environmental Impacts of Products and Materials to the International Panel for Sustainable Resource Manage, Paris: United Nations Environment Programme, p 34
Unganai L (2011) Climate change and its effects on agricultural productivity and food security: A case of Chiredzi district. In National Climate Change Workshop, Harare, Zimbabwe
UNDP (2015). Human Development Report. http://hdr.undp.org/. Accessed 05 Oct 2019
USACE (2016). Resilience Initiative Roadmap. Washington
USAID (2015) Ethiopia Pastoralist Areas Resilience Improvement and Market Expansion (PRIME) Project Impact Evaluation. Baseline Survey Report Volume 1: Main Report. A report prepared by TANGO International
Vogel K (2012) Review of the use of 'Theory of Change' in international development, DFID
World Bank (2003) World Development Report 2004 (Overview): Making Services Work for Poor People. World Bank.
World Bank (2013) Zimbabwe Economic Briefing. November 2013. The World Bank, Harare
World Bank (2014) Zimbabwe Economic Briefing June, 2014
World Food Programme (WFP) (2014) Zimbabwe: Results of exploratory food and nutrition security analysis. World Food Programme, VAME Unit, Harare, Zimbabwe
World Food Programme (WFP) (2015) Consolidated Approach to Reporting Indicators of Food Security (CARI) Guidelines, Second Edition. http://documents.wfp.org/stellent
World Food Programme (WFP) (2017) At the Root of Exodus: Food Security, conflict and international migration. May 2017, Rome
World Health Organization (WHO) (2017) Monitoring Health for the SDGs, Sustainable Development Goals. Geneva: World Health Organization; 2017. Licence: CC BY-NC-SA 3.0 IGO
ZimStat (2012) Census 2012, Zimbabwe National Report, ZIMSTATS, Harare
Zseleczky L, Yosef S (2014) Are Shocks Really Increasing? A selective review of the global frequency, severity, scope and impact of five types of shocks. In: 2020 Conference Paper 5. May 2014, Washington, DC

Chances of Conflict on Account of Large-Scale Influx of Climate Refugees in India

Devangana Gupta and Manish Sharma

Abstract Global mean sea level has risen by 21–24 cm compared to the preindustrial levels and they are expected to rise by half a metre in 2050. This has led to increased displacement of human populations. According to World Migration Report 2020, there were 28 million new internal displacements across 148 countries at the end of 2018. Sixty one percent of these displacements were triggered by disasters, and 39% by conflict and violence. In India, 1.5 million people are classified as internally displaced every year. Small islands in the Ganga–Brahmaputra delta in the Sundarbans are already disappearing into the sea. Out of the ten most climate affected countries listed by Global Climate Risk Index, 2019, three are India's neighbours: Nepal, Bangladesh, and Sri Lanka. In addition, there are many close island neighbours including Maldives, the island nation with the lowest elevation (1.5 m above sea level). In India, 350 million people live along the 7500-km long Indian coastline. Future climatic changes and increase of inflow from these neighbouring nations will add to the stress and strife the country is undergoing in view of migration of over 10 million people from the then East Pakistan (now Bangladesh) before and at the time of 1971 war. The problem might expand to political dimensions and could take communal turn considering that Maldives and Bangladesh are Islamic countries. The focus of the present chapter would be to look into the details of the above issues and try to figure out a solution.

Keywords Climate migration · Sea-level rise · India · Maldives · Bangladesh

1 Introduction

One of the fallouts of global warming is sea-level rise. Global mean sea-level rise from 1880 to 2009 was 21 cm and the annual trend of rise is approximately 3 mm per year, which brings it to 24 cm in 2020 (Church and White 2011). By 2100, global

D. Gupta (✉) · M. Sharma
Department of Gandhian and Peace Studies, Panjab University, Chandigarh, India

M. Sharma
e-mail: manish@pu.ac.in

M. Behnassi et al. (eds.), *Social-Ecological Systems (SES)*,
https://doi.org/10.1007/978-3-030-76247-6_9

mean surface temperatures will rise between 1.8 and 4° C and the mean sea level is predicted to rise by 50 to 100 cm due to the combined effect of thermal expansion of sea water and melting of polar ice caps and glaciers (IPCC 2014). Sea level has been fluctuating from 5 m above the present level during the last interglacial period (129,000 to 116,000 years ago) to 120 m below the present levels at the peak of last glacial age 20,000 years ago (Gornitz 2007; IPCC 2014).

Coupled with extreme weather events, floods and droughts, sea-level rise will leave several areas unliveable forcing people to migrate to other liveable places. India has a long coastline of 7517 km spanning nine maritime states and two Union Territories. Coastal districts host 14.2% of India's population. India has been identified as one of the countries which are most vulnerable to the impact of accelerated sea-level rise. Half of Asian cities with more than 1 million residents face multiple hazards such as floods, coastal flooding, and cyclones (IPCC 2014; India 2018).

Almost 85% of India's area is vulnerable to hazards. As much as 45.64 million hectares of land is subject to floods, and annual average flood damage during 1996–2005 was INR 47.45 billion (USD 753.2 Million). 1,238 Indian islands are vulnerable to loss of coastal wetlands, including mangroves, and saltwater intrusion in freshwater aquifers. Islands are highly susceptible to frequent and intense tropical cyclones, droughts, and tsunamis (India 2018).

India's coastal ecosystem is among the world's most diverse, and India is the second largest producer of fish in the world (India 2018). Such an ecosystem demands appropriate management practices and governance structures. As many as 106 coastal and marine sites have been identified in India as conservation or community reserves to increase participation of the local communities in governance (The Hindu 2019). The Internal Displacement Monitoring Centre (IDMC) has said that India will be among the frontline nations that will confront the human displacement due to natural calamities (IDMC 2017).

In December 2006, there were reports, published widely in the world media, of the submergence of an inhabited island due to climate change. Researchers reported that Lohachara Island, home to 10,000 people, in the Hooghly river delta in the Indian state of West Bengal, which had started flooding 20 years ago, had finally been entirely submerged. It was one of a number of vanishing islands in the delta. The loss of the islands and other coastal land in the delta left thousands of people homeless (Brown 2008).

Migration is as old as human civilisation. Population movements have always played a vital role in the structural transformation of societies and economies. However, unlike in the past when migration involved establishment of new habitations in fertile and virgin lands, migration today is characterised by movement of populations to other countries with people and cultures of their own. It, therefore, raises issues regarding loyalty, identity, development, and security.

The saga of large-scale migration in South Asia, often accompanied by violence, started with India's Independence from the British in 1947 when two nations – India and Pakistan – were carved out of British India on the basis of religion. Hindus migrated to India and Muslims to Pakistan, displacing 10 to 15 million people both ways (Talbot and Singh 2009; Khan 2017). The second wave of migration followed the 1950 occupation of Tibet by China during which 150,000 Refugees migrated to India, Nepal and Bhutan. (Mills et al. 2005; Dolma et al. 2006). The third major migration to India was in 1971 during the formation of Bangladesh. About 9 million refugees were driven out to West Bengal and Assam in India (Chen 1973).

The world has witnessed large migrations internally as well as externally. More than 76,000 people remain internally displaced in the North-East region and around 70,000 in the Kashmir part of India. Now climate change is displacing more people than other reasons of displacement like conflict or violence. Forced migration hinders development in four ways: increasing pressure on urban infrastructure and services; undermining economic growth; increasing the risk of conflict; and worsening health, educational and social indicators among migrants themselves (IOM 2019). The number of people affected by floods per year could be 160 million by the 2050s and 420 million by the 2100s (Nicholls and Lowe 2004). Large-scale population displacement will redraw the ethnic map of many countries, bringing previously separate groups into close proximity with each other and in competition for the same resources (Sindico 2005). Large population movements have been recognized by the UN Security Council as a potential threat to international peace and security, particularly with pre-existing ethnic and social tensions.

At present, the number of international migrants is estimated at 272 million, with nearly two-thirds being labour migrants. The proportion of migrants at 3.5% is small compared to the vast majority choosing to reside in the countries in which they were born. However, the number of international migrants has already surpassed some projections made for the year 2050 (IOM 2003), at 2.6% of population, accounted for 230 million (IOM 2019).

2 Climate Change and Migration

In the course of human evolution and settlement, large-scale migration is well known. There is archaeological evidence that human communities responded frequently to challenges posed by climate change. Large urban settlements were driven by climatic and environmental desiccation (Boko et al. 2006; McLeman and Smit 2006).

The Intergovernmental Panel on Climate Change (IPCC) in 1990 warned that the greatest impact of climate change will be on human migration. Coastal erosion, coastal flooding, agricultural disruptions etc. have displaced millions of people. Since then, analysts have tried to estimate the future flows of climate migrants (sometimes called climate refugees). The most widely repeated estimate is 200 million by 2050. In the mid-1990s, it was reported that up to 25 million people had been forced from their homes and off their land by such profound environmental pressures as pollution, land degradation, droughts, and natural disasters (Brown 2008).

Apart from climate drivers, non-climate drivers like public policies, population growth and community-level resilience also influence migration. According to Myers, 200 million people will be forced to move as a result of climate change. "When global warming takes hold," Myers argues, "there could be as many as 200 million people overtaken by disruptions of monsoon systems and other rainfall regimes, by droughts of unprecedented severity and duration, and by sea-level rise and coastal flooding".

Drivers of forced migration fall into two categories: climate drivers and non-climate drivers. Climate drivers can be classified into climate processes and climate events. Climate processes are slow changes like sea-level rise, salinization of land, desertification, and increasing water scarcity. Sea-level rise will make coastal areas and small-island states uninhabitable and land degradation will force the farmers to move out if agriculture cannot sustain livelihoods (De Wit and Stankiewicz 2006). This is relevant for India which has over 60% of population living in rural areas, depending on agriculture.

Climate events are sudden and dramatic hazards like floods, glacial lake outburst, storms and hurricanes. These force people off their land much more quickly and dramatically. Non-climate drivers, such as natural and man-made disasters, are equally important. A natural hazard only becomes a 'natural disaster' if a community is not prepared for its consequences. This could include lack of early warning system and poorly-built houses. Therefore, a community's vulnerability depends on exposure to climatic conditions and its adaptive capacity, that is, its ability to recover after the event is over (Brown 2008).

Other factors such as poverty, population growth, conflict and human security combine with climate change. It is therefore hard to predict the number of migrants globally. There could be between 25 million to 1 billion climate migrants by 2050, moving internally or across borders, on a permanent or temporary basis (Brown 2008).

The question now arises, can people displace by effects of climate change be called climate refugees? This is not just a matter of semantics. The label can have far reaching implications on the international community's obligation under the international law. Is it possible to interpret the word'refugee' to designate those escaping environmental stresses strictly in line with the refugee status according to the UN 1951 Convention and 1967 Protocol on refugees? The latter states that the term should be restricted to those escaping persecution on grounds of race, religion, nationality, or social grouping. Secondly, refugee status requires crossing of an international border. Internally displaced persons (IDP), who stay within their own borders, will not be able avail of the benefit. The question of definition is at present being hotly debated amongst international human rights lawyers. But so far, the concept of climate refugees has not been appreciated at the international level. There is no recognition in the 1951 UN Convention relating to Status of Refugees on the protection of climate refugees despite the fact that they are increasing worldwide. Efforts have been made in the 2015 Paris Agreement to protect climate migrants but it does not adequately address the need (Jolly and Ahmad 2019a).

A legal aspect of the refugee issue in India is that even after having hosted a number of refugees in the past, the country is not a signatory to the UN Refugee Convention of 1951 or the 1967 Protocol, which protect refugee rights. Apart from this, India also does not have a national refugee protection framework. However, it continues to grant asylum to a large number of refugees from neighboring States and respects mandate of UN High Commissioner on Refugees (UNHCR) for other nationals, mainly from Afghanistan and Myanmar.

3 Climate Change Vulnerability of South Asia

South Asia is recognized as being quite vulnerable to climate change. The region's varied geography combines with regional circulation patterns to create a diverse climate. The glaciated northern parts, which include the Himalayas, Karakoram, and Hindu Kush mountains, experience annual average temperatures at or below freezing, whereas much of the Indian subcontinent averages 25 to 30 °C (Mani et al. 2018). There are eight independent countries in South Asia namely India, Bangladesh, Afghanistan, Bhutan, Maldives, Nepal, Pakistan and Sri Lanka (Jolly and Ahmad 2019b). In 1985, the eight countries established the South Asian Association for Regional Cooperation (SAARC) to promote welfare of its peoples, improve quality of life, accelerate economic growth, and social and cultural development in the region.

In the past decade, nearly 700 million people, half of the region's population, were affected by climate-related disasters. Such weather patterns are expected to impact directly over 800 million by 2050 and will continue to burden South Asian countries (Fallesen et al. 2019). Climate Risk Index (CRI) positions India, and Sri Lanka in the highest risk category. Millions of people in South Asia will be forced to move due to blistering temperatures and crop failures unless steps are taken to reduce greenhouse gas emissions (Jolly and Ahmad 2019b) (Table 1).

Table 1 The climate risk index (CRI) for 10 most affected countries in 2018

Rank 2018	Country	CRI Score	Death toll	Deaths per 100 000 inhabitants	Absolute losses (in million US$ PPP)	Losses per unit GDP in %
1	Japan	5.50	1282	1.01	35, 839.34	0.64
2	Philippines	11.17	455	0.43	4, 547.27	0.48
3	Germany	13.83	1246	1.50	5,038.62	0.12
4	Madagascar	15.83	72	0.27	568.10	1.32
5	India	18.17	2081	0.16	37, 807.82	0.36
6	Sri Lanka	19.00	38	0.18	3,626.72	1.24
7	Kenya	19.67	113	0.24	708.39	0.40
8	Rwanda	21.17	88	0.73	93.21	0.34
9	Canada	21.83	103	0.28	2,282.17	0.12
10	Fiji	22.50	8	0.90	118.61	1.14

Source Eckstein et al. (2019)

3.1 Maldives at Risk of Submersion

Mohammed Nasheed, the former president of Maldives, made an impassioned plea to the world in 2012: "If carbon emissions continue at the rate they are climbing today, my country will be underwater in seven years" (Catoe 2012). There may have been an overreaction to the emerging discourse at that time, but the scenario may come true in another 80 years. The case of the Maldives, which faces the threat of total submergence with a rising sea level, symbolizes the face of climate disaster for the SAARC area. When Nasheed took over his post in 2008, he said that the country was considering buying land in neighbouring countries for settling its nationals as a preparation for the submergence (Ramesh 2008). The UN Global Environment Outlook (GEO-6) (Pereira et al. 2019) points to the alarming figure of 40 and 25 million population threatened by rising sea level in India and Bangladesh, respectively. The highest point on the 1200 islands that make up the Maldives is only 2.4 m above sea level, and 80% of the total land mass of the islands is only a metre above it. If the sea level rises by 60 cm, most of the republic's 200 inhabited islands will have to be abandoned (SMH 2012). This was substantiated also by Hassan earlier in 2010; even in a 10 cm-rise scenario (Hassan 2010). Being land scarce and low lying, the country is exposed to the risks of intensifying weather events such as damage caused by inundation, extreme winds, and flooding from storms.

4 India's Refugee Problem

Indian subcontinent experienced massive population movements across the international boundaries of India, Pakistan, Bangladesh, Nepal and Sri Lanka during seventeenth, eighteenth and nineteenth centuries when migration was much more common (Weiner 1993). India's Independence from the British in 1947 resulted in the partition of the country on religious grounds forcing 15 million people to migrate both ways in a matter of few months, accompanied by slaughter of thousands of men, women and children. Six to seven million Muslims moved from India to Pakistan and eight million Hindus and Sikhs moved from Pakistan to India (Weiner 1993).

The second wave of migration, this time from China, following the 1950 occupation of Tibet by China during which 150,000 refugees migrated to India, Nepal and Bhutan (Mills et al. 2005; Dolma et al. 2006).

The third major migration to India was in 1971 during the liberation struggle of Bangladesh when 9 million refugees were driven out to West Bengal and Assam in India (Chen 1973). A violent operation by the Pakistani army to suppress the demands of Bengali speaking East Pakistani population for regional autonomy resulted in the movement of massive influx of into India and provided the justification for the intervention of the Indian military to liberate East Pakistan, which was later rechristened as Bangladesh. So far, 35 to 40 million people have moved across international boundaries in India, Pakistan, Bangladesh, Nepal and Sri Lanka either as economic migrants or refugees (Weiner 1993).

The birth of Bangladesh is a tragedy rarely witnessed in modern times. Within a short span of one year in 1971, the people of Bengal (East Pakistan) faced natural disasters, repression, and civil war. But Bangladesh won its freedom and independence (Chen 1973).

Bangladeshi migrants settled in the border districts of West Bengal and Assam, which have thousands of kilometres of shared border with Bangladesh. Apart from border settlements, Bangladeshi migrants gravitated towards India's large urban centres such as Mumbai, New Delhi, and, in particular, Calcutta, which has had historical relationship with East Bengal as a dominant metropolis (Gillan 2002). Over the last decade Bangladeshi "infiltration" has been a prominent "national" campaign theme for the ruling nationalist Bharatiya Janata Party (BJP).

4.1 The Problem of Migration and Public Policies

The Indian parliament passed Citizen Amendment Act in December 2019 which created fear among the minorities. It amended the Citizenship Act, 1955 by providing a path to Indian citizenship for illegal migrants of Hindu, Sikh, Buddhist, Jain, Parsi, and Christian religious minorities, who had fled persecution from Pakistan, Bangladesh and Afghanistan before December 2014. Muslims from those countries were not given such eligibility. The Act was the first-time religion had been overtly used as a criterion for citizenship under Indian law.

The amendment may render millions of citizens stateless with the new clause of "illegal migrant" without any cut-off date, as most of the immigrants particularly from Eastern Bengal, came to India facing persecution and without any legal document of immigration (Bagchi 2017).

5 Possible Solutions to the Problem

The problem with migration arises when the migration is illegal and poses multi-layered threats to the state. Firstly, India's borders with its neighbouring countries, especially Bangladesh and Nepal, must be regulated. The border with Pakistan is already regulated. Secondly, migrants need to be strategically dealt with to reduce, if not eliminate, the level of threats they pose. Based on this, some broad policy recommendations may be made, keeping in mind the age-old relationship between the countries. They are enumerated below:

- India should become a signatory to the 1951 Refugee Convention, which protects refugee rights. India should also set up a national refugee protection framework.

- All past migrants should be given one-time citizenship with special identity cards and steps should be taken to regulate migration.
- Both the Indo-Bangladesh and Indo-Nepal borders must be managed. A separate Department of Border Management should be created. Effective border management, including regulation of the porous and open border, border fencing, effective utilisation of border guarding forces and control by immigration check posts (ICPs) will check illegal migration from both Bangladesh and Nepal or with any other neighbouring country.
- The porous and open nature of the border must be changed. There is no treaty or agreement to make the border between the any two countries open and porous. On the other hand, such a system developed gradually to suit the needs of both countries. This is now increasingly proving to be a security risk, particularly for India.
- Nepali migration to India has not been securitised. Given the state of affairs that exists today, it has to be securitised, particularly in the context of the growing Maoist threat to India and their cross-border linkages. The pattern and nature of Nepali migration to India need to be monitored and regulated.
- Although Bangladeshi migration has been securitised, action on the part of the government is lacking. Emotional considerations and petty political gains must not dictate policy making in India. The recommendations of a Group of Ministers in 2001 include the following observation: "the massive illegal immigration poses a grave danger to our security, social harmony and economic wellbeing."
- It is a fact that migrants to India, both Bangladeshi and Nepali, continue to nurture their original national identity. This makes them attached to their home country. While India should not have any problem on this count, the government has to take adequate preventive as well as curative measures, so that their original "identity", "obligation", "affiliation" and "attachment" do not go against India's national interests. In other words, the immigrants have to be increasingly made a part of the Indian mainstream. This could be done through various constitutional means including a one-time offer of citizenship, if our resources and law permit or may be via the Citizens Amendment Act 2019.
- In order to check migration from Bangladesh, India should assist in the economic development of that country. An enlarged aid package to Bangladesh is advocated. The logic is that with the economic development of Bangladesh, migration to India will decline. However, "social network" is the most important "pull factor" that attracts Bangladeshi migrants to India. Second, the logic that India should contribute to the economic development of Bangladesh in order to check migration needs to be carefully examined.
- In the case of Nepal, the 1950 Treaty needs to be revisited, particularly Articles VI and VII pertaining to reciprocal treatment meted to the nationals of one country in the territories of the other. This provision encourages and facilitates cross-border migration in an unregulated manner. If this is not in India's interests, it should be suitably amended to reflect present realities. It may be stated that the 1950 Treaty, as has been implemented, offers much greater benefits to Nepal than to India.

- Finally, in order to control illegal migration that brings with it other problems like smuggling of narcotics, arms and weapons and cross-border terrorism, there should be compulsory registration of all citizens and non-citizens living in India. All citizens should be given a multi-purpose national identity card and non-citizens should be issued identity cards of a different colour and design.

6 Conclusion

Rising sea level and increasing frequency and intensity of natural disasters are threatening coastal regions with displacement. Worldwide, 61% of displacements are triggered by disasters and 39% by conflict and violence. Small islands in the Ganga–Brahmaputra delta are disappearing into the sea forcing their inhabitants to shift to more viable places. The island state of Maldives is estimated to submerge into the sea by another 80 years. Large-scale cross-border migration is likely to happen, with consequences for peace and the society. To add to the problem, India is not a signatory to the 1951 Refugee Convention or the 1967 Protocol, which protect refugee rights. India also does not have a national refugee protection framework. However, it continues to grant asylum to a large number of refugees from neighbouring States and respects mandate of United Nations High Commissioner on Refugees (UNHCR) for other nationals, mainly from Afghanistan and Myanmar. Since India's neighbouring countries are among the top ten most vulnerable countries, the incidences of migration might increase over time due to increase in climate related extreme events. Therefore, there is a need to further regulate the borders and enhance preparedness for tackling the migrant problem which may include defining the term 'climate refugees' as per the national law as a first step.

References

Bagchi SS (2017) The great betrayal: potential statelessness after living decades in mother India. Vidyottama Sanatana Int J Hindu Sci Relig Stud 1:122–131

Boko MI, Niang A, Nyong C, Vogel A, Githeko M, Medany B, Osman-Elasha R, Brooks N (2006) Climate Change, Drought and Pastoralism in the Sahel. Discussion note for the World Initiative on Sustainable Pastoralism

Brown O (2008) Migration and climate change. International Organization for Migration, Geneva

Catoe L (2012) Endangered Island Nations Call for Global Action on Climate Change. The Guilfordian

Chen LC (1973) Disaster in Bangladesh. Oxford University Press, New York (1973)

Church JA, White NJ (2011) Sea-level rise from the late 19th to the early 21st century. Surv Geophys 32:585–602. https://doi.org/10.1007/s10712-011-9119-1

De Wit M, Stankiewicz J (2006) Changes in surface water supply across Africa with predicted climate change. Science 311:1917–1921

Dolma S, Singh S, Lohfeld L, Orbinski JJ, Mills EJ (2006) Dangerous journey: documenting the experience of Tibetan refugees. Am J Public Health 96:2061–2064

Eckstein D, Winges M, Künzel V, Schäfer L (2019) Global Climate Risk Index 2020 Who Suffers Most from Extreme Weather Events? Wether-Related Loss Events in 2018 and 1999 to 2018. Germanwatch, Bonn

Fallesen D, Khan H, Tehsin A, Abbhi A (2019) South Asia needs to act as one to fight climate change. https://blogs.worldbank.org/endpovertyinsouthasia/south-asia-needs-act-one-fight-climate-change. Accessed 27 Apr 2020

Gillan M (2002) Refugees or infiltrators? the bharatiya janata party and illegal migration from Bangladesh. Asian Stud Rev 26:73–95

Gornitz V (2007) NASA GISS: Science Briefs: Sea Level Rise, After the Ice Melted and Today. https://www.giss.nasa.gov/research/briefs/gornitz_09/. Accessed 25 Apr 2020

Hassan MW (2010) Climate Change in the Maldives. In: World Bank. https://www.worldbank.org/en/news/feature/2010/04/06/climate-change-in-the-maldives. Accessed 27 Apr 2020

IDMC (2017) Grid 2017. Global report on internal displacement. Internal Displacement Monitoring Center

India (2018) India's Second Biennial Update Report.

IOM (2019) World Migration Report 2020. International Organization for Migration, Geneva, Switzerland

IOM (2003) World Migration Report. International Organization for Migration, Geneva

IPCC (2014) Climate change 2014: Synthesis report. Contribution of Working Groups I, II and III to the fifth assessment report of the intergovernmental panel on climate change [core writing team, R. K. Pachauri and L. A. Meyer (eds)]. Intergovernmental Panel on Climate Change, Geneva, Switzerland

Jolly S, Ahmad N (2019a) Conceptualizing the Climate Change Migration in South Asia. In: Climate Refugees in South Asia. Springer Singapore, Singapore, pp 15–45

Jolly S, Ahmad N (2019b) Climate Refugees in South Asia: Protection Under International Legal Standards and State Practices in South Asia. Springer, Singapore

Khan Y (2017) The Great Partition: the Making of india and Pakistan. Yale University, New Haven, CT

Mani M, Bandyopadhyay S, Chonabayashi S, et al (2018) South Asia's Hotspots

McLeman R, Smit B (2006) Migration as an adaptation to climate change. Clim Change 76:31–53

Mills EJ, Singh S, Holtz TH, Chase RM, Dolma S, Santa-Barbara J, Orbinski JJ (2005) Prevalence of mental disorders and torture among Tibetan refugees: a systematic review. BMC Int Health Hum Rights 5:7. https://doi.org/10.1186/1472-698X-5-7

Nicholls RJ, Lowe JA (2004) Benefits of mitigation of climate change for coastal areas. Glob Environ Change 14:229–244

Pereira L, Asrar GR, Fisher LH, Hsu A, Nel J, Sitas N, Ward J, Vervoort J, Selomane O, Trisos C (2019) Bottom-up Initiatives and Participatory Approaches for Outlooks-Global Environment Outlook (GEO-6): Healthy Planet, Healthy People Chapter 23. Global Environment Outlook (GEO-6): Healthy Planet, Healthy People

Ramesh R (2008) The new president of the Maldives may buy land elsewhere because of climate change. The Guardian

Sindico F (2005) Ex-post and ex-ante (legal) approaches to climate change threats to the international community. N Z J Environ Law 9:209

SMH (2012) Where climate change threatens survival. In: Syd. Morning Her. https://www.smh.com.au/politics/federal/where-climate-change-threatens-survival-20120108-1pq4c.html. Accessed 27 Apr 2020

Talbot I, Singh G (2009) The partition of India. Cambridge University Press Cambridge, Cambridge

The Hindu (2019) 106 coastal sites picked for conservation. The Hindu

Weiner M (1993) Rejected peoples and unwanted migrants in South Asia. Econ Polit Wkly pp. 1737–1746

Key Knowledge Gaps at the Interface of Chemical Ecology and Ecosystem-Based Adaptation Planning: A Public Policy Perspective

Gopichandran Ramachandran

Abstract The review presents a snapshot of thrust areas at the interface of chemical ecology and ecosystem-based planning that have not received adequate attention, despite significant progress in the respective fields. It is important to take note of such calls for action, due to the growing emphasis on adaptation. Importantly, several facets of bio-resources management across the soil and plant continuum feature in this gaps framework. These include, amongst others, mineralization mediators in soils, age correlated changes in biochemical profiles with implications for tri-trophic interactions, and therefore productivity and related interpretation as signals of resilience. Recent advances in science communication that will enable target-oriented engagement with stakeholders are also referred to, as part of the public policy framework this review centers on.

Keywords Chemical ecology · Adaptation · Public policy · Stakeholder engagement

1 Introduction

Ecosystem-based adaptation planning should ensure sustained productivity. This implies the development and implementation of approaches that holistically consider phenological and related chemical ecological variations in the biochemical profiles of plants. Importantly, such variations could be natural or artificially induced, with individual and synergistic impacts on the adaptive ability of plants. National missions on climate-resilient agriculture in several countries could come across sustainability challenges, should the stated correlates not receive adequate attention. The principal objective of the present review is to highlight stated gaps in our knowledge about some important facets of chemical ecology with special reference to the applications of such knowledge to the development and implementation of strategically important climate change adaptation actions. This is also timely because of the impending launch of the

G. Ramachandran (✉)
NTPC School of Business, Noida, Uttar Pradesh, India
e-mail: gopichandran@nsb.ac.in

M. Behnassi et al. (eds.), *Social-Ecological Systems (SES)*,
https://doi.org/10.1007/978-3-030-76247-6_10

sixth assessment report of the Intergovernmental Panel on Climate Change (IPCC) and a large number of adaptation plans already available in the public domain.

The methodology for the review presented is guided by questions about the significance conferred on aspects of chemical ecology in design of productivity-related adaptation strategies. It extends into calls for such integration in order to enhance the robustness of related approaches.

The case in point is the adaptive abilities of crop plants, weeds and trees interpreted within the larger context of chemical environment they are part of. This has to inevitably be so at the interface of soils mediated by microbial conversions, and therefore access to nutrients and herbivory as a dominant perturbation. Changes in biochemical profiles of plants are known to be on account of phenology too, and it is therefore important to differentiate natural and artificially induced variations (Pugilisi et al. 2006; Lee Dyer et al. 2019). The inextricable impacts of the changing composition of greenhouse gases compound our analysis. The latter is further complicated through demands on resilience imposed by ultraviolet-B radiations. The present chapter brings to the attention of science technology and innovation policy makers important domains of knowledge pertaining to the above stated, that have to be addressed on priority because they have reportedly not received adequate attention (Knutti 2019; Neufeldt et al. 2018; Persson 2019; Adger et al. 2018). The references discussed above are quite invaluable because of the extensive spread and depth of sourcing of evidences they present. Importantly, most of them have been recently published; thereby their temporal relevance for decision support is quite profound.

The fundamental premise is of such biological realities as age correlated biochemical profiles, quantitative thresholds of mediators of change, recalcitrance of materials and delayed and rapid induced resistance (Beck 1965; Janzen 1968; Blum 1981; Haukioja 1991; Anathakrishnan 1993; Ananthakrishan and Gopichandran 1993). They continue to be nebulous at best, because of the rapidly changing knowledge landscape about continually and co-evolving quantitative and qualitative profiles of plants and several elements of tri-trophic interactions, potentially attributable to mutually reinforcing impacts of climate change. It will be useful to, therefore, take stock of recent calls for action, amongst several others, in the fields of microbial ecology, micro arthropod dynamics, and biochemical profiles of plants with implications for natural enemy complexes. These obviously have a bearing on the overall productivity of crops or forests, as the case may be. Adaptation planning cannot oversimplify sustainability only through the quantum of crop yield.

Models interpret impacts across large temporal and World Bank (2019) spatial scales and local levels to deliver adaptation benefits. Beveridge et al. (2018) present a detailed analysis of elements common to both contexts and the scope to enrich them for greater real-life gains at the local level. In fact, they indicate the possibility of mal-adaptation if adequate evidence-based safeguards are not ensured. Chemical ecological correlates are yet to gain their visibility in this analysis too. Spatial distribution patterns of crops are a case in point (Leng and Huang 2017). Several publications elaborated on the scope for such value adds even in the Indian context (Singh et al. 2015; Behnassi et al. 2014; Vijayan and Vishwanathan 2018). Productivity correlates cannot be isolated from physical and chemical ecological parameters. Learnings from the interplay of these parameters should define policies on cultivation practices. It is, therefore, important to ask if we are adequately information-empowered to define spatially and temporally robust adaptation strategies, especially from a public policy point of view. The present statement of gaps is intended to be a snapshot at best, with two mutually reinforcing strands, across the stated areas, with implications for a robust systems approach. This is to reinforce the scientific basis of decision making across policy, plans, programmes and project continuum. More than twenty such areas are highlighted in the following.

The *first* segment is about some aspects of chemical ecology, with implications for adaptation strategies that target ambiguities in resilience. The *second* strand is about some specific public policy interfaces that can make use of insights from the suggested areas. These could feed into the much-neglected stakeholder engagement dynamics, central to participatory bio-resources conservation and management. Provisions of Article 6 of the Convention could be exploited to hybridize knowledge. The 2019 United Nations Report titled *Climate action and support trends* takes note of persistent gaps in individual and institutional capacity of governments to better mainstream climate considerations in national planning and budgeting. This is true especially of adaptation measures (UN CC Secretariat 2019). The latter specifically pertains to the science and science communication that sets the context to transform policy intents to reality. Additional insights about the relevance of chemical ecological aspects stated and the felt need to embed them in policies that guide action are discussed in the following. This is dovetailed with the need to rationalize communication strategies as part of public policy frameworks.

2 Discussion

Debra (2016) stresses, adaptive measures are not adequately mainstreamed in the growing body of adaptation decisions and that physiological and ecological defense costs in small timeframes have to be addressed on priority. It is essential to identify and evaluate synergies between multiple adaptation options with a clear understanding of their limits and limitations. Ruth et al. (2019) elaborate on five hypotheses with implications for costs of adaptation determined further by natural defense mechanisms of biological systems. Secondary compounds, their derivatives and related dynamics of carbon and nitrogen offer signals characteristic of their respective landscapes. These call for a deeper understanding of the interplay of selective pressures including competition sensing and interspecific interactions across domains. These are essentially bi-directional to multi-directional in nature and are little understood in terms of climate change impacts.

Neher and Mary (2019) cite significant literature on the role of micro-arthropods and set a context for related insights about food chains they govern in soils. The importance of comminution with implications for nutrients cycling cannot be overemphasized. However, despite significant progress in this domain, the authors rightly highlight paucity of knowledge about the ecological functions of such micro-arthropods across significantly large spaces and the dynamics of multi-channel omnivory. Importantly, they relate it to soil health and related pathways of nutrient access. This is also true of the occurrence and distribution of all microorganisms with measures of specific activity in soils (Hoyle 2013).

The specific thrust is on the threshold of microbial populations that impact soil functions. Individual and synergistic impacts, especially of metabolites in the root zone, need to be addressed systematically with particular reference to allelopathy. This is reiterated recently by Cavicchioli et al. (2019) with the highlight that role of microbes is not explicitly taken into account in policy making. This is especially true of management of land resources and their rehabilitation through afforestation and select agriculture crop management. Recalcitrant substances and their fates determine biochemical dynamics within this micro-milieu with implications for nitrogen compounds in particular. Paucity of information about long-term changes in soils, bloom cycles and exudates that influence microbe-mineral association may reduce the level of confidence on related adaptation planning. Climate change correlates further enhance complexity.

Bonanomi et al. (2005) and Francisco et al. (2019) draw attention to the fact that plant soil feedbacks are temperature sensitive due to which, it is not possible to generalize trends across spatially heterogenous and temporally variable landscape. Microclimate conditions influence pathways in enzyme mediated conversions in soils that in turn determine successions and resultant nutrient limitations. Enhanced CO2 and its impacts are relatively less known with respect to fungi-mediated nutritional and defense traits of plants (Haukioja 1991; Frew and Jodi 2018). Classen et al. (2015) and Soares et al. (2019) reinforce these links with considerations of above-ground interactions implicating salinity, waterlogging and drought-stress on the concentration of nutrients in edible plant parts. This is due to the type, duration and intensity of stress that could determine breeding preparedness for select nutrients and related farm management practices to sustain production.

Jacktel et al. (2019) focus on the palatability of host leaves as a function of exposure to heat stress due to heat waves or thermal shocks. These could alter pest dynamics in forest systems too, further correlatable to the profiles of resistance exhibited by the trees. These could in turn be correlated with nutrient access enabled by microbes in soils. Cornellisen (2011) highlighted species-specific community structure changes, especially in the tropics, and resulting tri-trophic interactions. Larson et al. (2019) draw attention to insect hybridization and related temperature effects imposed by climate change with reference to variations within species and their geographic boundaries. Interestingly, a citizens' science perspective is proposed to capture large samples for study.

Craig et al. (2014) define the links between climate change and ozone depletion and emerging related dilemmas. Recent efforts to prevent backsliding on ozone layer protection, call for detailed analysis of risks and benefits of compliance. Biological effects of ultraviolet-B radiations and concomitant impacts on materials across land and ocean surfaces are well documented by the Montreal Protocol. The most important takeaway from these investigations is the urgent need to investigate individual and synergistic impacts of both phenomena, namely climate change and ozone depletion. Such differentiation and knowledge that guide adaptation action are few and far between. Thais et al. (2018) report the consequences of exposure to ultraviolet on the anti-oxidant capacity of plant systems and its correlation with injury susceptibility. Raquel et al. (2008), more than a decade ago, highlighted the fact that changing carbon dioxide levels affect disease resilience. Robson et al. (2019) clearly highlight adverse effects on plant growth. Tropical areas experience relatively higher levels of UV exposure and, therefore, adaptive abilities of bio-resources should be carefully calibrated on this account also.

Detailed insights from such pointers as above are expected to modulate practices by farmers through targeted engagement strategies. Leng and Huang 2017, Aggarwal et al. (2019) indicated that farmers will do well to adopt location-specific climate resilient varieties, planting schedules, nutrient, fertilizer and water management approaches to stay less affected by inclement weather systems. Such targeted interventions are conspicuous by their meagre presence, further compounded by only incremental adaptation measures. The UNEP adaptation gap report (2014) UN Environment (2019) corroborates the above.

Frameworks of stakeholder engagement evolve in response to their specific information and capacity building needs (Samaddar 2019; Gardner 2009). Recent thrusts on the science of science communication highlight the need to address controversies in science on priority. It is essential to remove ambiguities about factors that limit performance of practices or technologies; to enhance willingness of communities to accept them (Bucchi 2019). Article 51 A(h) of the Indian Constitution presents a unique opportunity through a call for citizens to imbibe and practice scientific temper. It can be interpreted for the scope to understand forms and functions of logical frameworks on cause-effect relationships and differentiate natural and artificially induced variations (Gopichandran R. 2013).

An India specific climate change analysis (Mehajan et al. 2019) refers to the increased intensity and frequency of heat waves. Scenarios regarding adaptive mechanisms of plants would therefore change significantly. Empirical evidences regarding such changes are critical especially because of the large latitudinal spread of the country. An interesting public policy interface is about information that can be provided by national meteorological services to guide adaptation across various sectors of economy. This is true also of the energy sector and related infrastructure exposed to the vagaries of weather. Their vulnerabilities should be assessed in greater detail. Mehta et al. (2019) draw attention to the fact that climate assessments are fraught with uncertainties about scale and intensities of impacts across sectors, including socio-economic parameters. Ecological uncertainty, knowledge and epistemic uncertainties, and political economy uncertainties determine the precision with which such ambiguities could be tackled.

Sandra (2011) reviewed a large number of European Union projects and called for systematic efforts to upscale mitigation and adaptation benefits on the basis of ecosystem-based approaches. Even a decade ago, inadequate quantitative data impeded assessments, especially with reference to the cost effectiveness of projects to garner more finance. Such ambiguities continue about differences between ecosystem-based approaches to mitigation and adaptation, with implications for resource extraction policies that further determine landscapes management and appropriate remediation strategies.

Jyotiraj (2016) also elaborated on the inter-sectoral and sector specific vulnerabilities. The energy sector, in particular, had significant interface with water, agriculture, forestry, and infrastructure systems. These are increasingly embedded in planning exercises. India's 12th 5-year plan, the subsequent "Draft national energy policy 2017", and the program on optimizing energy mix, proposed by the Indian Ministry of Power emphasized sustainability (NITI Ayog 2017; Ministry of Power, Government of India 2018).

Rebecca et al. (2016) and van der Geest et al. (2019) pointed out that the assessment of impacts is difficult because of reinforcing drivers over long-time scales and uncertainties. These are not adequately addressed with respect to the source of perturbations – especially so with reference to ecosystem services. European Commission's COACCH project (CO-designing the Assessment of Climate CHange costs (COACCH 2018) discussed several gaps in impact analysis and related policy challenges pertaining to biophysical and socio-economic tipping points. Such parameters as climate shocks, and mitigation/amelioration management styles could influence the biochemical basis of adaptation and, therefore, the need for multi-sectoral perspective. Ryan and Bustos (2019) discussed six policy implementation related barriers with respect to six Latin American countries.

India made a comprehensive assessment of its ecosystems and biodiversity within an economic perspective. The interim report as a working document (GIZ 2015) presented several case examples representing forest, inland wetland, coastal and marine systems. These substantiated important economic consequences of biodiversity loss and impacts on declining ecosystem services. This followed its scoping efforts timed with the 11th Conference of Parties of the Convention on Biological Diversity (CBD). It called for concerted efforts to evaluate several ecosystem services to define economic incentives for conservation. MoEF, Government of India (2019) discussed the BIOFIN that followed in 2019 Rattani (2018). It focused on several financial solutions to enlist the private sector through its corporate social responsibility (CSR) initiatives. Sustainable use planning is at the core of this initiative, aligned with Aichi targets thereby an interplay of provisioning, payments, regulatory, cultural and supporting services. The UN Environment GIZ (2019) elaborated on ecosystem-based adaptation governance centered on social, institutional and economic foundation across all levels to determine and foster nature-based solutions in addition to adaptation accountability for coherence.

The Centre for Science and Environment in 2018 summarized the scale and diversity of impacts worldwide with implications for India. It traced India's initiatives through the fourth assessment period and in response to the Bali action plan, logically extending to the Cancun and Paris agreements. Synergies across missions and programs are quite evident with insights on community, technological and institutional mechanisms to meet the stated goals. Four important components of emissions on agriculture in particular set the context for robust management of outcomes through integrated water, soil health and knowledge thrusts. The UN Environment,

CBD and GIZ in 2019 presented voluntary guidelines for design and implementation of the ecosystem-based adaptation macro-level framework integrating six sectors as a primer for policy makers. The latest call from the World Commission on Adaptation and the World Resources Institute in 2019 focuses on the triple dividend framework, duly emphasizing the fact that knowledge gaps retard progress.

Action on Climate Today (Virginie et al. 2019) presented a case for private sector engagement for adaptation action. The private sector could support with budget and skills as part of environmental stewardship. This could further strengthen governance and, therefore, a need for stronger foundations of narratives. The recent World Economic Forum Risk report (2019), Valerio et al. (2018), Barbara et al. (2019), and the Emissions Gap Report (2019) of the UN Environment provide the necessary impetus for private sector engagement by highlighting the likelihood of extreme weather events in addition to the consequences of failures on the mitigation and adaptation. It duly recognizes the importance of economic, environmental, geopolitical, societal and technological verticals in this context. The World Bank in 2019, stressed on the need for rating system of adaptation and resilience action plans to derive mutually reinforcing benefits from finances and programmatic approaches. It further highlights five important consequences of non-adaptation and, therefore the importance of a needs-driven adaptation plan.

3 Conclusion

Recent literature pertaining to several facets of chemical ecology of direct relevance to ecosystems-based approaches to adaptation highlight areas of investigations that have not received adequate attention. These have to be addressed on priority especially due to the perceived growing nexus between climate change and resultant impacts at the local level in particular. On the other hand, they set the context to differentiate consequences of other natural and artificially induced variations that influence biochemical and related productivity profiles and their ecology as a whole. Holistic understandings of the science of such phenomena should be suitably embedded in adaptation strategies to sustain the momentum for change as part of integrated resilience. Public policy applications of science and its applications call for a detailed understanding of the limits and limitations of knowledge systems. The present snapshot is, therefore, meant to help focus appropriately on the stated knowledge gaps to reduce uncertainties in decision making. Interestingly, these gaps appear to have persisted despite advances in measurements and interpretation of qualitative and quantitative profiles of plants and other elements of their biotic environment. The scientific and policy assessment reports of the UNFCCC, the Montreal Protocol and the Convention on Biological Diversity in particular reinforce the stated context to evolve and apply deeper insights for local level resilience gains. The references cited substantiate this premise.

Conflict of Interest The authors declare no conflict of interest.

References

Adger NW, Shardul AM, Monirul QM, Cecilia CKO, Juan, PRP, Barry S, Kiyoshi T, Brenna E, Samuel F, James F, Simone G, Simon JN , Richard JTK, Irene L, David CM, Tristan DP, Arun S, Shukla PR, Joel BS, Tim R, Cynthia R, Katharine V, Johanna W (2018) Assessment of adaptation practices, options, constraints and capacity https://www.ipcc.ch/site/assets/uploads/2018/02/ar4-wg2-chapter17-1.pdf

Ananthakrishnan TN Bionomics of thrips 1993. Annu.Rev.Entomol 38: 71–92.

Ananthakrishnan TN, Gopichandran R (1993) Chemical Ecology in Thrips – Host Plant Interactions, 125p Oxford & IBH, New Delhi

Aggarwal P, Vyas S, Thornton P, Campbell BM (2019) How much does climate change add to the challenge of feeding the planet this century? Environ Res Lett 14(4):043001. https://doi.org/10.1088/1748-9326/aafa3e

Barbara B, Alex C, Angela F, Rob M, Chavi M, Rowena T, Cooper W (2019) Global Landscape of Climate Finance https://climatepolicyinitiative.org/wp-content/uploads/2019/11/2019-Global-Landscape-of-Climate-Finance.pdf

Beck SD (1965) Resistance of plants to insects. Annu Rev Entomol 10:27–32

Behnassi M, Muteng'e MS, Gopichandan R, Shelat KN, (eds.) (2014) Vulnerability of agriculture, water and fisheries to climate change. Towards sustainable adaptation strategies. (336p) Dordrecht: Springer

Beveridge L, Whitfield S, Challinor A (2018) Crop modelling: towards locally relevant and climate-informed adaptation. Clim Change 147(3–4):475–489. https://doi.org/10.1007/s10584-018-2160-z, https://link.springer.com/content/pdf/10.1007/s10584-018-2160-z.pdf

Blum MS (1981) Chemical Defences in Arthropods. (540p) Cambridge: Academic Press

Bonanomi G, Giannino F, Mazzoleni S (2005) Negative plant/soil feedback and species coexistence. Oikos 111:311–321. https://www.academia.edu/17753812/Negative_plant-soil_feedback_and_species_coexistence

Bucchi M (2019) Facing the challenges of science communication 2.0: quality, credibility and expertise. EFSA J 17(S1):e170702 https://doi.org/10.2903/j.efsa.2019.e170702 https://efsa.onlinelibrary.wiley.com/doi/pdf/10.2903/j.efsa.2019.e170702

Cavicchioli R, William JR, Kenneth NT, Farooq ALRB, Matthew B, Michael JB, Antje B, Philip WBTC, Thomas WC, Roberto D, Christine MF, Jef H, David AH, Janet KJ, David MK, Britt K, David BMW, Jennifer BHM, Mary AM, Victoria JO, David SR, Justin VR, Virginia IR, Brajesh KS, Lisa YS, Frank JS, Matthew BS, Madeleine JHVO, Scott CW, Eric A, Nicole SW (2019) Scientists' warning to humanity: microorganisms and climate change. Nature Rev. Microbiol. 17(9):569–586 https://www.nature.com/articles/s41579-019-0222-5.pdf

Rattani V (2018) Coping with climate change: An analysis of India's NAPCC.Centre for Science and Environment, India http://www.indiaenvironmentportal.org.in/files/file/coping-climate-change-NAPCC.pdf

Classen ATMKS, Jeremiah AH, Gregory SN (2015) Direct and indirect effects of climate change on soil microbial and soil microbial-plant interactions: what lies ahead? Ecosphere 6(8):1–21 https://www.researchgate.net/publication/281453577_Direct_and_indirect_effects_of_climate_change_on_soil_microbial_and_soil_microbial-plant_interactions_What_lies_ahead

Cornelissen T (2011) Climate Change and Its Effects on Terrestrial Insects and Herbivory Patterns. Neotropical Entomology 40(2):155–163 https://www.researchgate.net/publication/51138701_Climate_Change_and_Its_Effects_on_Terrestrial_Insects_and_Herbivory_Patterns

Craig EW, Richard GZ, Robyn L, Sasha M (2014) Solar ultraviolet radiation in a changing climate. Nature Clim. Change 4(6):434–441 https://doi.org/10.1038/nclimate2225 https://www.researchgate.net/publication/264696919_Solar_ultraviolet_radiation_in_a_changing_climate https://www.nature.com/articles/s41598-017-18710-2

Debra D (2016) Gaps in agricultural climate adaptation research. Nature Clim. Change 6:433–435 www.nature.com/natureclimatechange https://www.nature.com/articles/nclimate3007.pdf?draft=marketing

Francisco I, PJosé AM, Josep P, Peter BRRD, BAurora G, David AW, Wim HP (2019) Climate change effects on plant-soil feedbacks and consequences for biodiversity and functioning of terrestrial ecosystems https://advances.sciencemag.org/content/advances/5/11/eaaz1834.full.pdf

Frew A, Jodi NP (2018) Mycorrhizal-mediated plant–herbivore interactions in a high CO2 world. Funct. Ecol. British Ecol. Soc. 33:1376–1385 https://besjournals.onlinelibrary.wiley.com/doi/epdf/10.1111/1365-2435.13347

Gardner, J, Dowd, A-M, Mason C, Ashworth P (2009) A framework for stakeholder engagement on climate adaptation. CSIRO Clim. Adapt. Flagship Working paper 3:1–31 http://www.csiro.au/resources/CAF-working-papers.html. https://research.csiro.au/climate/wp-content/uploads/sites/54/2016/03/3_CAF_WorkingPaper03_pdf-Standard.pdf

GIZ (2015) The economics of eco – systems and bio diversity, India Initiative interim report Working document https://www.giz.de/en/downloads/giz2015-en-ecosystems-biodiversity-interim-report-india.pdf

GIZ (2019) Governance for Ecosystem-based Adaptation Understanding the diversity of actors & quality of arrangements https://www.adaptationcommunity.net/wp-content/uploads/2019/09/giz2019-en-eba-governance-study-low-res.pdf

Gopichandran R (2013) Opinion: enable and manifest scientific temper in tangible development oriented outcomes: go beyond rhetoric. J Sci Temper 1(3 & 4): 198–207

Government of India ministry of Power, Central Electricity Authority 2018 Report of the Government of India, Ministry of Power, Central Electricity Authority, Committee on optimal energy mix in power generation on medium and long term basis 01–01–2018 https://powermin.nic.in/sites/default/files/webform/notices/Report_of_the_Committee_on_optimal_energy_mix_in_power_generation_on_medium_and%20long_term_basis.pdf Optimal Power mix

Haukioja E (1991) Induction of defences in trees. Annu Rev Entomol 36:25–42

Hoyle F (2013) Managing soil organic matter, A practical guide. GRDC Project Code: KDI00023 https://grdc.com.au/__data/assets/pdf_file/0029/107696/grdc-guide-managingsoilorganicmatter-pdf.pdf.pdf

Jacktel H, Julia K, Bastien C (2019) Responses of forest insect pests to climate change: not so simple. Curr Opin Insect Sci 35:103–108 https://reader.elsevier.com/reader/sd/pii/S2214574519300227?token=BE09DFDAF156CFB60120CF39537E9F91C5482D07D7BB23E2C12E96B5524F4040A8D6863A29869EEDF4A95D4F0BEEB80C

Janzen DH (1968) Host plants as islands in evolutionary and contemporary times. Amer Nat 102:592–595

COACCH (2018) The Economic Cost of Climate Change in Europe: Synthesis Report on State of Knowledge and Key Research Gaps. Policy brief by the COACCH project. Editors: Paul W, Jenny T, Katriona M. Published May 2018

Jyotiraj P (2016) Review of Current and Planned Adaptation Action in India CARIAA Working Paper #10 CARRIA https://www.iisd.org/sites/default/files/publications/idl-55866-india.pdf

Knutti (2019) Closing the Knowledge-Action Gap in Climate Change https://www.cell.com/one-earth/pdf/S2590-3322(19)30028-4.pdf

Larson EL, Robin MT, Scott A (2019) TInsect Hybridization and Climate Change Front. Ecol. Evol. https://doi.org/10.3389/fevo.2019.00348 https://www.frontiersin.org/articles/10.3389/fevo.2019.00348/full

Dyer LA et al (2019) Modern approaches to study plant–insect interactions in chemical ecology https://doi.org/10.1038/s41570-018-0009-7 https://www.researchgate.net/publication/325365847_Modern_approaches_to_study_plant-insect_interactions_in_chemical_ecology

Leng G, Huang M (2017) Crop yield response to climate change with crop spatial distribution pattern. Nat Sci Rep 7:1463.10 p

Mehajan RK, Abha T, Shreekant G (2019) Towards effective climate services: Indian context. Curr Sci 117(8):1274 https://www.currentscience.ac.in/Volumes/117/08/1274.pdf

Mehta L, Shilpi S, Hans NA, Alankar, SB, Upasona G, Vijay KV (2019) Climate change and uncertainty from 'above' and 'below': perspectives from India. Reg Environ Change (2019) 19:1533–1547 https://doi.org/10.1007/s10113-019-01479-7 https://link.springer.com/content/pdf/10.1007/s10113-019-01479-7.pdf

MoEFCC (2019) Biodiversity finance plan working document https://www.biodiversityfinance.net/sites/default/files/content/knowledge_products/Biodiversity%20Finance%20Plan%20Report%20Updated%20and%20Final%20%28Digital%20Presence%20-%20Low%20resolution%29%2008-07-2019.pdf

Neher DA, Mary EB (2019) Soil microarthropods and soil health: intersection of decomposition and pest suppression in agroecosystems. Insects 10(12):414 https://doi.org/10.3390/insects10120414

NITI Aayog, Government of India 2017 Draft National Energy Policy, 106p https://niti.gov.in/writereaddata/files/new_initiatives/NEP-ID_27.06.2017.pdf

Neufeldt, H, Sanchez MGOA, Knudsen CMS, Dorkenoo KEJ (eds.) (2018) The Adaptation Gap Report 2018. United Nations Environment Programme (UNEP), Nairobi, Kenya. United Nations Environment Programme https://unepdtu.org/wp-content/uploads/2019/04/agr-final-version-2018.pdf

Pugilisi MP, Jennifer MS, Koty HS, Raphael RW, Valerie JP (2006) Marine Chemical Ecology. Natural Prod Rep 23(2):153–80 https://doi.org/10.1039/b404735b https://www.researchgate.net/publication/7207634_Marine_Chemical_Ecology

Persson A (2019) Global adaptation governance: An emerging but contested domain WIREs Clim Change 10:e618 https://doi.org/10.1002/wcc.618

Raquel G, Emília H, Wagner B (2008) Sci. agric. (Piracicaba, Braz.) 65: no.spe Piracicaba https://doi.org/10.1590/S0103-90162008000700015 http://www.scielo.br/scielo.php?script=sci_arttext&pid=S0103-90162008000700015

Rebecca KR, Brett AB, Laura ED, Fleur JFM (2016) Incorporating climate change into ecosystem service assessments and decisions: a review. Glob. Change Biol 23(1): https://doi.org/10.1111/gcb.13457 https://www.researchgate.net/publication/306041356_Incorporating_climate_change_into_ecosystem_service_assessments_and_decisions_A_review/link/57d11a5d08ae601b39a0d555/download

Robson MT, Pedro JA, Agnieszka KB, Paul WB, Craig CB, Gareth IJ, Titta KK, Justyna Ł, Javier MA, Luis OM, Susanne N, Marta P, Neha R, Filip V, Marcel AK, Jansen MA (2019) A perspective on ecologically relevant plant-UV research and its practical application. Photochem Photobiol Sci 18:970 https://pubs.rsc.org/en/content/articlelanding/2019/pp/c8pp00526e#!divAbstract

Ruth S, Dana U, Lukas YW, Helge BB (2019) Microbe-driven chemical ecology: past, present and future. ISME J 13(11): https://doi.org/10.1038/s41396-019-0469-x https://www.researchgate.net/publication/334351605_Microbe-driven_chemical_ecology_past_present_and_future

Ryan D, Bustos E (2019) Knowledge gaps and climate adaptation policy: a comparative analysis of six Latin American countries. Clim Policy 19(10): 1297–1309 https://www.tandfonline.com/doi/abs/10.1080/14693062.2019.1661819?needAccess=true#aHR0cHM6Ly93d3cudGFuZGZvbmxpbmUuY29tL2RvaS9wZGYvMTAuMTA4MC8xNDY5MzA2Mi4yMDE5LjE2NjE4MTk/bmVlZEFjY2Vzcz10cnVlQEBAMA==

Samaddar S, Ayaribilla AJ, Oteng-Ababio M, Dayour F, Yokomatsu M (2019) Stakeholders' perceptions on effective community participation in climate change adaptation. In: Sarkar A, Sensarma S, vanLoon G (eds) Sustainable Solutions for Food Security. pp 355–379 Springer, Cham https://www.researchgate.net/publication/330475983_Stakeholders'_Perceptions_on_Effective_Community_Participation_in_Climate_Change_Adaptation_Combating_Climate_Change_by_Adaptation

Sandra N, Gerardo A, Pam B, Sarah B, McKenna D, Ana F-L, Holger G, Michele S (2011) Assessment of the potential of ecosystem-based approaches to climate change adaptation and mitigation in Europe. Final report to the European Commission, DG Environment, Contract no. 070307/2010/580412/SER/B2, Ecologic institute and Environmental Change Institute, Oxford University Centre for the Environment https://ec.europa.eu/environment/nature/climatechange/pdf/EbA_EBM_CC_FinalReport.pdf

Singh AK, Dagar JC, Arunchalam A, Gopichandran R (eds) (2015) Climate Change Modelling, Planning and Policy for Agriculture. Springer India 245p.

Soares JC, Carla SS, Susana MPC, Manuela M, Marta WV (2019) Preserving the nutritional quality of crop plants under a changing climate: importance and strategies https://link.springer.com/content/pdf/10.1007/s11104-019-04229-0.pdf

Thais HR, Andrea S, Antonella C, Eric GC, Annamaria R, Lorenzo G (2018) Physiological effects of short acute UVB treatments in Chenopodium quinoa Willd. Sci. Rep. 8:371 https://doi.org/10.1038/s41598-017-18710-2 https://www.nature.com/articles/s41598-017-18710-2.pdf

UNEP (2014) The Adaptation Gap Report United Nations Environment Programme (UNEP), Nairobi http://wedocs.unep.org/bitstream/handle/20.500.11822/9331/-Adaptation_gap_report_a_prel.pdf?sequence=2&isAllowed=y

UN CC Secretariat 2019 Climate action and support trends https://unfccc.int/sites/default/files/resource/Climate_Action_Support_Trends_2019.pdf

UN Environment 2019 Emissions Gap Report https://wedocs.unep.org/bitstream/handle/20.500.11822/30797/EGR2019.pdf?sequence=1&isAllowed=y

UN Environment GIZ 2019 CBD Technical Series No. 93 Voluntary guidelines for the design and effective implementation of ecosystem-based approaches to climate change adaptation and disaster risk reduction and supplementary information https://www.cbd.int/doc/publications/cbd-ts-93-en.pdf

Valerio M, Bella T, Federico M (2018) Understanding and Increasing Finance for Climate Adaptation in Developing Countries https://www.international-climate-initiative.com/fileadmin/Dokumente/2019/20190225_Understanding-and-Increasing-Finance-for-Climate-Adaptation-in-Developing-Countries.pdf

van der Geest K et al (2019) The impacts of climate change on ecosystem services and resulting losses and damages to people and society. In: Mechler R, Bouwer L, Schinko T, Surminski S, Linnerooth-Bayer J (eds) Loss and Damage from Climate Change. Climate Risk Management, Policy and Governance. Springer, Cham https://doi.org/10.1007/978-3-319-72026-5_9, https://www.researchgate.net/publication/329277938_The_Impacts_of_Climate_Change_on_Ecosystem_Services_and_Resulting_Losses_and_Damages_to_People_and_Society_Concepts_Methods_and_Policy_Options

Vijayan I, Vishwanathan PK (2018) india's initiative on climate resilient agriculture - a preliminary assessment. Int J Pure Appl Math 118(9):491–497 https://acadpubl.eu/jsi/2018-118-7-9/articles/9/44.pdf

Virginie F, Caroline F, Vidya S, Vandana N, Sunil A, Naman G, Luca P (2019) Engaging the private sector in financing adaptation to climate change: Learning from practice. ACT http://www.acclimatise.uk.com/wp-content/uploads/2019/02/ACT-Private-Sector-paper_final_web-res.pdf

WEF 2019 The Global Risks Report 2019 14th Edition http://www3.weforum.org/docs/WEF_Global_Risks_Report_2019.pdf

World Bank 2019 Action Plan on Climate Change Adaptation and Resilience MANAGING RISKS FOR A MORE RESILIENT FUTURE http://documents.worldbank.org/curated/en/519821547481031999/The-World-Bank-Groups-Action-Plan-on-Climate-Change-Adaptation-and-Resilience-Managing-Risks-for-a-More-Resilient-Future.pdf

World Commission on Adaptation and WRI 2019 Adapt Now: A Global Call for Leadership on Climate Resilience. https://cdn.gca.org/assets/2019-09/GlobalCommission_Report_FINAL.pdf

When Environmental Inequalities Lead to Social Inequalities!

Moulay Hicham El Amrani

Abstract The state of natural environment often impacts human society. The degradation of Earth's ecosystems due to anthropogenic dynamics has so far benefited economically only a minor proportion of humanity while making the majority more vulnerable. Those who consume and pollute less are often those who suffer more; a situation of inequality which is increasingly growing through the deepening of gaps between the rich and the poor. Decision making processes, in a desperate attempt to set the right balance when the political will exists, are often unable to find efficient and long-lasting solutions for these emerging issues, focusing instead on managing their related implications (poverty, displacement, conflicts, etc.). Therefore, the fight against inequalities is still at an early stage and achieving equality often requires breaking the wall and the adoption of unconventional conceptual, strategic, and operational frameworks. The viability and resilience of social-ecological system require that the ecological and social equality and equity should go in the same direction, unlike current dynamics. Thus, is it currently possible to develop in the future a trend where individuals, communities, and countries reduce, on a voluntary basis, their unsustainable environmental practices – such as the overuse and degradation of nonrenewable natural resources – on behalf of a viable and resilient social-ecological system where environmental and social equality is ensured? The answer to this dilemma may evoke the need to make transformative changes and enhance both the efficiency and effectiveness of our actions. Currently, the prevailing strategies are limited to the protection and survival of Earth's ecosystems, often with insufficient outcomes. This chapter analyzes these dynamics and linkages while focusing of their social dimensions and environmental justice. The analysis value the need to achieve a balance which has the potential to enhance both the environmental and social sustainability.

Keywords Ecological crises · Inequalities · Environmental justice · Sustainable development

M. H. El Amrani (✉)
Faculty of Law, Economics, and Social Sciences of Agadir, Ibn Zohr University, Agadir, Morocco
e-mail: elamrani@mondeannuaires.com

M. Behnassi et al. (eds.), *Social-Ecological Systems (SES)*,
https://doi.org/10.1007/978-3-030-76247-6_11

1 Introduction

Today, more than ever, the world is increasingly interconnected. This interconnection covers the relations between countries, societies, economies, social organizations, cultures, ecosystems' dynamics, etc. As a result, it creates a state of global interdependence. In this context, ecological conditions have a large-scale impact on the various aspects of the life of individuals and communities. The degradation of Earth and its ecosystems is linked to many reasons such as the increased use of non-renewable resources and the generation of huge quantities of pollutants and emissions by productive activities. This leads to a myriad of ecological problems such as global warming, biodiversity loss, and the degradation of many terrestrial and marine ecosystems. Current development models have positively impacted the economic situation of a part of world population. However, the negative externalities are increasingly affecting both ecological and human systems (i.e., negative impacts on humans due to pollutions, emerging diseases, and consumption of unhealthy products), thus reducing the quality of life for a large part of the humanity across the globe.

Moreover, environmental risks are not evenly distributed and those who have no economic power are most vulnerable with a marginal influence on decision-making processes. Such a situation creates environmental inequalities with various impacts. Environmental inequalities are built on several factors, mainly the lack of internalization of negative externalities by their creator. For instance, when a factory is settled in a specific area, the social and economic benefits are unequally distributed in favor of affluent communities and businesses, whereas the environmental risks and effects are disproportionately concentrated between vulnerable communities and groups.

These dynamics constitute emerging challenges for the policy-making and, consequently, require new conceptual, strategic, and operational frameworks to overcome related effects. However, the fight to create equal opportunities, where each individual should benefit from sufficient conditions to enhance well-being and human development in a healthy environment, seems still long.

This chapter seeks to shed light on the social impact of environment and environmental justice. The encounter between these dimensions can be a source of complexity, given the divergence of actors' interest and their power relations. A coherent and humanist position of these dimensions should open avenues for reflection in order to reach efficiency in preserving the environment and the human development in a balance, without compromising the expectations of the actors.

2 Environmental Inequalities: Underlying Drivers and Responsibilities

2.1 Social and Political Aspects of Environmental Inequalities

Many countries around the world are facing great pressure on the environment and natural resources. This increases the level of pollution and, consequently, exacerbates the delicate existing environmental balances. In the absence of efficient political measures to enforce the capacities of countries to access and benefit from natural resources, the poor will be deprived from exercising their rights and will always suffer from inequality.

The weak understanding and awareness of policy makers of the social-environmental interdependence generate many problems and risks such as: structural poverty, disempowerment, exclusion from decision-making processes, income inequality, and lack of agency among citizens and communities (UNDP 2011). This situation creates multiple forms of vulnerability and complicated the linkages between people and their environment and, consequently, generates an injustice even between them.

So, in the current context where there are attempts to adapt the territories to environmental change in order to mitigate noted injustices, it is necessary to understand the interactions between the social, political, and environmental dynamics. It is also a question of determining criteria of the appropriate indicators to analyze them and highlight appropriate solutions. Firstly, it is useful to consider the multidimensional aspect of the environment, which covers both natural and social components and dynamics. Yet, the way in which societies currently interact with the environment, perceive and exploit it, is not the best way. In other words, it is our way of using this environment that creates inequalities.

Secondly, the spatial approach can allow us to cross social and environmental inequalities in order to highlight the phenomena of environmental discrimination. As a result, we will have a mapping of environmental and social inequalities. This situation may be the result of an unbalanced political management of resources, which highlights contrasts between areas that have been able to combine social and environmental benefits and others that have accumulated only environmental disadvantages. As a consequence, it will lead to the formation of marginalized populations. This inequality can be observed both at the local and regional levels in the same country, and at the global level between the North/South and South/South. The Western countries were generally able to efficiently control and preserve the environment and natural resources with a capacity to achieve food self-sufficiency and minimize natural and health risks. Additionally, these countries continue to exploit developing countries' resources through outsourcing or relocation of polluting productive activities in order to maintain a domestic environmental balance. On the other hand, the Global South is always penalized by the mismanagement of its environment and natural resources and the transition towards development models generally harmful to the environmental balance.

To address this situation of injustice, a number of voices are raised in favor of those who suffer from ecological or environmental segregation, through a new concept of 'ecological debt', inspired by a financial model of developing countries toward developed countries, especially in the early 1990s, by NGOs such as the Instituto de Ecologica Poluticia (IEP).[1] This concept of 'ecological debt' was later developed to the extent that an alternative debt treaty was signed. Admittedly, this initiative has not so far yielded the expected benefits, but at least it has emphasized an important element and raised the awareness of the international community on environmental injustice-induced challenges.

Indeed, a policy of compensation can neither ensure environmental equity, nor correct the injustices caused now and in the future. This policy of compensation for environmental damage raises ethical, economic, and financial issues. Ethical, because one wonders whether the policy of privatization of environmental goods adopted by most developing countries since the beginning of the 1990s can be offset by the transferred funds, technology or social and/or economic aid granted to such countries. For the economic and the financial aspects, the compensation refers to the payment made by private corporations for the negative impacts of using and damaging natural resources. This will, in turn, negatively affects the financial balance of these corporations, therefore making their competitiveness at risk, especially in poor countries. Consequently, many countries prefer not to invest in environmental protection.

2.2 *Who Is Responsible for the Environmental Justice?*

Today, the environmental crisis concerns everyone since its impacts affect the whole humanity (i.e., climate change). However, such impacts are unequally distributed. At the level of the same country, it is the most vulnerable and disadvantaged social groups who often live in degrading environments. This microscopic (local) schema also extends to the (global) macroscopic level. In other words, this polarity between rich and disfavored people at the national level is found between developed and developing countries as noted by Jamieson (2002:297): "the poor – or the most disadvantaged – suffer disproportionately environmental pollution that is produced by society as a whole".

These ecological inequalities lead to the emergence of certain active movements with the objective to provide workable solutions. During the 1980s, an environmental justice movement was established in developed countries with the aim to fight against inequalities and to ensure access to decision-making processes. Unfortunately, this dynamic was lacking in developing countries, because of their conception of the situation which was considered as a conspiracy by developed countries to prevent

[1] IEP is an NGO working in the field of the environment since 1987.

them from developing (Larrère 2009). However, this conspiracy theory does not change the reality of things.

It is important to note that environmental inequalities concern the distribution of environmental benefits and problems, while environmental justice is intended to be the fair distribution of benefits and costs. In other words, environmental justice entails the distribution of present or future environmental policies related charges in a context marked by the attitude of some countries to withdraw from their responsibilities towards the environment (Brazil and USA as example) with the desire to avoid bear unilaterally the costs. The following statement made by the President George Bush, during his presidential campaign in 2000, is a clear example about this mindset: "The United States took care of cleaning the air as the Kyoto Protocol wanted to force them to do, while China and India were exempted" (Gardiner 2004).

On the other hand, developing countries are suffering from old and massive industrialization, which is taking advantage in favor of developed countries.[2] This massive industrialization causes air pollution, loss of biodiversity, climate change, etc. It is the duty of the developed and highly industrialized countries to take responsibility for the situation. They must also take the necessary measures to amend the imbalance and prevent it from further deterioration. Far from this duality, it seems that each party has a share of responsibility, which must be assumed in order to safeguard the environment and ensure a healthy future for the next generations. For this to happen, solutions must be considered in a multidimensional and multilateral way within the framework of an equitable cooperation. Several scenarios are possible, including the distribution of the reduction of greenhouse gas emissions targets based on equity and fair share:

- Apply the equalizing principle at the international level, that is, allocate efforts to reduce greenhouse gases equally among all countries;
- Apply a principle based on inequalities, according to the level of development. Thus, targets for reducing emissions can be set according to per capita GDP. This scenario proposes to adopt an effort according to the degree of development of the country;
- Have to respect a *vital minimum* (Shue 2001), excluding any possibility for poor countries to impose a minimum volume of emissions that does not allow them to meet their basic needs.

These scenarios will have varying impacts on reality. From allowing carbon emitting countries not to reduce their emissions, to imposing constraints on the poorest and richest countries at the same time. This would certainly hurt the poor countries and limit their development potential. Thus, setting 1990 as reference date in the Kyoto Protocol allowed developed countries to preserve their historic advantage while discharging themselves from any responsibility over the current climate and environmental crisis due to decades of massive exploitation of natural resources during the period preceding such reference date. In the same time, such reference date blocks or constrains the future development process of other countries. Far from this

[2]Especially the biggest polluters of the world: China and the United States of America.

dimension, the acceptance and recognition of developed countries' historic responsibility does not mean being able to determine the level of ecological debt (Emelianoff 2006) towards developing countries.

3 Links Between Natural Disasters and Environmental Inequalities!

3.1 Natural Disasters Impacts' on Humans

The massive exploitation of natural resources exhausts all the wealth of the countries in question, cause environmental (deforestation, climate change, biodiversity, etc.) and social tragedies (impoverishment and marginalization of local communities). In this perspective, Sukhdev (2008) set out the close link between poverty, the degradation of ecosystems, and the loss of biodiversity. This situation destroys any possible economic development and completely aborts the environmental Kuznets curve. This curve links the level of economic development (measured by per capita GDP) to the environmental degradation and creates an automated link between the pursuit of economic development (increase in living standards) and decreases in the environmental condition: environmental degradation increases with the increase in per capita income before reaching a peak, then decreases. Figure 1 shows clearly the theorem of this curve.

Unfortunately, the real situation is contrary to the expectations. In fact, the volume of waste and greenhouse gas emissions increases with the economic growth. In our opinion, for this theory to be valid, a redistributive policy must be adapted.

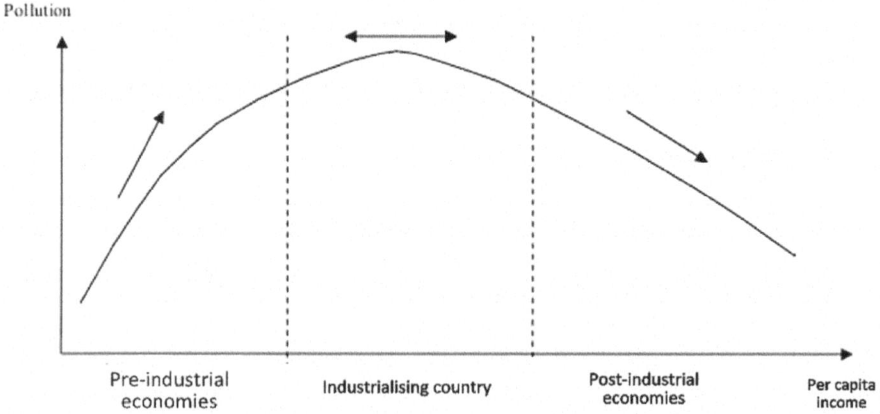

Fig. 1 The environmental curve of Kuznets Source: Meunié (2004)

Table 1 Rich and poor countries facing natural disasters

Income category	Number of disasters	Population (millions)	GDP per capita	Numbers of dead	Total cost as% of GDP
High income	1 476	828	23 021	75 425	0,007
Low income	1 533	869	1 345	907 810	0,55

Source:Stromberg (2007)

From an economic perspective, the economic growth does not automatically mean slowing down environmental degradation and pollution. In China for instance, which is currently experiencing an environmental catastrophe, the unsustainable degree of pollution exceeds in some cities, such as Shijiazhuang, the 74 μg per cubic meter in 2018, while the threshold recommended by the World Organization of Health (WHO) is set at 2.5 μg per cubic meter (Reuters, 10th May 2018). The dimension of environmental inequality can also have an impact on the policy: for example, the United Nations Environment Program (UNEP) (UNEP 2007) identified 17 major environmental-induced conflicts that occurred between 1990 and 2002 due to the will to control natural resources.

In short, poor governance, loss of biodiversity, deforestation, pollutions, carbon emissions, and damage to the natural balance seem to be the main drivers of environmental inequalities, with all the consequences for the security of human beings and their development. As a result, the populations of the least developed countries are considered the ones who suffer the most from such inequalities (Table 1).

Moreover, it was found as well that low-income countries are more vulnerable to natural disasters than high-income countries. To better illustrate this inequality even in terms of human loss, Table 2 shows the consequences of inequality by comparing the ten richest countries and the ten poorest countries globally for the period 1974–2003. Even if the richest countries – like Luxembourg, Denmark, Belgium, Swiss, etc. – have natural disasters, the number of victims is always very low in comparison with the poorest countries – like Malawi, Somalia, Ethiopia, etc. This situation could be explained by various reasons like the insufficient adaptive capacities of people and the weakness of infrastructure that exposes populations to increased risks of disasters and failed in protecting them from related impacts.

3.2 *Mitigation of Environmental Inequalities in Developed and Developing Countries*

Environmental inequalities reflect the difference in access to use valued resources among individuals or social groups. Such inequalities may be social, economic, gender or territorial. Many movements are trying to reduce the negative impacts

Table 2 Human impact of natural disasters in the ten richest and ten poorest countries

Countries	GDP per capita in $, 2002à	Annual number of victims/100,000 inhabitants 1974–2003	Countries	GDP per capita $, 2002	Annual number of victims/100,000 inhabitants 1974–2003
Luxembourg	44 000	0	Somalia	550	2 701
U.S.A	37 600	59	Sierra Leone	580	155
Norway	31 800	5	Burundi	600	674
Swiss	31 700	2	Congo RD	610	114
Ireland	30 500	4	Tanzania	630	1 531
Canada	29 400	72	Malawi	670	8 748
Belgium	29 000	2	Afghanistan	700	1 120
Denmark	29 000	0	Eritrea	740	6 402
Japan	28 000	182	Ethiopia	750	5 259
Austria	27 700	29	Madagascar	760	2 090

Source:Guha-Sapir et al. (2004)

of inequalities such as: local movements denouncing the situation deemed unfair with environmental conditions; and transnational movements which integrate the sustainable development perspective and the ambition to achieve social equity in their agendas and actions. With regard to local movements, it has emerged in developed and developing countries. Their main concerns are to denounce diverse and unfair social, political and environmental situations. The 1980s coincided with the development of US environmental justice movements, whose main objective was to fight the construction of polluting factories in areas populated by African-Americans (Ghorra-Gobin (2005). In this context, the concept of environmental racism has emerged due to the work of Chavis,[3] including his report on *Toxic Waste and Race in the United States* (1987). The conclusions of this report ended with the introduction of a Federal regulation and the obligation to take environmental justice into account in 1994 at the level of the US Environmental Protection Agency (EPA). Environmental justice, frenzied, in a particular geographical context of the United States where cities are highly urbanized and unequal, thus creating divergent social settings. The issue of environmental justice is also present at the level of developing countries, but with less structured forms than in the United States. For instance, the indigenous[4] movements or the movement of landless[5] in Brazil, which claim a justice of resource use and preservation of the environment.

[3]Benjamin Franklin Chavis, born January 22, 1948 in Carolina, is an American leader of civil rights. He was an assistant to Martin Luther King.

[4]Brazilian environmentalist movement.

[5]The Movement of the Landless is a peasant organisation born in Brazil in 1980s. Today, it is considered as one of the most important social movements in Latin America.

The claims of these movements take different forms with a tendency to increase coordination between their actions and to reorient their guidelines; thus, instead of remaining stuck in the defense of fauna and flora, they have also integrated the right to access to a healthy environment and natural resources (water, land, forest resources…). This last dimension is integrated given the political aspects of the issue and the oligarchic control of resources by a small group of people, politicians or economic actors. Such a control makes it difficult to voice the demands of protest movements, and explains the use of the coordination strategy at the transnational level by such movements to build relationships with international NGOs in order to be able to take efficient actions at the local level.

In Europe, local movements are more active than in developing countries, with the involvement of political institutions in the fight against social and environmental inequalities: i.e., the case of United Kingdom is illustrative (Agyeman and Evans 2004) compared to France where the interest in the issue of inequality is limited and recent (Emelianoff 2006) even if it is gaining an increasing interest among researchers, civil society actors, media, and politicians. In this regard, Faburel and Roy (2006) had identified four aspects of environmental inequalities, which are: territorial inequalities; inequalities of access to urbanity and living environment; inequalities in relation to nuisances and risks; and inequalities in the capacity of action and interpellation of the public power for the transformation of the living environment.

In developing countries, local movements, which are generable unable to make great changes in the political system or at least to influence it, try to organize themselves into networks or joining programs of major global NGOs in order to have an influential impact. For example, the protest movement against the construction of the Sardar Sarovar dam[6] in India was supported by the International River Network,[7] involving both local and international activists, mobilizing the scientific community, lobbying through the press and submitting petition to the World Bank. It was a subject of controversy for a long time. This foundation stone was laid on April 5th, 1961 but because of the dispute about the project itself and the sharing water method between states, Gujarat, Maharashtra, and Madhya Pradesh raised this case to the Narmada Water Dispute Tribunal, which was created in 1969. The tribunal gave its judgment 19 years after (on 1988). But, since 1985 it was also subject of protests from social activists because of its negative social and ecological impacts.

Under the pressure, the World Bank was forced to review the project, and at the same time the government canceled the loan authorized by the World Bank on Mach 31th, 1993. After several years of much deliberation, however, the Supreme

[6]The Sardar Sarovar Dam is a gravity dam on the Narmada river near Navagam, Gujarat in India. Four Indian states, Gujarat, Madhya Pradesh, Maharashtra and Rajasthan, receive water and electricity supplied from the dam. The foundation stone of the project was laid out on 5 April 1961. The project took form in 1979 as part of a development scheme funded by the World Bank through their International Bank for Reconstruction and Development, to increase irrigation and produce hydroelectricity.

[7]It is an NGO created since 1985 as an actor in the protection of the rivers and the right of the rowers to take advantage of these resources.

Table 3 Typology of movements for environmental justice

Concerned places	Political social context	Marker of injustices	Environmental design	Environmental indicators	Emblematic cases
U.S.A	Democratic political system strong socio-spatial inequalities, urban segregation	Urban industrial risks	Living environment	State of health of the population	Warren county
Emerging country	Democratic society very strong socio-economic disparities, urban segregation	Industrial and urban risks question of access to resources	Living environment access to resources (economic, cultural, identity)	State of health of the population Lack of rights poverty exclusion of protected and/or appropriate areas	Thor chemical in South Africa; Bhopal in India Chicho Mendes in Brazil; Chipko Movement in India; Narmada Bachao Andolan in India
Poorest country	High poverty little political relay	Question of access to resources exclusion of protected areas	Access to resources	Few data on the environment	Ogoni in Nigeria
Non-democratic country	Lack of freedom of expression	Nascent ecological challenge	Localized pollution problems	Not generally communicated to the general public	China

Source:Blanchon et al. (2009)

Court allowed the construction of the dam, provided certain conditions are met. The foremost condition placed by the Court was that all those displaced by the increase in height of five meters be satisfactorily rehabilitated and that the process be repeated for every five meters increase in height (The Indian Express, September 17th 2017).

Another example illustrating the strategic alliance policy between local and transnational movements is the resistance to the concession of public water services to private companies in Bolivia, notably the city of Cochabamba (De Gouvello and Founrier 2002). Thereby, and with the strength of their organization, they could force powerful transnational companies to withdraw from the market, even after the signature of the concession contract.

Table 3 summarizes primarily a classification of movements for environmental justice across the globe.

By reference to these data, we notice that the marker of injustice depends on political and social context, and that the experience can be different from one country to another. So, environmental movements try to adapt their actions with reality while exploring international linkage to force changes inside.

4 Conclusion

Can we achieve economic development by using natural resources while simultaneously protecting the environment and reducing social inequalities? In principle, the two processes must go side by side to preserve the future of humanity. Unfortunately, our daily live practices show the opposite. The needs of the world's population are constantly increasing, which means that the use of available natural resources will also increase, reaching levels beyond the sustainability thresholds with all the devastating consequences that we can imagine. In addition, many developed countries continue to directly or indirectly exploiting the resources of poorest countries, thus causing more damage to environment and deepening environmental and social inequalities, especially in the Global South. This is a contradictory reality because the more developing countries are in need of financial means. The more they receive from developed countries, the greater is the risk of overuse or exploitation of their natural resources. Thus, the worst polluters are, at the same time, the best payers. Yet, we note that public environmental policies in many developing countries are still failing to preserve the environment or indirectly protect the livelihood and health of vulnerable populations. To reverse such a situation, it would be wise to integrate an approach linking ecological and social policies in order to ensure an ecologically efficient and socially neutral condition for human societies.

References

Julian A, Bo E (2004) Just sustainability: the emerging discourse of environmental justice in Britain? Geograph J **170**(2): 155–164

Blanchon D, Moreau S, Veyret Y (2009) Comprendre et construire la justice environnementale. Armand Colin, Annales de Géographie, 1 n° 665–666

Chavis BF (1987) Toxic wastes and race in the United States, A National Report on the Racial and Socio-Economic Characteristics of Communities with Hazardous Waste Sites, Commission for Racial Justice. United Church of Christ, Exhibit 13

De Gouvello B, Fournier JM (2002) Résistances locales aux « privatisations » des services de l'eau: les cas de tucuman (argentine) et cochabamba (Bolivie). Presses de Sciences Po (P.F.N.S.P.). Autrepart, 1 n° 21

Emelianoff C (2006) Connaître ou reconnaître les inégalités environnementales?", Grégum - Université du Maine, ESO - UMR 6590 CNRS

Executive order 59. Fed. Reg. 7629 (1994) Clinton 11/02/1994

Faburel G, Roy A (2006) Les inégalités environnementales, L'environnement en France, Paris, Institut français de l'environnement

Faburel G, Gueymard S (2008) Inégalités environnementales en région Île-de-France: le rôle structurant des facteurs négatifs de l'environnement et des choix politiques afférents», Espace populations sociétés, 2008/1l, 159–172
Gardiner SM (2004) Ethics and global climate change. Ethics 114(3):555–600. https://doi.org/10.1086/382247
Ghorra-Gobin C (2005) De la ville à l'urban Sprawl La question métropolitaine aux États-Unis. Cercles 13
Guha D, Hargitt D, Hoyois P (2004) Thirty years of natural disasters: 1974–2003: The numbers. Press universitaires de Louvain, Centre for research on the Epidemiology of disasters
Larrère C (2009) La justice environnementale, Multitudes /1 (n° 36), 156–162 https://www.epa.gov/environmentaljustice/learn-about-environmental-justice.. Accessed 24 May 2020
Jamieson D (2002) Global Environmental Justice, *in* Jamieson D. Oxford, Clarendon Press, Morality's Progress
Meunié A (2014) Controverses autour de la courbe environnementale de Kuznets. Centre d'économie du développement, IFREDE-GRES- Univ. Bordeaux: IV, DT/107/2004
Reuters (10 May 2018) China's Shijiazhuang to take air pollution fight to crematoriums, public toilets. https://www.reuters.com/article/us-china-pollution-hebei/chinas-shijiazhuang-to-take-air-pollution-fight-to-crematoriums-public-toilets-idUSKBN1IB16H. Accessed 6 Jan 2021
Shue H (2001) Climate, Companion to Environmental Philosophy. Malden DJ (Eds). Blackwell Pubs, Mass
Stromberg D (2007) Natural disasters, economic development, and humanitarian aid. J Econ Perspectives **21:** 199–222
Sukhdev P (2008) L'économie des écosystèmes et de la biodiversité, Communautés européennes: Welzel, Hardt, Wesseling
The Indian Express (September 17, 2017) A short history of the Sardar Sarovar Dam on river Narmada. https://indianexpress.com/article/research/a-short-history-of-the-sardar-sarovar-dam-on-river-narmada-4847807/
UNEP (2007) Annuel report, DCP/0926/NA, in. https://wedocs.unep.org/rest/bitstreams/11459/retrieve. Accessed 01/04/2018
UNDP (2011) Human development report: sustainability and equity: a better future for All. Palgrave MacMillan, New York

Approaching Sonic Devices from a Multicriteria Sustainable Approach to Enhance the Resilience of Social-Ecological Systems

Gueiros Teixeira Suzana, Jules Ghislain Slama, Luiz Pinguelli Rosa, and Carlos Eduardo Milagres Pereira

Abstract In the *Art of War*, Sun Tzu has quoted: *If you know the enemy and know yourself, you need not fear the result of a hundred battles. If you know yourself but not the enemy, for every victory gained you will also suffer a defeat. If you know neither the enemy nor yourself, you will succumb in every battle.* Based on those sayings, this chapter aims to highlight the aspects of noise and vibration by reference to their impacts on living beings, how damaged might they be, depending on physical and psychophysical conditions of beings and environment; and to question, based on a multicriteria sustainable approach, the concepts of noise repellants, sonic non-lethal weapons, and other devices generating the soundscape scenarios over human health and other non-human species. This research is the result of many insights developed from different approaches brought together for the creation of a new research group addressing vibration, sustainable management, and strategic planning, with special emphasis on sonic devices under a defense approach.

Keywords Sonic devices · Noise · Sustainable management · Multicriteria · Defense

---

G. T. Suzana (✉)
Federal University of Rio de Janeiro, Associate Professor Level 1, Technology Center, SAGE-COPPE, Rio de Janeiro, Brazil
e-mail: gueiros@poli.ufrj.br

J. G. Slama
Former Titular Professor at Mechanical Engineering Program – PEM - COPPE - Current Professor at the Urban Engineering Program, Federal University of Rio de Janeiro, Rio de Janeiro, Brazil

L. P. Rosa
Federal University of Rio de Janeiro – COPPE, Rio de Janeiro, Brazil
e-mail: lpr@adc.coppe.ufrj.br

C. E. M. Pereira
Former Military Police of Rio de Janeiro Officer - Commander, Rio de Janeiro, Brazil

M. Behnassi et al. (eds.), *Social-Ecological Systems (SES)*,
https://doi.org/10.1007/978-3-030-76247-6_12

1 Introduction

"To feel a thing is to be affected by that thing".

Goodman (2010:94).

This chapter is a result of the questionings developed by an interdisciplinary research group regarding noise and vibration and their associated threats and benefits, especially in the context of the design and use of sonic devices tailored for defense, such as sonic repel products and non-lethal sonic weapons. In the XXI century, human societies are progressively evolving under a mandatory sustainability approach since the economic, social, and environmental impacts of product innovation – in its very conception - must be increasingly mitigated. Under noise and vibration impacts, externalities originated from sonic sources may be evaluated considering their physical properties, under a system approach containing three major parts: source, path and the receiver – as learned from the acoustical specialist Beraneck on his methodological approach (Beraneck and Vér 2006) – but also, when addressing impacts from noise and vibration measurements we propose reviews regarding environmental values, in a complex and multi-attribute insight of product engineering, and a sustainable management approach regarding its use phase impacts under scientific metrology approach considerations.

Either for safety matters - intentionally and (or) unintentionally – when a sonic source is capable of emitting harmful vibration and (or) noise towards a receptor – either towards a living being or a structure - there are many scientific considerations taking place. Harmful factors must be analyzed under scientific and moral principles, since sonic impacts from mechanical and acoustical vibrations become dangerous to human and environmental health exposure, with the causing of displacement of various organs, pulmonary hemorrhage, changes in the heart rate, among other health disruption data, and for such it is used as a weapon (Stefan et al. 2017). We propose a multi-criteria approach investigation, in order to provide the product engineering area – and the many other stakeholders which partner with the subject, including strategic planning actors, associated with Defense market – an appropriate scientific review regarding sonic products innovation, use phase impacts, and consequences over: public health; defense (health, safety and security) and legislation; environmental values; bioethics and biopolicy; psychoacoustics and physio acoustics; metrology and standards; and biomimetics – the mimicking of nature addressing the innovation process, ecodesign, Life Cycle Assessment (LCA), Product Category Rules (PCRs), Environmental Product Declaration (EPD), and Human Factors. These are the trends of this research group schematically shown through a cognitive map in Fig. 1.

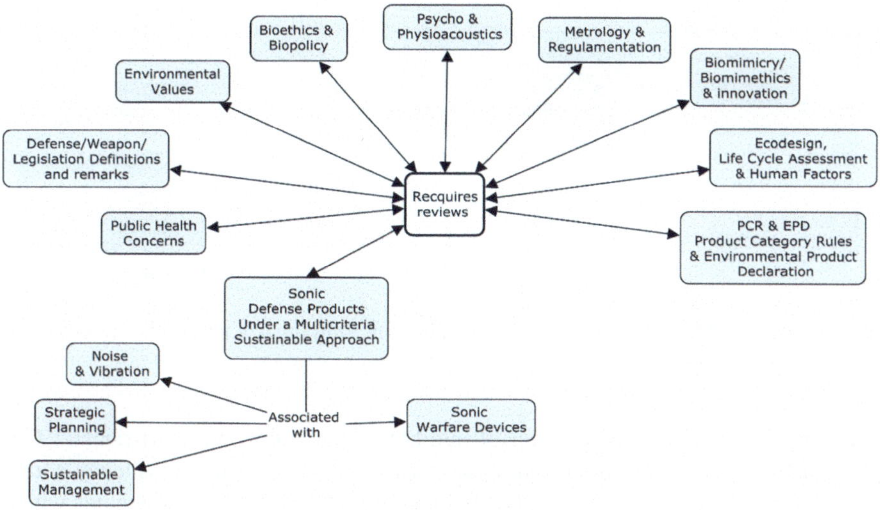

Fig. 1 Cognitive map proposed by the Research Group on Vibration, Sustainable Management and Strategic Planning with the Focus on Sonic Defense Products Guidance

2 Preliminary Assumptions and Concepts

> *"Humans interact with the acoustic environment in many complex ways. The human hearing system not only consists of the ear but also encompasses conduction and mediation via bones, flesh, and body cavities. Acoustic energy has the ability to the posing of profound influence in an individual's brain waves, respiratory cycles, nervous system, muscle function, heart rate and glandular function"* Davies (1992:25).

According to Mansfield (2005), "vibration is a mechanical movement that oscillates about a fixed (often a reference) point. It is a form of mechanical wave and, like all waves, it transfers energy". Despite the human perception of vibration and/or audible sound waves or noise, it might affect human health pervasively and destructively. Human response to vibration is complex in nature since there are simultaneous components taking place. According to Beraneck and Vér (2006:1), a "sound wave is any disturbance that is propagated in an elastic medium, which may be a gas, liquid or solid. Ultrasonic, sonic, and infrasonic waves are included in this definition".

Both the exposure to vibration and the hearing mechanism in all living beings are biologically, physiologically and, anatomically differentiated as well as with the nature of vibration, exposure and response (Mansfield 2005). The sound pressure level and its response differ from one living species to another, and vibration might have at a determined intensity and frequency a different effect amongst tissues of the same being. Everything in the known world has a resonant frequency; by such we understand a physical phenomenon in which an external force or a system of vibrating forces stimulates another system around it such as to vibrate with greater amplitude at a specified frequency of operation.

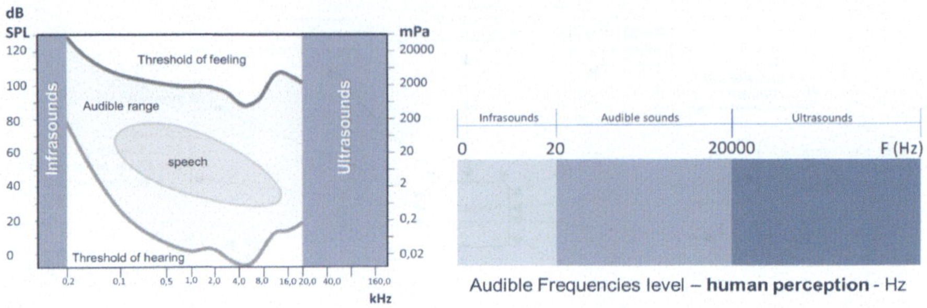

Fig. 2 Limits of human perception – threshold of hearing – threshold of pain– (dB – Decibel/SPL – Sound Pressure Level)

The range of human perception of audible sound according to Broner (1977) can be differentiated into three main categories: subsonic or low frequency sound which is defined in the range of 20 Hz to about 500 Hz; midrange frequencies which inhabit the realm of 500 Hz to 6 kHz (6000 Hz); and high frequency sound defined in the remaining 6 to 20 kHz. Figure 2 shows the limits of human perception; frequencies above 20 kHz are considered ultrasounds and those frequencies beneath 20 Hz are infrasounds. According to (Davies Alex Bfa Honours 1994), the ear is a very easy structure to attack. Physiological changes in the body may start to occur at higher sound pressure levels, and pain becomes evident when the middle ear system mechanically displaces beyond its normal operational limits, at an increased sound intensity at determined frequencies, it is reached the threshold of pain, which differs among animals.

Vibrations are a dynamic phenomenon found in life. In human beings, under the purpose of determining the risk of harmful exposure, it is important to identify the sources despite the human awareness of related exposure and harm. To approach the harmful effects of vibration on humans, Stefan et al. (2017:140) claims that "vibrations have complex harmful effects on human and on the built environment, affecting health of the human body, the quality of the human work, the physical and mental comfort the resistance of the building components, etc.".

Despite having the biological condition to perceive noise, species can be harmed by vibrations, depending on the intensity level, and exposure conditions, that is the reason why scientific standards have placed limited time – dose –for human exposure. The excessive or focused amplification of a determined range of waveforms directed to some parts of the human body are prove to cause overwhelming nausea, loosening of the bodily organs, and ultimately death. According to ISO 2631-1:2001, there are four physical factors of fundamental importance to determine human risk exposure and response to vibration: intensity, frequency, direction, and time of exposure to the source. There are three criteria for assessing the harmful effect of vibration on humans: decrease of work efficiency, health hazard, and comfort limit. For the purpose of this research, we have mainly considered the first two. Frequency plays a strong influence in the body's response, and high frequency vibrations – which transmitted through tissues – may pose health hazards under determined conditions of exposure. Among the harmful effects on human health, research data has additionally observed: changes in tendons, muscles, bones and joints, headaches, balance, walking impacts, and other associated effects on nervous system. Figure 3 shows the limiting levels of infrasonic exposure and Table 1 displays low frequency noise effects.

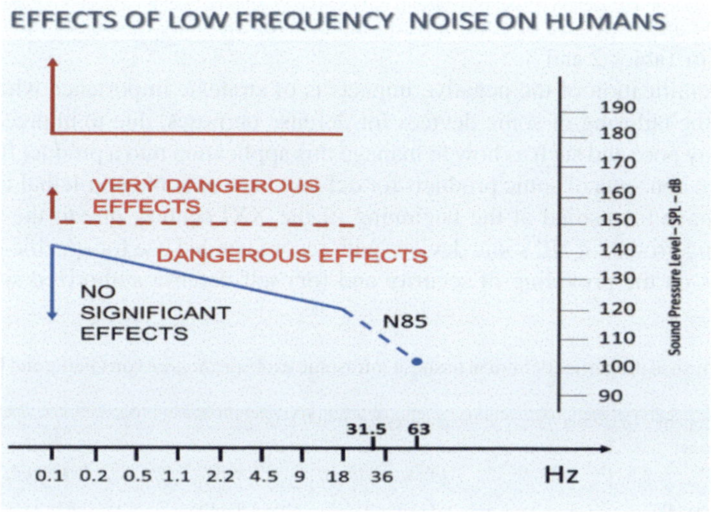

Fig. 3 Limiting levels for infrasonic effects *Source* Broner (1977)

Table 1. Low frequency noise effects observed*Source* Mohr et al. (1965)

TOLERANCE DATA	
EXPOSURE Frequency HZ _Intensity dB	OBSERVED BEHAVIOUR
0-50 HZ	Chest wall vibration, gag sensations
Up to 145 dB	Respiratory rhythm changes, post-exposure fatigue, voluntary tolerance not exceeded
50 to 100 Hz Up to 154 dB	Headache, choking, visual blurring and fatigue, voluntary tolerance limit reached
DISCRETE FREQUENCIES	TOLERANCE LIMITS SYMPTONS
100 HZ at 153 dB	Mild nausea, giddiness, sub-costal discomfort, cutaneous flushing
60HZ at 154 dB	Coughing, severe substernal pressure
70Hz at 150 dB	Chocking respiration, pain on swallowing, giddiness

The human body and its tissues when under exposure of vibration are harm according to the vibration amplitude and pressure level – which will vary according to body biomechanics in different locations of the body. Vibration plays a dominant role in creating ill health in the case there is lack of environmental awareness regarding its source, path, biological conditions, and exposure. Living tissues are affected by vibrations either through positive effects on health (like healing), or negatively through inflicting the harm. Those conditions result from a combination of intensity and exposure of certain frequencies over the body tissues of the receptor as shown in Tables 2 and 3.

The identification of the negative impacts is of strategic importance when associated to the tailoring of sonic devices for defense purposes, due to unprecedented harm it may pose and such as how to manage this application into a product function. The design concepts of sonic products for defense such as sonic non-lethal weapons – NLW- have intensified at the beginning of the XXI century due to the product engineering growth of NL sonic devices with recommended use for specific military operations on the providing of security and (or) self-defense authorized sites and,

Table 2 Physical properties of human tissue at infrasonic exposure *Source* von Gierke and Brammer (2002)

Physical Properties of Human Tissue at Frequencies below 100 kHz			
Bone, compact			
	Tissue, soft	Fresh	Embalmed, dry
Density, gm/cm3	1–1.2	1.93–1.98	1.87
Young's modulus, dyne/cm2	7.5 × 104	2.26 × 1011	1.84 x 1011
Volume compressibility, dyne/cm2	2.6 × 1010	…	1.3 × 1011
Shear elasticity, dyne/cm2	2.5 × 104	…	7.1 × 1010
Shear viscosity, dyne-sec/cm2	1.5 × 102	…	…
Sound velocity, cm/sec	1.5–1.6 × 105	3.36 × 105	…
Acoustic impedance, dyne-sec/cm3	1.7 × 105	6 × 105	6 x 105
Tensile strength, dyne/cm2	…	9.75 × 108	1.05 × 109
learing strength, dyne/cm2, parallel	…	4.9 × 108	…
Shearing strength, dyne/cm2, perpendicular	…	1.16 × 109	5.55 × 108

Table 3 Additional data on frequency of vibration (Hz)/effects on human body *Source* Broner (1977)

Frequency of Vibration and its Effects on Human Body	
Frequency	Effects
BELOW 1 HZ	Motion sickness
3,5 TO 6,0 HZ	Alerting effect
4,0 TO 10 HZ	Chest and Abdomen pain
AROUND 5,0 HZ	Degrades manual actions
7,0 TO 2O HZ	Communication problem
8,0 to 10 HZ`	Back Ache
10,0 TO 20 HZ	Intestine and bladder pain
10,0 TO 30 HZ	Degrades Manual and Visual Controls
10,0 TO 90 HZ	Degrades Visual Actions

by law enforcement staff. Under the scientific assumption that sound is a vibration which is differently perceived among species, other sonic devices, such as animal repels, have been conceived, manufactured, and adopted by civilians, thus entering the regular market consumer products. The inclusion of pest repels – fully available to consumer access – among other categories of consumer products with the same function, offered as being of lesser environmental impact, alongside with other alternative military sonic products engineered to address defense within the debilitating function. The following cited scientists, experts in the investigation of soundscape scenarios regarding sonic harm and sonic assaults, highlight the need of additional scientific research in these areas:

- "This human range of perception frequencies constitute only a small bandwidth of the full waveform spectrum that we exist within… It is clearly apparent that the human organism is in an extremely delicate state of equilibrium with the sonic environment and any profound disturbance of this system will have subsequent ramifications on the individual" (Goodman 2011).
- "Research into the effects that these three bandwidths of frequencies have on humans was only institutionalized, in any sense of the word, at the beginning of the Twentieth century. Ultra-sonic and infrasonic phenomena, which is not consciously perceived by our ears, in order to detect the ways that waveforms (a term connoting all the frequencies within the infrasonic, sonic, and ultrasonic ranges) have been harnessed as tools and weapons to demarcate physiological, spatial, and psychological territories" (Heys 2011:10).
- "The acoustic effects on the body are more complex. Research has concluded that with low frequency sound in the region of 50–100 Hz at levels of 150 dB or more, intolerable sensations in the chest and thoracic region can be produced – even with the ears protected" (Davies 1994).

Soundscape is the human perception dependable on the socio-cultural background and the psychological dimension with the acoustic environment in context. This terminology was according to Schafer (Schafer 1993) was first acknowledged by Michael Southworth when addressing the acoustic environment perceived by humans. Our approach deals with the soundscape security proposed concept, under which additional concerns are considered such as the inclusion of safety, sustainability, source type, and its mobility and ecological attributes which may as well be classified as of an illicit nature and (or) terror. The concept of sound as a resource reaches across a broad range of applications, among them, a resource for weaponneers. When migrating from the limits of an anthropocentric approach towards a post-human one (Lindgren et al. 2018), considering a shift towards an improved ethical position, environmental and sustainable aspects can be improved by a metrology review. Under a biocentric or a post-human review, accordingly to Naess (1994), it is claimed that "man may be the measure of all things in the sense that only human being has a measuring rod, but what he measures he may find to be greater than himself and his survival". The consequences of a sonic siege will appear on other populations which may be unintentionally on the path of man-made sonic devices in harmful use and (or) abuse. The extension of the affected population and the consequences to the local ecology health may be negatively increased.

This highlights that we aim to embrace concerns to the fact that soundscape scenarios pose different responses to species, according to physical properties of sound and vibration waves, path of propagation and biological characteristics of the receptor. From a consumer society perspective, when in the process of product engineering of sonic devices for defense – either for personal, public or crowd control – the fostering of bioethical aspects should be mandatory on preliminary phases of testing and of use. Product awareness upgrade to a bioethical consumer education should consider testing, animal and human health suffering and the additional scientific extension of intentionally or unintentionally effects that poses on the surrounding species, which may be affected. The embracement of animal differences for product engineering regarding sonic devices may not only be looked from an anthropocentric aspect, but from the animal welfare attributes under a sustainable environmental management approach.

3 Into the Pathways of the Research Group

"When it comes to environmental happiness, the choice is simple: Thrive or die". Delaney and Madigan (2014).

Upon publication of *Beyond Sustainability: A Thriving Environment,* in which Delaney and Madigan (2014) developed interesting insights regarding human actions on the planet, many questions have arisen including the idea proclaimed for many years by pro-environment extremists who advocates that *we are all doomed unless we drastically change our behavior*. In this perspective, environmental ethics for Delaney and Madigan (2014:128) is consider as a "thinking which makes us accountable for our actions, in that we do not act merely from instinct, and can thus be held morally responsible and accountable for what we do. For most ethical thinkers, animals and other forms of life have no moral standing, but rather can be used as means to an end for human happiness". The vision, which is mostly based on the human anthropocentric nature and their decision-making processes regarding environment, has progressively been challenged with the encompassing of arguments for additional understandings such as: the imperative condition of fragility of life on earth with the values in favor of environmental thrivability and, the assumption that happiness is a thriving environment which leads us to the understanding of the concept of environmental happiness.

Although the research on the sustainable management approach in the context of sonic devices conceived for defense means may appear as a contradiction, it is perhaps for such very same reason that many scientists highlight the need to add this aspect to the product engineering of such products. This encompasses its very own function, its use phase impacts, by whom, where, under what circumstances, the bioethical concerns of the use, biopolitical aspects and bio-diplomacy agreements when in trade, and foreign affairs related aspects regarding the externalization costs that might occur with unprecedented risks of damage to population, are as well to be reviewed.

The risks of such products to become accessible to civilian consumers throughout open markets, in a context marked by the lack of stakeholders awareness and of third party risks - when dealing with extended magnitude of impacts associated to either the licit or illicit use of these products - have motivated the current insights, due to the asymmetrical use of power, moral and ethics and ruinous effects; considering that product engineering in this case, deals with debilitation and the possibility of the posing of suffer, health debilitation and collateral damage. Regardless of how one product may offer a better alternative in terms of harm (what kind of harm, if the harm itself is required to be an attribute), in comparison with others with the same function and, in the same category of impacts, the SNL weapons or sonic sources used as weapon, are a risk due to the chances of being used under abuse. There is a diversity of sonic defense products, some considered non-lethal devices, some not even considered as weapons; nevertheless, in despite of not being classified as a weapon, sonic sources may pose harm which in certain sites and conditions, may become a matter of national security.

The following topics were selected by the research group as a mandatory multicriteria approach to ensure the appropriate sustainable overview of the product conception and the review of its life-cycle approach when considering their availability in the market, and the associated risks from a diversity of ethical backgrounds relevant to local consumers.

3.1 Public Health Concerns

"Population Health is a Public Good" McMichael (2001).

Noise and vibration affect both poor and wealthy people alike. Among the psychological dimensions associated with the impact, some may pose harm on the local site health with the risk of extended consequences of global impacts, depending on the population affected. Environmental impacts on health are investigated and classified in terms of local, regional and global scales. The infrasound environmental impact is observed in natural events, but it can as well be originated from anthropogenic sources. Once inaudible frequency for human perception, the impact may affect human body not only individually but also collectively due to the propagation path of the source. The use of infrasound detection and frequency ranges which are inaudible to humans, have been investigated for many decades to determine their effects on human body and the strategic application of those sources, including applications for self-defense, crowd-control defense and specific military use. Sonic sources as repels might be posing danger to local environment health and on the disruption of human health without that population awareness of its consequential damages.

The collapse of local wellbeing health might as well be a result of local exposure conditions to sonic sources. Intentionally or non-intentionally, those sources placed on sites may be harmful for population, thus compromising the public health with

Table 4 Sound source and effects *Source* Altmann (1999)

SOUND SOURCE	EFFECTS	TARGETS
Infrasound	*May affect labyrinths, vertigo, imbalance, etc. resonances in inner organs, e.g., heart with effects up to death*	*Riot control*
Infrasound from non-linear superposition of two ultrasound beams	*Intolerable sensations*	*Riot control*
Infrasound	*Incapacitation, disorientation, nausea, vomiting, bowel spasms, effect ceases when generator is turned off, not lingering physical damage*	*Crowd riot control, Psychological operations*
Very low frequency noise	*Disorientation, vomiting fits, bowel spasms, uncontrollable defecation Anti-personnel: resonances in body cavities causing disturbances in organs, visual blurring, nausea – temporary discomfort to death*	*Enemy troops*
Infrasound – tuned low frequency, high intensity	*Anti-material: embrittlement or fatigue of metals, thermal damage or delamination of composites, against buildings: shattering of windows, localized earthquakes*	-

debilitation and/or severe illness due to many impacts – such as progressive organs damage, the compromise of respiratory and visual system – already scientifically proved and listed on Table 3. According to Mcmichael (2001), "the changes in the social, economic, biological and ecological conditions of living are the ones that shape the patterns of health and disease in population"; therefore, the author urges to investigate health under the frame of sonic products. Once introduced in the marketplace, these products might negatively interfere with local health patterns and create illness in affected areas. Risk exposure and conditions under which sonic devices are being use require additional technological tools and products in order to detect, identify, and monitor if an individual or a group of individuals are under siege. There is a demand to undertake an investigation about the sonic sources that might have the potential to pose harm and the introduction of a soundscape security policy on sites/ receptors that may be consider as targeted areas. There might be either stable, local or fixed sources, as well as moving sources.

Sonic sources originated from anthropogenic actions – under the intent of affecting a population – require from security and law enforcement actors the understanding that the source physical characteristics generates a potential harm, which may affect both the physical and behavioral conditions of affected receptors as the prevailing ecological soundscape characteristics and ecological features of the environment where it is affected. Population affected by sonic attacks represent a group which is in disadvantage when considering the biological vulnerability to noise and vibration damage from its exposure conditions as well as the ignorance by the population of both the existence, perception and extension of its harm, once they become incapable of immediate defense response due to debilitation level it causes. According to Naess (1994), "man should learn to behave in an ecologically sound manner simply because this is necessary for human survival".

Among the subjects investigated by this research group there is the need of a sonic disease risk mapping which has the potential to assist local authorities in identifying, monitoring, avoiding and/or managing the risk of sources and population exposure to related harms. Local data may offer paths to identifying the disruption on health and the posing of illness caused by sonic sources exposure. Through mapping, transparency may be provided, such as of local vulnerability and health indicators for epidemiologists. These inputs may help policy makers to determine

Table 5 Non-lethal sonic devices for law enforcement and defense purposes *Source* USAF Institute for National Security Studies USAF Academy, Colorado (1997) and Art 36 of CCW (CCW (2018)).

NON-LETHAL SONIC WEAPONS	
ACOUSTIC BEAM – HIGH POWER - LOW FREQUENCY BEAM	Envisioned to be a piston-driven or detonation driven pulser which forces compressed air into tubes generating low frequency wave. Creates a blast wave.
ACOUSTIC BULLETS – HIGH POWER, VERY LOW FREQUENCY WAVES	Results in blunt object trauma from waves generated in front of the target. effects range from discomfort to death.
ACOUSTIC CURDLER UNIT	Device plugged into an hps* sound system to produce blatting noise. irritates disperse rioters and had a decibel range just below that of the danger level to the human ear.
ACOUSTIC DEFERENCE TONES	Devices which can project a voice or other sound to a particular location. the resulting sound can only be heard at that location.
ACOUSTIC DOPPLER EFFECT ALARM	Alarm system based on any movement in the area between a transmitter and a receiver.
ACOUSTIC HIGH INTENSITY SOUND	Loud music use with the intention to drive away the receptor, also known as polysound.
ACOUSTIC HPS-1 SOUND SYSTEM	A 350 watt sound system with an audible voice range of 2 1/2 miles.
ACOUSTIC, INFRASOUND	Very low-frequency sound which can travel long distances and easily penetrate most buildings and vehicles. transmission of long wavelength sound creates biophysical effects; nausea, loss of bowels, disorientation, vomiting, potential internal organ damage or death may occur. by 1972 an infrasound generator had been built in France which generated waves at 7 hertz. when activated it made the people in range sick for hours.
ACOUSTIC SQUAWK BOX	Crowd dispersal weapon. this directional device emits two ultrasonic frequencies which when mixed in the human ear become intolerable. it produces giddiness, nausea or fainting. the beam is so small that is can be directed at specific individuals in a riot situation.
ACOUSTIC TELESHOT	Cartridge projecting a powerful sonic device delivered by a 12-gauge shotgun.
ACOUSTIC & OPTICAL, AIR BURST SIMULATOR	A diversionary device normally used to simulate the air burst of artillery rounds during infantry training.
ACOUSTIC & OPTICAL COD-WEIGHT	A heavy, (2 pounds or greater) weight to which a diversionary device is attached to allow it to be thrown through window screens, window glass, bushes and similar materials
ACOUSTIC & OPTICAL, DIVERSIONARY DEVICE	A hand thrown pyrotechnic device which emits a loud bang and dazzling light when ignited. engineered to create a sensory overload which temporarily causes confusion and an inability to effectively respond to a tactical team's actions.
ACOUSTIC & OPTICAL, FLASH STICK	A stick or pole to which a diversionary device is affixed, allowing it to be precisely placed and held during ignition. often used for exact insertion through chain link fences, windows, heavy brush and so forth
ACOUSTIC & OPTICAL, PAINTER'S POLE	An extendible pole to which a diversionary device is affixed allowing it to be precisely placed and held during ignition. often used for supporting second story entries from beneath.
ACOUSTIC & OPTICAL, PHOTIC DRIVER	Crowd control device based on ultrasound and flashing infrared lights which penetrate closed human eyelids. potential for epileptic fits because of the stroboscopic flashing effect.
ACOUSTIC & OPTICAL, PSYCHO-CORRECTION	Russian technology that involves influencing subjects visually or aurally with imbedded subliminal messages.
LONG RANGE ACOUSTIC DEVICE LRAD	Acoustic hailing device or sound cannon, operates in the audible range and issue high-energy acoustic beams to communicate with, warn and potentially disorient or disable a person. such devices can reportedly produce harmful, pain-inducing tones over some distance, damage the human ear and cause permanent hearing loss (short distances).

Table 6. Principles of deep ecology movement *Source* http://www.deepecology.org/platform.htm

PRINCIPLES OF THE DEEP ECOLOGY MOVEMENT	
1. The well-being and flourishing of human and nonhuman life on Earth have value in themselves. These values are independent of the usefulness of the nonhuman world for human purposes;	*5. The flourishing of human life and cultures is compatible with a substantial decrease of the human population. The flourishing of nonhuman life requires such a decrease;*
2. Richness and diversity of life forms contribute to the realization of these values and are also values in themselves;	*6. Policies must therefore be changed. The changes in policies affect basic economic, technological and ideological structures. The resulting state of affairs will be deeply different from the present;*
3. Humans have no right to interfere destructively with nonhuman life except for purposes of satisfying vital needs;	*7. The ideological change is mainly that of appreciating life quality (dwelling in situations of inherent worth) rather than adhering to an increasingly higher standard of living. There will be a profound awareness of the difference between big and great;*
4. Present human interference with the nonhuman world is excessive, and the situation is rapidly worsening	*Those who describe to the foregoing points have an obligation directly or indirectly to try to implement the necessary changes.*

appropriate response mechanisms regarding sonic toxicology and the design of a soundscape security policy. The consequences on debilitated health will call for attention on Policy makers, accordingly to Morrison (1995) the challenge to policy makers "is to design policies that carry the needed signal without provoking unacceptably high levels of conflict." The research group's assumption is that depending on the characteristics of the affected ecosystems and of exposure conditions to sonic impacts, local resilience might have the risk to deal with the progressive decay on public health. Scarcity of data over sonic sources and affected populations might be a contributing factor to the further compromising local health and safety, security and biosecurity of species.

3.2 Defense and Legislation

"History has shown more than once that the fates of the greatest empires have been decided by the health of their peoples or their armies" (Virchow 1848, cited on Herzlich 2013).

The scale of today's environmental problems requires human societies to develop sustainable solutions when dealing with them. As mentioned earlier, noise and vibration can cause both reversible and irreversible damage for public health. The deterioration of public health is a matter of security, and by extension a matter of Defense, which is understood as the action of protecting populations from or resisting to threat. A debilitated population is an undermined group within its capacity of empowerment. The enhancement of Defense capabilities is conducted by science, technological innovations and organizations with the power to address them a secure, livable and environment to prosper. The use of sonic devices either for personal safety and (or) security purpose, requires the review on Health and Safety concerns in order to provide stakeholders a scientific approach on the consequential health impacts from exposure, since sonic attacks or threats may limit conditions of local ecological health in complex ways, that might, eventually be of inaccessible acknowledgment to decision making process, due to the ignorance of damage extension of affected population. Under such ignorance, unasked questions regarding compliance may be of legal and safety concern. From the point of view of concerned victims, they might have ignored; the vulnerability has often made those under siege defenseless. Engineering and scientific operators should be capable of tracking such dynamics in order to redesign safety standards for such devices and appropriate compliance.

The regulation of both non-lethal-sonic devices adopted for Defense and sonic devices used as pest deterrents, despite of their availability in international markets, should undergo, according to our research group proposal, a mandatory review by reference to a sustainable management approach regarding the consequences and risks associated to their use. This should be done while questioning the related bioethical issues, which have strongly increased in the XXI century, due to the actions of many state and non-state actors who have worked on the reframing of the product engineering concepts and functions under the sustainable multi-criteria attributes.

Regarding the corporate role in protecting health safety and environment, Ashford (2013) said: "to put it another way, health, safety and environmental concerns must be constructed expansively to include both physical and economic security".

The nature of sonic-non-lethal product engineering function, its concepts and access by operators of the products, and their introduction to the civilian markets, are required of appropriate tracking and high awareness level from every involved stakeholder. That is justified because of the externalities it may generate when in use and the risks of illicit use by non-authorized personnel. Adequate use and/or even the prohibition process of the tailoring of sonic devices to be used for Defense require an absolute awareness in the Intelligence and Defense sector in order to remain in absolute control of the process, of determining the alternatives and where, how and whom to make use. In addition, the lack of cognitive awareness of dangers among human populations, it may become of asymmetric condition once it may cause harm and long-term illness without a clear understanding on the damage extension, despite the labeling of the products as non-lethal devices. This may be highlight in the raising of questions regarding the forms of an ecological shift, the way to shape production systems, and the nature of impacts of such products when based on the level of possible risks and/or abuse. According to Ashford (2016), "views of regulation on industry are conflicting". Despite of their impact, sonic devices and/or non-lethal sonic devices designed for Defense purposes and repels, may be consider as market opportunities.

When dealing with Defense-tailored products – repels and non-lethal sonic devices – product's function itself requires the presence of certain rates of environmental and health risk considered under acceptable limits. For this reason, regulation, standardization and compliance process, communication of damage risk, and the stakeholders' awareness are required to place such questions: to whom and where it is acceptable the allowance of its use whom are the ones who may profit from? Could this ever be considered under sustainable management approach? In this perspective, the ethical and bioethical considerations regarding the debilitating function – which should be transparent under a product life cycle approach – will offer the product the needed accuracy of information in terms of economic, social and environmental aspects. The need to provide appropriate regulatory environment under sustainable management should be required, especially regarding the role of product innovation in dealing with debilitating functions.

Regulating the tailoring and use of such devices at present times is a somewhat a challenging process. The mainstreaming of an ecological approach in the process of product engineering and technological innovation must be guided by basic principles of unity. According to Morrison (1995) the latter "can be represented in the term of an ecological circle which includes the practice of a moral ecology, that is: the lived awareness that actions have profound consequences on the environmental health". That said, questions are placed such as where those products will be tested and how, where will they remain in use, by whom, to whom, and the extent of their negative externalities, which must be assessed from a cost–benefit perspective - in order to justify them as an avoidance of a greater harm by dissuasion of a supposed enemy or attack. Trade-offs among Defense's needs regarding the design of sonic non-lethal devices and their every application must be consider by reference to these aspects: by whom, where, against whom, and under what circumstances. This is in line with Ashford's assumption: "Technology is more than hardware and sustainability is more than pollution prevention" (Ashford 2013:5). In addition to this, the concept of 'sustainability' requires a broad review when considering the placement of a technological innovation in the market whose functions generate harms and risks for both human and other populations, given the connections between species, which affect ecological systems. In fostering a transition to sustainability, technological innovation should be oriented by the need to boost economic wealth while developing products have own the potential to minimize, eliminate, and mitigate adverse effects on health, safety, and the environment.

3.2.1 Non-lethal Sonic Weapons, Devices and Repellants – The Concept

> *"A transition to sustainability will necessarily involve changes to the technological basis of corporate activities"* (Ashford 2013).

According to registers, it was around the early sixties when a set of varied weapons technologies began collectively to have the classification as *non-lethal*. With technological innovation and product innovation designed for Defense purposes, there was a growth in demand for weapons engineered without lethal consequences, thus raising many questions regarding reliability in safety and security associated to uncertain markets, especially when those products have the intent to debilitate the receptor. As mentioned in Scott (2010) and Littfield (2016), throughout the twentieth and twenty-first century, there has been a vast amount of research collected and interest gained in the use of non-lethal weapons (NLW), which are intend to immobilize or impair targets without causing permanent or severe damage to the human body. Nevertheless, it also states that NLW shall not be required to have a zero probability of producing fatalities or permanent injuries. In that case, depending on the source characteristics, receptor and exposure conditions, there might be lethality. In the United States, the definition and allowance of use of weapon is legally define under various federal, state, and local laws, which vary by jurisdiction. Accordingly, a weapon is generally something used to injure, defeat, or destroy, and may cover many types of instruments, such as a blackjack, slingshot, metal knuckles, dagger, knife, pistol, revolver, or any other firearm, razor with an unguarded blade, and any metal pipe or bar used or intended to be used in a club, among others. By that definition, the intention of defense by dissuasion and possibility of harm and the use of the local resources as to accomplish those means might as well be consider as a weapon; therefore, under defense circumstances, an animal may be considered as a weapon if used against something or someone. The extension of weapon definition also calls for psychological effects, such as laughter (Cambridge Dictionary 2019), which in some situations can be used as a tool for offending and inflicting harm. We are under a progressively complex and multi-criteria approach when determining what kind of noise – when there is the perception of sound – or infrasound and ultrasound vibrations, and under what physical and psychological circumstances we are able to determine the related harm, based on the level of abuse, the damage extension for an individual's and/or a group's health.

These products have been under research and development in a few countries and have been the subject of both interesting and speculative discussions (Art. 36 of the CCW, 2018). This is because of their silent nature of perception (of a sonic attack under population ignorance of harm) and the lack of cognitive awareness of the attack among affected populations regarding the associated damage and the unknown condition if they are even being under exposure and the alternatives of defense. In the discussion paper for the Convention on Certain Conventional Weapons (CCW) (2018), among other observations there was the fact that the use of acoustic weapons raises the question of our orientation towards technologies that target the human senses and our capability of resilience. Additional considerations have called attention to the ethical, health, legal, and environmental concerns about the acceptability and desirability of acoustic violence – sound as a technique of authority and control – especially frequencies off the limits of the human audible range. In other words, the source of harm, by not being perceived and identified or comprehensible for those experiencing it, cancels the right of defense from the subject who may have been under exposure and compromises the resilience of subjects' health to the recovering of its siege.

The US Department of Defense Directive 3000.3, *Policy for Non-Lethal Weapons*, adopted in 1996, defines NLWs as: to be used with the intention to provide commanders with weapons that "are explicitly designed and primarily employed with the goal to incapacitate personnel or material, while minimizing fatalities, permanent injury to personnel, and undesired damage to property and the environment" (US Dep. of Defense 2002). This definition represents a broad range of products, including sonic devices, which may be consider – depending on the product and receptor – as sonic weapons. The expression 'while minimizing fatalities and permanent injury', nevertheless, might be a questionable statement. The risk of abusive use may cause permanent damage to the receptor. When in need of use, a NLS device will not guarantee the inexistence of a permanent health harm for the victim. The argument used to introduce some of those products into the non-military market as personal defense gadgets, has empowered civil population to use them when the governmental permission exists, but without a guarantee of additional safety.

The Geneva Report (CCW 2018) has stated: "given the well-documented health impacts of weapon noise on humans, consideration of acoustic devices also raises the question of whether political measures should be taken at the international level to better protect both civilians and soldiers from weapons that cause noise-induced hearing loss". Indeed, this is a case of public health, safety and security concern, especially when considering that hearing loss is not the only damage that such devices may cause. The risks involving product technology capability of harm has considerable sustainability arguments, such as risks affecting civilian population and even animals' rights due to animal testing. It should be highlighted again here the difficulty to control the context of exposure to a weapon, especially with the risk of using such a weapon incorrectly (Davison and Lewer 2006). The Table 4 presents a list of damages in association with the sonic source while the Table 5 highlights some of the non-lethal sonic weapons currently in use.

Consumer products for civilians – such as sonic pest repellants, sonic dog repellants, and other living species excluding humans – have been consider profitable compared to others with similar functions. Current sonic repellents technologies for pests and mosquitoes for instance, have been effective to benefit consumers by avoiding one of the most significant threats to human and veterinary health that may have resulted by mosquitoes, due to their capacity of being vectors of debilitating diseases (Norris and Coats 2017). Some repels have already been in market for consumer use for a long time while holding the label as not being harmful to human health and pets compared to chemical ones with the same function. Sonic devices, as animal and pest repel, are weapons for those species not for humans, although the circumstances under which the pet repellent is used may raise some bioethical concerns. Figure 4 lists some of the sonic products and some of the questionings regarding product and technological innovation in association with risk assessment.

Regarding the military and law enforcement use of non-lethal sonic weapons, according to the International Committee of the Red Cross (ICRC 2002:18-5), “at first sight, these weapons seem very attractive. Is it not more in keeping with the principle of humane treatment to put someone out of action temporarily than to kill them with conventional weapons? Closer examination reveals, however, that the use and development of each proposed non-lethal weapon requires serious scrutiny”. There are consequential local health damages. Among the platforms of its illicit use there can be listed: i. out of an appropriate formal, legal, and/or accredited use, and under unauthorized military or conditions such as the military theater operations

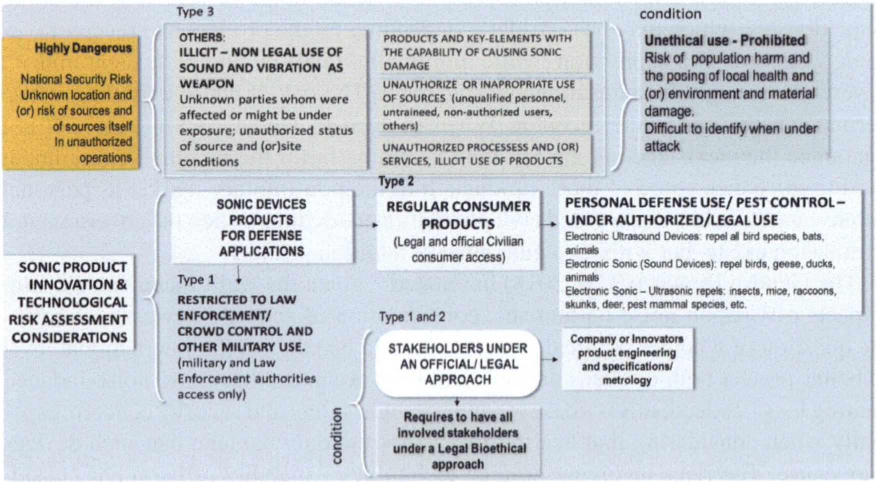

Fig. 4 Sonic Devices technological innovation and consumer products under a risk assessment approach *Source* Developed by the authors

(TO[1]); ii. off the previously authorized sites and subjects and off targets/or countries which are legally authorized to its use, iii. off the legal trade agreements - since we are also dealing with the trading of military equipment, and (or) areas which are required previous authorization and trained, authorized personnel and operation; iv. off the reach of authorized accredited laboratories which may operate with illicit testing. Under those conditions it may cause unprecedented abuse and harm with the compromising of the resilience of recovering on the affected area and of population health if illicitly used and security.

3.3 Environmental Values, Bioethics, and Biopolicy

> *"How can one turn a threatening gunman into a retching bundle of nerves, suffering simultaneously from bowel spams and a loss of courage before surrendering to the police? Simply use infrasound on him"* (Altmann 2001 quoted by Goodman 2010).

Sonic non-lethal weapons in use for military/law enforcement scenarios and sonic repellants engineered for animal and pests used by civilian population, although broadly different, have a common approach when considered from the perspective of their functions and the risks of abuse due to the governing regulatory aspects on their use. Understanding that the use of infrasonic frequencies devices, according to Goodman (2010), affect the brain and nervous system; for instance, frequencies, such as of 7 Hz, may coincide with theta rhythms, which are capable to induce moods of fear and anger. Because every organ within the human body has a resonant frequency, it is so delicate to deal with such a harm. The nature of sonic devices for defense may arguably be questioned: what are the sanctions of abuse, once it is identified? Because of their function to debilitate the targets coupled with the ignorance of the latter of both act and collateral effects are of generating unnecessary suffering, the abuse of sonic sources as weapons are capable of affecting local resilience, stricter rules and sanctions should be applied. Therefore, those products have been considered of efficient application in military confrontations. However, when considering a sustainable management approach, the practice of moral ecology may be added, which means that lived awareness that actions have profound consequences that we both can and must consider and act upon. The characteristics of the environment in which we develop our biological and mental being and the quality of our life is affected beyond its perception of damage. The fight against the destructive activity of humans for the planet has its own approaches. Among them, the principles of the deep ecology (Table 6) and their core assumption that the nonhuman life on earth has an intrinsic value (Naess 1994).

According to Kellert (1993), human identity and personal fulfillment somehow depend on our relationship to nature. The human need for nature links not just to

[1]It is defined as a formal geographic combatant commander for the conduct or support of specific military operations. TOs, according to the United States Department of Definitions (2019), are usually of significant size, allowing for operations in-depth and over-extended periods of time.

the material exploitation of the environment but also to the influence of the natural work on our emotional, cognitive, aesthetic and even spiritual development. Human intelligence is bound to the presence of animals. They are the means by which cognition takes its first shape and they are the instruments for imagining abstract ideas and qualities. Animals share our common planet as well as the growth and development of the human species. When in use, sonic deterrents and/or repel are created to negatively affect and weaken biological systems under vibration sources at frequencies with sufficient amplitudes and with appropriate intensity to debilitate. The animal instinct provides itself with the ability of alertness and defense under the perception of attach; nevertheless, if those biological abilities are undermined by technological innovations or site conditions, bioethical questions ought to be raised.

3.4 Metrology and Regulation

"Sound deserves its own pollution category" (Ron Pellegrino quoted on Davies 1994)

Rule-making shall inevitably investigate what will be the effects on human and non-human species which are not the main goal or focus of attack. Nevertheless, the conditions under which this will happen ought to be review to ensure the avoidance of unnecessary suffering of the targets. According to Broner (1977), in order to avoid the damage to human species due to sonic devices during the impact assessment, the use of cadavers, experimental animals, or simulations for most studies on mechanical injury is common. This is the reason why the acquisition of impact related data has been under scrutiny. This is the reason as well why metrology, regulation, and standardization in this area require as well a review of the prevailing scientific methods, which should remain subject to a periodic review. The invisible aspects of data acquisition have to be under audit and tracking by reference to the ethical principles of the scientific methods used; thus, supporting the innovation process, especially when considering the nature of the product's function, and more importantly under what values.

It would be a superficial approach to introduce sonic defense function products and devices in the market without considering the complex scenario and the criteria when in testing and use. The scientific approach of the measurements regarding noise and vibration under a sustainable management review will require its development under bioethical principles. The standards for quality and control of product manufacturing and testing, in this case, will need to assure interested stakeholders about the types of risks being generated and the guarantee of accountability and certainty of the testing methods and measurements being performed. The metrology traceability for sonic devices must bridge with advances made in biology and the non-maleficence principle, alongside with the determination of what technical information will be of relevance for stakeholders awareness, those adjustments on the process of testing and measurements are required to perform under previous established environmental, bioethical, and biopolitical strategies, as approached in the previous section: Broner

(1977:13) "caution is necessary in future research because artifacts produced by faulty experimental procedures can suggest genuine psychological or physiological effects. Therefore, a lethal infrasonic weapon appears unlikely, and such reports as the one which suggested that a sound gun had been developed but had proved unusable as it would have killed everything within four miles.

In this perspective, we may question: what conditions of engineering and use, a non-lethal sonic source may be considered lethal or (and) of torture? Many questions will raise concomitantly with the authorization of its use for civil consumers, as the ones already being argue within law enforcement and military spaces.

Questionable aspects evoke, such as those regarding the testing. Some research projects are currently under development or evaluation addressing acoustical bioeffects. It is thus of interest to know which vibration frequencies have the largest effects, not only on humans, but on non-human species too, and what magnitudes are under tolerable acceptance by reference to a sustainable management approach. Sonic-induced stress is considered to be difficult to quantify – 'stress' can be defined here as a perturbation to homeostasis – with accuracy since the response over population exposure is difficult to be tracked. Some studies (i.e., effect of noise on marine) have found a tractable ground by investigating repeatedly the chronic sounds under a population exposure. Regarding non-human species, and due to anthropogenic sonic distress and its effects on the ocean environment, for instance, research concluded that long-term population effects will have the greatest impact on marine mammal species (Erbe et al. 2018). These effects include transformations such as the risk of permanent threshold shift and an increase in occurrence of pathological stress. It was proposed that the homeostasis could have been restored based on an adaptive stress response. Nevertheless, depending on the frequency of disturbance, out of the normal physiological response range, or persistent, the stress response can be pathological.

The measuring of sonic impacts on non-human species is extremely difficult to realize and it remains arguable, especially when approached under bioethical considerations and by reference to the harm and suffering inflicted upon these species. According to Satterfield and Kalof (2012:29), "If one limits the concept of value to the degree which an item contributes to an objective or condition in a system, then we can see how natural scientists use the concept of value all the time to talk about casual relationships between different parts of a system… sustainability values may be more important than efficiency values around and below threshold limits". Metrology must address the necessary adjustments not to mask damage while providing communication of the dangers, otherwise communities might be at a critical risk, under siege, under a condition which are out of human cognitive inputs of awareness, neither perceived or understood and of compromised system resilience.

3.5 Ecodesign, LCA and Human Factors, Biomimicry and Innovation

> *"Nature´s inventions have always inspired human achievement and have led to effective algorithms, methods, materials, processes, structures, tools, mechanisms, and systems"* (Cohen 2006:6).

Nature results from a Divine process. Mankind is considered to be inspired by God – or Creation – when in the process of creation. What is Divine within us? Despite of any type of belief, or incredulity, the word religion has its etymological meaning of *religare*, derived from the *Latin* meaning tie, bind; other philosophers have made it *relegere* with the meaning of *reading over again*. What is the reframing process of function we can recreate within a sustainable process of design? We may call it ecodesign. Many definitions have added to the concepts of ecodesign, among them: "A Methodological Systematic Assessment of the process of product system development which are economically and ecologically sustainable. The product life cycle should be associated to the product development process" (Keoleian et al. 1993) and, as a process that reduces the environmental burden of a product considering its entire life cycle (Suzana 2018). The importance of ecodesign is the questioning of the function (either a product, a process or a service) under its environmental, ecological aspects of its externalities.

How can we reframe what urges for defense or attack? It is perhaps more than complex and impossible question to answer, since we are under a constant process of creation, and creation requires its own forces. The bionic, biomimetics or biomimicry with the meaning of bios – life and mimicking or mimesis – to imitate, is a scientific approach where the *man-made* process of creation will urge to look forward and be inspired by the sacredness of life. This is the reason why we shall question, in the process of creating the purposes for *man-made* creations, the use of those functions in man-made creation. How does nature perform in such functions and how it regenerates – if it is ever capable – in a sustainable way, and if Mankind should ever dare to risk the testing if the resilience capability becomes under progressive extermination by external sources. From such we can understand the delicate nature of sonic devices and non-lethal weapons approach and the type 3 sources defined in this framework; those whom are in use by humans through the knowledge of technology of sound and vibration, to inflict harm upon the innocent. The responsibility of scientific knowledge in this area is such that it cannot exist without approach of the Sacred, once it affects life. What aspects of defense innovation products shall be mimic from nature when in the process of engineering; can we conceive it fair? What key elements shall we search? What are the key elements that define a sonic product and its harm? How will they operate, under what conditions of the system? The area of Human Factors Engineering has evolved from such interface, with the aim of reducing risks regarding user interface, while providing improvement on the many processes involved in a system, and it is mandatory on the highlighting of discussions, the many aspects regarding user interface, such as risks and perception of risk, product interface with the clear understanding of operations and functions by operator among many other human factors engineering attributes involving sonic devices for defense (Fig. 4).

3.6 *Life Cycle Assessment (LCA), Product Category Rules and Environmental Product Declaration*

In the process of product engineering (either a product, a process or a service), once the directives of product function and its attributes have been defined, there are many alternative processes, material inputs, and phases which, under a sustainable management approach, may have to be tracked and reported. Among the tools that provide such transparency, the LCA is the one that is formally in use under environmental standardized methods to provide the society with correct quantifiable impacts regarding a product's function and externalities for environment, health, and resources. The methods are formalized under ISO Standards. According to the Environmental Protection Agency, "LCA enables the estimation of the cumulative environmental impacts resulting from all stages in the product life cycle" (EPA (2006)). There is not only Environmental Assessment, but also Social Life Cycle Assessment, which evolved from the Corporate Social Responsibility Statements, enabling the identification of Social Hotspots in association with the site and processes involved. ISO Standards 14040 and 14044 have given support to the improvements in transparency; among many statements, it includes "the intended application (What); the reason for carrying out the study (Why); The intended audience (for whom); whether the results are intended for comparative assertions intended to be disclose to the public (how)" (Simonen 2014). Figure 5, adapted from ISO 1440 and Simonen (2014), shows LCA strengths, framework and application.

Product Category Rules (PCR), according to ISO 14025:2006, is a set of specific rules, requirements and guidelines for developing Type III Environmental Declarations for one or more product categories. A group of products that can fulfill equivalent functions (Ingwersen 2013; 2019). This is the reason why there is a need for a

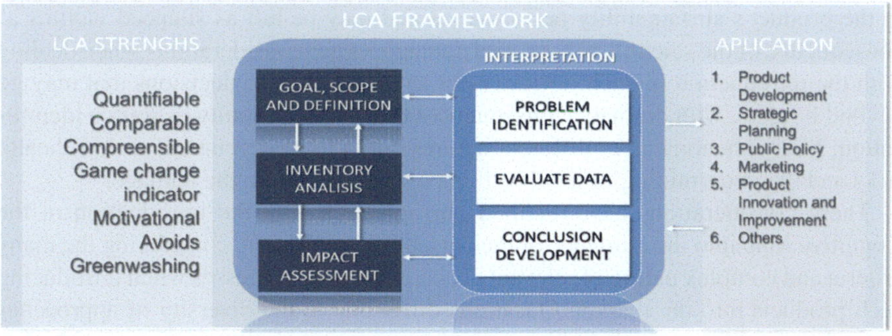

Fig. 5 LCA strengths, framework and application *Source* ISO 14040 and Simonen (2014)

periodic review of the types of sonic products and their use, when considering their function of repel, debilitate, and/or the possibility of alternative function of product key-elements, based on product redesign, which may intentionally or unintentionally inflict harm on population without previous legal authorization. PCR has its working active group from EPA and American Committee of Life Cycle Assessment (ACLCA), where it remains under permanent review. Scientists involved are in a Mission of permanently improving a better life on this planet based on tools, such as LCA, regarding product reviews under the traceability of their function materialized into products, services, and processes. Every PCR is required to have a Program Operator under a certain background of qualification in order to propose formulate and approve PCRs. PCR is more than a label, it is based on a product's full life cycle; it can as well influence and be under the influence of a decision-making process, policies regarding trade and purchasing decisions. Based on this, it periodically requires the monitoring of product innovation and the mapping of security key-elements that need to be under surveillance in the product engineering process. Further discussions on this subject are required while involving all concerned stakeholders. Among the issues to be discussed, the profile of Program Operators in this framework should be evoked, considering that not only those should have absolute knowledge of their products, but they must be inquired about their environmental values.

4 Conclusions

> *"Scientists often disagree on a lot of things, but that group was a great example of why it's important to trace the money behind the science"* (Billot 2019:266).

The insertion of technological innovation into products in the area of sonic devices for Defense, requires the creation of a sonic ecological intelligence and a strategic planning when considering its use in further applications, under authorized military use. There is as well the need to take into account the many alternatives dealing with different scenarios from diverse stakeholders and shareholders when under a risk approach. Consumers cannot process the amount of relevant information related to the product's sustainability performance and may be left as silenced victims as consequence of the scientifically proved damage. Other considerations when dealing with the introduction of such devices refer to the impacts of decisions that may go beyond borders, with dramatic consequences due to the difficulty of source identification. Many phenomena of different natures ought to be accounted, such as health risks and the undermining of local resilience to recover from the damage.

These considerations have justified our approach with the introduction of the Cognitive Map into the guidance of the current research team, considering the many criteria and complex nature of relevant decision-making processes when introducing such products into the market. Questions differ due to the diversity of approaches and values of different stakeholders, which are invest with the power to decide. For Ashford (2013*)*, "a technological innovation can be characterized by its motivating force, by its type, and by its nature". When looking into the sonic devices' function – which are address for defense means – what are the other alternatives? What are the motivating forces to introduce them in the market and/or its use to certain populations? When compared to others, does it appear as a technological shift that will be harmless and more effective when meeting the function of defense? There is also the bioethical pillar of questionings the sonic devices' function from a sustainable management approach, which will remain as driving forces behind every technological change in this current century. Since we are dealing with the function of debilitating which may generate a permanent damage, there are serious issues regarding the choices available for the decision-making process. Is resilience consider a positive attribute under sonic distress? How will industry comply or respond to health, safety, and environmental regulations and pressures? In the case of countries already under poverty suffer and social agony, where environmental laws are lenient and health issues are already critical, shall Defense – under authorized and legal situation – make use of them? And if so, under what circumstances?

The environmental exposures and extended damage associated with type 3 source (Fig. 4) in affected physical area and population, may be extremely difficult to avoid, monitor, and control; such is the risk which may be faced when dealing with sonic sources, which, "although not lethal, would be very painful and would probably disable a man for an appreciable length of time" (Davies 1994). From the Insurance stakeholder view, it will be mandatory to refer to the multicriteria decision making model. According to Linkov et al. (2018), "the insurance company's perspective, innovations from an emerging technology may provide substantial benefits in unexpected ways". From the ethical perspective, there is more to be add. Under scientifically proved damage, the use of sonic sources designed for debilitating function may be capable of slowly killing while causing suffering for the affected population by its siege. Research in improved metrology investigations to identify sonic attacks are required concomitantly with the avoidance of suffering from sonic attacks and their harmful consequences. The insertion of products with debilitating function in the market, associated with environmental and health risks, should only be conceivable with the previous communication and acknowledgements of their potential damage. Any confidentiality around such innovations and their associated risks and damages, should be incriminated, once the authorization of the product does not correspond to the required public awareness that remain under inability of reciprocate response and defense. Moral corruption of those who approved the use of such products as profitable goods should be review under the attempt of criminal sonic torture risks and humanitarian abuse.

References

Jürgen A (1999) Acoustic Weapons — A Prospective Assessment: Sources, Propagation, and Effects of Strong Sound. Experimentelle Physik III Universität Dortmund, Germany. http://scienceandglobalsecurity.org/archive/sgs09altmann.pdf

Nakashima A, Bob C (October 2006) Defense R&D Canada – Toronto Technical Report DRDC Toronto TR (2006) -218 DRDC Toronto TR 2006–218
Article 36 (2018). Discussion paper for the Convention on Certain Conventional Weapons. (CCW). Geneva, November
Ashford NA, Heaton GR, Priest (1979) W.C. Ehs Regulations and T.I. In: Technological innovation for a dynamic economy; Hill CT, Utterback JM, (eds.) Pergamon Press, NY, pp 161–221
Ashford N (2013) Building a Sustainable Future. MIT. file:////SERVER/Current%20Work/DOROTHY/PUBS/proceedings/VOL5/html/ash_usa.htm[26/4/2013 2:54:09 μμ]
Ashford N, Hall RP. The Importance of Regulation Induced Innovation for Sustainable Development. ISSN 2071-1050
Axelsson O, Guastavino C, Payne SR (eds) (2020) Soundscape Assessment. Lausanne Frontiers Media SA. 103389/978-2-88963-329-6
Beraneck L, Vér IL (2006) Noise and Vibration Control Engineer – Principles and Applications. John Wiley and Sons
Billot R (2019) Exposure. Atria Books, NY
Broner N (1977) The effects of low frequency noise on people- a review? J. Sound Vibration (I 978) 58(4), 483–500. Department of Physics, Chelsea College, University of London, London S W6 5PR, England, Received 15 November 1977
CCW (2018) Convention on Certain Conventional Weapons (CCW), Geneva
Cohen YB (2006) Biomimetics. Taylor & Francis, Biologically Inspired Technologies, p 2006
Davies Alex Bfa Honours (1994) Acoustic Trauma: Bioeffects of Sound, Applied Biophysics, Aether Research Laboratory
Davison N, Lewer Nick BRADFORD (2006) Bradford Non-Lethal Weapons Research Project (BNLWRP). Occasional paper No. 1. The Early History of Non-Lethal Weapons. Bradford: University of Bradford, Department of Peace Studies. https://ipcan.org/wp-content/uploads/2017/06/Research-Report-of-the-University-of-Bradford-on-Non-Lethal-Weapons-mars-2006.pdf
Delaney T, Madigan T (2014) Beyond Sustainability: A Thriving Environment. McFarland & Company. Inc., Publishers
Department of Defense (2002) Non-Lethal Weapons and Equipment Review: A Research Guide for Civil Law Enforcement and Corrections. National Security Research, Inc. July 2002
Environmental Protection Agency (EPA) (2006). EPA/600/R-06/060 – Life Cycle Assessment: Principles and Practice
Erbe C, Sara DJ, Dunlop RA (2018). Effects of Noise on Marine Mammals. https://doi.org/10.1007/978-14939-8574-6_10
Daphne F-L, Peter N, Bor-Kai H, Matheu K (2016) Biomimicry: A Path to stainable Innovation
Mohr GC, Cole JN, Guild E, Von Gierke HE (1965) Aerospace Medicine 36,8 17–824. Effects of low frequency and infrasonic noise on man
Munda G (2008) Social Multi-Criteria Evaluation for a Sustainable Economy
Goodman S (2010) Sonic Warfare: Sound, Affect, and the Ecology of Fear. MIT Press
Suzana GT (2018) Technological Innovation & Ecodesign _ Beyond Compliance. Lecture Slides Lecture on Birla Institute of Technology- Pantheon
Herzlich A (July 2013) The Meaning of Illness. History of Diseases. Edited by Mark and H. Auge. Routledge.
Heys T (2011) Sonic, Infrasonic, and Ultrasonic Frequencies: The Utilization of Waveforms as Weapons, Apparatus for Psychological Manipulation, and as Instruments of Physiological Influence by Industrial, Entertainment, and Military Organizations. Thesis for the Degree of Doctor of Philosophy, Liverpool John Moores University. March 2011
Hull MS, Bowman D, Linkov I, Trump J, Trump BD (2018) Chapter 9 - Nanotechnology Risk Management: An Insurance Industry Perspective, pp 195_223. Nanotechnology, Environmental Health and Safety – Risks, Regulation and Management – Third Edition – ELSEVIER
ICRC (2002) International Committee of the Red Cross Unit for Relations with Armed and Security Forces 19 Avenue de la Paix 1202 Geneva, Switzerland

Ingwersen W (2013 and 2019) – Guidance For Product Category Rule Development. 2013 Version Public Draft on ACLCA – Florida 2013, and Current updates 2019, from ACLCA PCR Group. Subramanian, V. Editors. Product Category Rule Guidance Development Initiative

Institute for National Security Studies (1997) – INSS. US AIR FORCE ACADEMY COLORADO. Non-lethal Weapons Terms and References – RJ Bunker Editor

Institute for Non-Lethal Defense Technologies Applied Research Laboratory. The Pennsylvania State University Park, PA 16804–3000 (February 2001) Crowd Behavior, Crowd Control, and the Use of Non-Lethal Weapons. Approved for Public Release, JNLWD, 20 Feb 2004

ISO 14040/ 14044, 14025 – ISO Reviews and Online Previews of Their Standards. https:www.iso.org/obp/ui/#home.2006.

ISO 2631–5.2018- (2018). Mechanical Vibration And and Shock — Evaluation of Human Exposure To Whole-Body Vibration — Part 5: Method For Evaluation Of Vibration Containing Multiple Shocks

Kellert SR (1993) The Biophilia Hypothesis. Island Press. ISBN 1- 55963-147-3

Kellert SR (1983) Affective, cognitive and evaluative perceptions of animals. In: Altnman, Wohlwill J (eds.) Behaviour and the natural environment – chapter 7. Plenun. NY

Keoleian GA (1993) The application of life cycle assessment to design. J Clean Prod 1(3–4):143–149

Lindgren NO, Ohman J (2018) A Posthuman Approach to Human-Animal Relationships. Environmental Education Research

Linkov I, Trump B, Cummings C, Galaisti SE (2018) Synthetic biology: perspectives on risk analysis, governance, communication, and ELSI. In: Comparative, collaborative, and integrative risk governance for emerging technologies. https://doi.org/10.1007/s10669-018-9686-5.2018

Littlefield R (2016) The psychoacoustic effect of infrasonic, sonic and ultrasonic frequencies within non-lethal military warfare techniques. Exploring the use of audio to influence humans physically and psychologically as a means of non-lethal warfare methods throughout both the 20th and the 21st century. http://Littlefield.co

Living Building Challenge 3.1. (2016) A Visionary Path to a Regenerative Future. International Living Future Institute

Neil M (2005) Human Response to Vibration. ISBN 0–415–28239-X 1. Vibration—Physiological effect. I. Title. QP82.2. V5M36 2004 612′.01445–dc22 2004054491. CRC Press LLC.2017

McMichael T (2001) Human Frontiers. Past Patterns, Uncertain Futures. Cambridge Press, Environments and Disease

Mihai Budescu M, Movilă M, Gheorghe A (2017) The harmful Effects of Vibrations on Human

Morrison. R, (1995) Ecological Democracy. Sound End Press. Boston, MA

Naess A (1994) ARNE. A Defense of the Deep Ecology Movement Earth. 1– Reprinted from Environmental Ethics, vol 6

Non-Lethal Weapons (NLW) REFERENCE BOOK (2012). Joint Non-Lethal Weapons Directorate. Supreme Court

Norris EJ, Coats JR (2017) Current and future repellent technologies: the potential of spatial repellents and their place in mosquito-borne disease control. Int J Environ Res Public Health

Skopec R (2018) The frey effect of microwave sonic weapons. Drug Des Int. Prop Int J 2(3). September ISSN: 2637-4706

Satterfield T, Kalof L (2012) Environmental values: an introduction – relativistic and axiomatic traditions in the study of environmental values. In: Linda K, Terre S (eds) Environmental values, Earthscan Reader Series

Schafer M (1993) The Soundscape. The Tuning of the World. Destiny Books. Rochester, Vermont

Scott, Richard, Major (2010) Nonlethal Weapons and the Common Operating Environment. April 2010. US Army Publication.

Shivakumara BS (2010) Study of Vibration and Its Effect on Health of the Motorcycle L. Rider. Online Journal of Health and Allied Sciences.

Simonen K (2014) Life cycle assessment – ryan smith editor. Routledge Taylor and Francis Group

Stefan I, Budesco M, Movla M (2017) The Harmful Effects of Vibrations on Human. Buletinul Institutului Politehnic

UNITED NATIONS OFFICE FOR DISARMAMENT AFFAIRS (2019) United Nations Institute for Disarmament Research – Hypersonic Weapons - A Study Prepared on the Recommendation of the Secretary-General's Advisory Board on Disarmament Matters

Vallee M (2017) The science of listening in bioacoustics research: sensing the animal sounds. theory, culture & society. SAGE

Agni V-A (2013) Biopolitics: A Shift from Anthropocentric to Biocentric Values in Business. Biopolitics International Organization

Von Gierke HE, Brammer AJ (2002) UConn Health Center. Effects of Shock and Vibration on Humans January 2002

Mansfield NJ (2005) Human Response to Vibration. CRC Press. ISBN 0-415-28239-X

Augé M, Herzlich C (1995) The Meaning of Illness. Harwood Academic Publishers GmbH

Altmann J (2001) Acoustic weapons - a prospective assessment. Sci Glob Secur 9(3):165–234

The Anatomy of Uncertainty and Implications for Preventive Environmental Management

Gopichandran Ramachandran, Mohamed Behnassi, and Gireesh Chandra Tripathi

Abstract That it is important to internalize externalities and reduce environmental burdens is an understatement. Even if this is an imperative, it is important to ask if it has after all been possible to reduce burdens commensurate with growing developmental aspirations and material progress. Have compliance and logical reduction of burdens evolved through mutually reinforcing efforts? Or, is it a case of willful non-compliance or lack of drivers to mediate concerted actions, notwithstanding the immense variety of environmental protection tools deployed world over? Some questions that dominate the interface of science and public policy to enable preventive management pertain to the profit motives of business and their tendencies to overlook the call of science for mitigation and adaptation. On the other hand, scores of businesses around the world have demonstrated the feasibility of successfully securing economic and environmental gains while fulfilling commitments to comply. The present chapter addresses these questions from the perspective of uncertainty, central to resolving dilemmas in collective management of externalities through a public policy outlook.

Keywords Scientific uncertainties · Industry initiatives · Financial institutions · Externalities · Preventive environmental management

G. Ramachandran · G. C. Tripathi
NTPC School of Business, Noida, UP, India
e-mail: gopichandran@nsb.ac.in

G. C. Tripathi
e-mail: gireesh.tripathi@nsb.ac.in

M. Behnassi (✉)
College of Law, Economics and Social Sciences of Agadir, Ibn Zohr University; Center for Research on Environment, Human Security and Governance (CERES), Agadir, Morocco
e-mail: behnassi@gmail.com; m.behnassi@uiz.ac.ma

M. Behnassi et al. (eds.), *Social-Ecological Systems (SES)*,
https://doi.org/10.1007/978-3-030-76247-6_13

1 Introduction

Industry-led sustainability is an interesting framework for integrated management of environmental externalities.

Preventive management through at-source reduction of waste streams, alternative uses through eco-industrial development frameworks, material substitution, good operating practices, and complementing end-of-pipe interventions are predominantly applicable within the four walls of the industry with significant mitigation outcomes. Importantly, environmental impacts of emissions and other wastes are felt far away too, across latitudes and longitudes. These call for focused remediation/adaptation efforts by such sites at the local and ecosystem levels (UN 2017). Information leads indicate that the industry is actively engaged with multilateral and financial institutions, guided by governments, and is responding to market enablers in tackling the impacts of climate change (Government of India 2017, 2018, 2019). This covers sector and operations specific interventions and at sites beyond their physical boundaries. Therefore, it is imperative to sustain the momentum through enablers based on a deeper understanding of their individual and synergistic impacts specific to sectors, locations, and challenges.

Some questions about the sustainability of the momentum of cleaner transitions relate to forms and functions of uncertainties in science-based decision making.

Is it likely that uncertainty perpetrates bias and is almost a ploy for inaction? Are indices of environmental performance persuasive enough to inspire businesses to dive deep into the real causes of process-related perturbations? Is there a growing demand from citizens who experience the impacts of externalities to increasingly invoke legal provisions and force industries and businesses to comply? These could also be related to the principles of perceptions about science and, therefore, preparedness of stakeholders to comprehend solutions as viable over immediate, medium- and long-term horizons. Interestingly, such perceptions appear to pervade all levels of organizations, including individuals, communities, businesses, financial institutions, and regulators. Challenges posed by climate change and the recent COVID-19 pandemic are typical cases in point. Two strands of facts reveal the complexity of the context.

Empirical evidences derived through deep scientific investigations point to unabated onslaughts due to environmental externalities.

The NASA (2021) recently provided indisputable empirical evidences about artificially-induced perturbations on climate and ecological systems. These pertain to the steady rise in acidity of surface ocean waters and the actual increase in the quantum of carbon dioxide delivered into oceans on a yearly basis. Rise in sea levels is documented categorically as the phenomenon of diminishing ice sheets. The Global Precipitation Measurement Mission and the NASA's Soil Moisture Active Passive Mission reinforce deep insights in this regard. Interestingly, Hepburn et al. (2020) and Shaddick et al. (2020) cite evidences of sharp falls in the levels of carbon dioxide correlated with lower coal and oil consumption due to COVID-19 related lockdown world over. The American Geophysical Union (2020) corroborated significant reduction of pollutant onslaughts due to the lockdown and argued that emissions regulations could become more stringent, duly recognizing other mutually reinforcing natural processes. The latter was seen in the case of surface ozone levels. Importantly, several Fortune 500 firms responded to the changing dynamics of inadvertent reduction of externalities (Kelly 2020). They argued that the COVID-19 experience has awakened them to risks they were not prepared for. It was, therefore, essential to build back better with environmental safeguards and stimulus funds to accelerate transitions to a net-zero emissions economy and invest in a cleaner and resilient energy infrastructure. This included a special focus on training workforces to imbibe climate-efficient practices.

Industry initiatives and financial institutions around the world join the chorus for enhanced mitigation and adaptation outcomes.

They mandate climate stress tests and climate risk disclosure to protect capital markets (CERES June 2020; MSCI 2020). This was highlighted by the CERES through a comprehensive review of losses incurred by economies due to climate change related challenges reaching about 05–20% of the GDP. The Institute for Energy Economics and Financial Analysis (Buckley 2019; Buckley et al. 2019) highlights coal divestment and related restricting policy instruments supported by investors and initiatives based on science-based targets and the Carbon Disclosure Project (CDP) which indicate climate and energy efficient initiatives by hundreds of firms in several sectors. The Climate Analysis Indicators Tool (World Resources Institute) is also used in many cases. In April 2020, G20 nations signed fiscal measures worth trillions of dollars for clean physical infrastructure investment, including grid modernization and carbon capture and storage (CCS) technology. These are to complement ecosystem resilience and regeneration, including restoration of carbon-rich habitats and climate-friendly agriculture as part of clean R&D spending.

It is well known that in 2006, the United Nations launched a set of six investment principles to incorporate environmental, social and governance (ESG) matters into investment practice, to assess the impact of the sustainability and ethical practices of companies as a function of financial performance. This is built on the premise that shareholders' demands escalate for companies that fail to adapt. One of the best ways to evaluate these metrics is through ESG integration, as recently highlighted by Demine (2020) and MSCI (2021) through the India ESG Leaders Index and Wills Towers Watson (2020). Hundreds of money managers collectively oversee trillion in assets, aligned with the principles for responsible investment across such aspects as socially responsible and impact investing. **Strategic, substantive, and impactful ESG programmes** and e**nvironmental stewardship** are initiatives which could determine the scope for e**nergy diversification, carbon footprint reduction,** and n**atural resource sustainability** through cleaner renewable energy and products. These are substantiated through leads presented by, for instance, the American Council on Renewable Energy, REPSOL, American Electric Power, Edison International, TATA Power, Uniper, Kansai Electric Power Group and Orsted.

2 Signals of Promise Emerge from Other Multilateral Initiatives and Additional Industry – Specific Action

The Kigali Amendment of the Montreal Protocol embeds energy efficiency enhancement of refrigeration and air conditioning appliances and control of leakages as part of its transitions with enormous implications for climate and ozone layer protection. Several countries have developed their respective cooling action plans for this purpose. The energy security, access, and sustainability tool of the World Energy Council enables a seamless integration of business and environmental protection goals. Some of the other best examples of industry environment leadership include Pivot goals, EP 100, RE 100, The Energy Charter Treaty, The Task Force on Climate Related Financial Disclosures that linked financial and reputational risks with climate inefficiency, The Business Environmental Leadership Council, The Green Power Partnership, The Corporate Climate Tracker, The Carbon Leadership Framework, The Oil and Gas Climate Initiative, The Methane Guiding Principles and The Mineral Methane Initiative that facilitate appropriate documentation and reporting. The UN Global Compact's and the International Labor Organization (ILO)'s principles are also adopted by firms with a commitment to protect, respect, and remedy externalities.

Yet another interesting facet of environmental management has been the roles played by platforms such as: the CarbonBrief which maps world's coal power plants; the Urgewald research project; and the Banktrack and clean energy transitions templates of the International Energy Agency, for the energy sector in particular. The latter also relates to energy efficiency enhancement and technology collaboration programmes. Can these be taken as promising signals of a positive trend towards comprehensive environmental protection or will these be just transient? How long it will take before such initiatives become evident in a significantly large scale even from clusters of small and medium enterprises? and what could be the implications of uncertainty that influence decisions to act?

3 Co-evolution of the Uncertainty Contours and Related Response

This co-evolution is probably reflective of nature and human endeavors replete with diversity and resilience characteristic of all constituents. This is evident from the rich body of literature referred to in this chapter. We commence with a landmark publication by Pouyat (1999) leading up to the recent the COVID-19 pandemic and post-COVID-19 transitions; including the most recent efforts of the Asian Development Bank which position courts as conscience keepers for amelioration (ADB 2020). We also draw upon positions defined by representatives of the energy sector and with reference to the growing impetus of the ESG framework.

Pouyat (1999) presents a comprehensive review of the continually evolving challenge due to uncertainty in the interface of science and policy for preventive environmental action, spanning four decades. The author prompts a conclusion that science-based solutions are not assertive or comprehensive enough and, therefore, the public tends to stay confused about possible outcomes of adopting such solutions. Is it therefore possible to invoke the fundamental premise that science will stand only for truth and be accordingly embedded in policy making? This is especially so of planetary boundaries and resilience, including environmental fate of pollutants and the urgent need for comprehensive abatement. Inadequate assertion from science may give way to legal and regulatory remedies; yet based on the limits and limitations of available knowledge. Businesses are implicated in underplaying the importance of science and appear to rely on political influence to evade responsibility. This generates a call to analyze the motivation and barriers that divide science and policy intended to translate mitigation intent to reality.

Almost fifteen years later, Fischhoff and Scheufelec (2019) highlight the persistent dilemma of uncertainties and the need to communicate facets of such uncertainty in policy analysis. This is especially true at the confluence of science, health, and cultural literacy in a rapidly changing communication landscape. Tools of communication that also check veracity of messages are ever more invasive, prompting an analysis of the consequences of hyper-transparency. This creates an additional challenge related to the present-day ability of stakeholders to comprehend the reach of disruptions across geographies, evidenced in the case of climate change and ozone layer protection, with implications of technology transitions and sustainable development. This reinforces the context of science communication and evidence-based policy making framework.

The IPCC in 2014 presented a risk management framework focusing on the linkages with uncertainty for policy choices (Kunreuther et al. 2014; Sharma and Ravindranath 2019). It acknowledged the growing range and number of stakeholders, including governance forums, which are interested by and influencing climate policy choices. Related policy instruments are equally diverse in the area of emissions, market instruments for fossil fuel-based energy systems and renewable energy systems, and within extraction and end-use applications across sectors. Public policies could drive investments and behavioral responses aligned with expectations.

One of the best analyses of the preparedness of industry to protect was presented earlier by Nadakavukaren and Cottier (2012) with special reference to the 'responsibility to protect' framework. It centers on the tendency to consider individual interest ahead of common interests and the misconception of moral pressure against legally permissible preference for inaction. A fourth facet of decision making was highlighted by Winterfeldt (2013) according to which beliefs and values appear to be at the core of perceptions and, therefore, the groundswell for environmental protection. This probably builds on the cultural literacy elements stated above. Winterfeldt (2013) raises the concern that scientists do not understand the needs of decision makers and the much felt need to bridge the gap through science and technology panels. The Montreal Protocol exemplified this aspect of communication through its Technology and Economic Assessment Panels and the Technology Options Committees. Fischoff and Davis (2015) argued that decision makers look for signals without delving deep into science and tend to choose from fixed options. It is, therefore, essential to create a wide array of options to choose from with a clear statement of uncertainties. Tools to characterize and assess such uncertainties have to also be provided to decision makers. This is probably best exemplified by the International Energy Agency (IEA 2020a) and the IEA Renewable energy markets update (IEA 2020b) with reference to countries grappling with energy mix options and the tendency to rush, and others without a comprehensive assessment of enabling circumstances. This is true of the interplay of regulations, market instruments, and institutional mechanisms across all areas of energy and industrial applications.

Oreskes (2015) captured the dynamics of resistance to change by arguing that doubt and uncertainty are probably exploited as defensive tactics by the industry to challenge the scientific evidence. This is based on the stance of a wide variety of industrial sectors such as tobacco, growth hormones in food, and fossil fuels. This is probably closely aligned with the earlier stated premise that business interests are opaque to the larger goals of preventive management. Oreskes (2015) escalates the argument to cultural persistence of doubt and the tendency of industry to exploit a resultant vulnerability of the end users of its products and services. This appears to border on the willful ignorance and tendency to challenge consensus by highlighting dissent. Can it also be a case of playing on the divide and the processes of making an unmaking of ignorance? Science too appears to be probabilistic at times and not deterministic enough. Therefore, scientists have to redouble their efforts to resolve uncertainties, thus strengthening social consensus (Yung et al. 2019).

It is in this context that one of the most recent trends about the groundswell of public outcries was highlighted by Setzer and Byrnes (2019). Human rights- and science-centered appeals are growing in scale and form, calling for focused mitigation and adaptation actions to tackle climate change related challenges. Communities also look for impacts beyond the court room in real life. More than 300 cases in a score and more countries prove this point. This corroborates the view presented by Pouyat (1999) that legal and regulatory remedies will prevail should industry evade environmental responsibilities. The meta-precautionary principle (Patterson and Mclean 2019) substantiates the synthesis of knowledge and harm conditions as a function of the plausible and actual severity of threat.

A detailed analysis of the typology of industry response to the call for compliance was presented by Hove et al. (2002) with reference to the oil and gas sector. They differentiated the higher-priority consequential strategy of business for self-interest, the willingness to accept responsibility, and procedural strategy to let go of some economic interest due to ethical considerations. While the burden of proof about alternatives is imposed on the society, the industry also tends to acknowledge social responsibility after it is ready to act upon its commitment to ameliorate. This is seen as an opportunity to gain reputation through concerted efforts as dictated by the government. Hove et al. (2019) present a three-strand continuum wherein industry views its business pursuits as larger than acknowledging its externalities; on the premise that the society prods industry in its pursuits for profits. This differs from the other perspective in which private companies modify their businesses to limit negative externalities or compete/cooperate with the society and valorize to form value through an ethical commitment. On the other hand, it is equally important for the government under all these circumstances to arrive at a consensus with the industry and enhance its stake on the basis of underlying values and options for changes. This invokes the process-consequence framework and its interface with motivations for proactive amelioration and ethical reasoning. The authors further point to the possibility that proactive companies may be prepared to incur costs due to an ethical stand and probably convert the ethical advantage to a competitive advantage through innovations.

4 The Centrality of Science, Political Posturing, and Public Opinion that Influences Industry Action Sustains

This is best exemplified by CDP's rankings that measure transparency with respect to understanding the impacts of climate change for instance and mitigation efforts (ADEC Innovations ESG Solutions 2017). The CDP introduced transparency assessments through four steps along with the scope to integrate public responses and tackle reputational risks. Talbot and Boiral (2015), however, showed that there is no linear relationship between credibility and disclosures made; this is due to the ineffectiveness of assurances to remedy vis-à-vis quality of data. The ethics related issues continue.

The Oil and Gas Preparer Forum's initiatives were examined by the World Business Council for Sustainable Development recently (WBSCD) with special reference to implementing the recommendations of the Task Force on Climate Related Financial Disclosure (WBSCD 2018). While it highlighted the seamless transition to governance and risk management from only reducing the volume of greenhouse gases, it also indicated the need to disclose relevant information that can be easily compared. This is due to the call of stakeholders according to which climate change considerations should be integrated into governance and are verifiable. This imposes elements of ethics, including procedural and distributional injustices, attributable to the actual intentions of actions and outcomes of intended actions (Dahlmann et al. 2019).

The implications of regulations on energy transitions were recently examined by S&P Global Ratings (2019) ; S&P Global Ratings (2019) through its Oil and Gas ESG industry report card highlighted the need to address safety and community impacts on priority in addition to other aspects such as social cohesion and consumer behavior risks. Climate inaction is implicated in these circumstances, despite the pervasive influence of the long arm of law. Climate litigation as referred to earlier (ADB 2020), especially in the Asia and Pacific regions, has been growing in forms and functions. Larger and increasingly intensive impacts compound the outcome of externalities. This enhances the vulnerability of communities and fragile ecosystems they depend on. Lawsuits accordingly invoke the provisioning, regulatory, supporting, and cultural functions of ecosystems.

The economics of ecosystems and biodiversity framework correlates these with enormous implications for the quality of life. The Constitution of India, for instance, inspires stewardship through an interplay of provisions in Articles 19 (1) (g), 21, 48A, 51 A (g) and (h) while 32 and 226 empower citizens to litigate in public interest. These resonate with the typology of routes adopted by affected communities worldwide with reference to human and natural rights, constitutional, statutory, and policy rights and climate change impacts management specific focal points (Gitanjali 2017). These encompass cases faced by the energy sector in particular through two thrusts: extraction and use of fossil energy substrates; and pollution impacts related to such processes. The latter is compounded further when the greenhouse-gas perspective is invoked about transboundary harm.

Yet another valuable insight presented by the ADB (2020) is about the lack of attention to biodiversity and eco-system resilience in climate change related litigation. This is a quite profound deficiency within analytical frameworks that should actually relate causes and effects of perturbations. Chemical ecological profiles of bio-resources and ecosystems exhibit rapid and delayed induced responses to changes even at the scale of the micro-environment. These mediate biochemical transformations and environmental fate of chemicals in particular, with implications for the quality of water and soil resources. Litigations on these aspects of environmental management should prompt ever more stringent norms and compliance.

A special emphasis is on the interplay of regulations, market instruments, and institutional mechanisms duly recognizing limits and limitations of mitigation and adaptation options. This also acknowledges the continually evolving nature of frontiers of knowledge and the need to adapt appropriately. The seamlessness of these elements of public policy is presented in Fig. 1. The ADB (2020) most aptly highlights the scope for climate vigilance related justice and the emerging ray of hope through 'better capitalism' and the World Economic Forum's Great Reset Initiative by businesses. A typical case in point has been highlighted by Coeckelbergh (2020) with reference to artificial intelligence and climate change impacts management.

Disclosure of information is, therefore, a crucial element in determining the veracity of options. It is important to ensure reliable information about abatement by companies in a timely manner, to create and strengthen credibility perceptions with consumers and investors in particular. This could also relate to a critical mass of information about sustainability to properly inform markets and enable performance benchmarking. It is equally important to also ask if the proportion of companies that report and those that don't will influence the seriousness with which they are viewed by the stakeholders. Governments too should meet administrative costs of promulgation and enforcement and risk of implementation failure. It was further inferred that some companies could face costs that may not be outweighed with benefits of compliance. This creates challenges by which the business case for corporate reporting could be elusive. Individual leadership, however, could drive higher levels of compliance (Artuso 2020).

A comprehensive review of sustainability reporting (KPMG 2020) demonstrates the growing spread and depth of parameters presented by a growing number of firms across the globe. They recognize the raising attention climate change issues

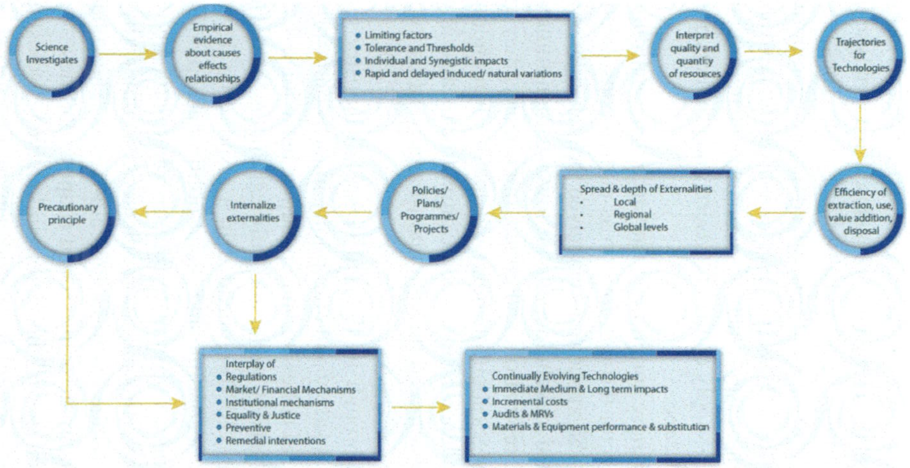

Fig. 1 A science-policy continuum presenting some of the lacunae to be addressed

have received and the potential for such aspects as labour, inclusivity, and justice to co-dominate soon. They have to, however, choose appropriate methodologies, documentation, and interpretation approaches to substantiate their claim. This dilemma persists even as the ESG framework has apparently provided the much-needed impetus to become increasingly environmentally efficient. It remains open to interpretation because ESG ratings and reporting boundaries with special reference to risk assessments and managerial plans vary significantly as a function of metrics and indicators.

The Transition Pathway Initiative (Dietz et al. 2020) recently argued that claims of net-zero made by oil and gas companies considered in its analyses are not substantiated enough with respect to intensity and absolute targets. These could vary further with respect to baselines, absolute and normalized quantities, reporting periods, and data consolidation protocols. This is especially so with reference to alternative energy sources tapped as in some cases. Such assessments get more complicated when carbon performance of the supply chain, end-users, and products are included in the overall management landscape.

The IPIEC et al. (2020) have committed to advance their environmental and social performance across the sector through four cross-cutting consolidation approaches. They acknowledge society's expectations perceived through eight reporting channels which connect with related stakeholders about lower-carbon futures with a special emphasis on governance, reputation, and ethics. Importantly, they also take note of the need to address challenges about quantum and quality of amelioration related data that will inspire the confidence of investors. Commitments also appear to pressure firms to deliver; further reinforced through audit-based verification and third-party commentaries.

S&P Ratings (2020) through its Oil and Gas ESG industry report card highlighted the need to address safety and community impacts on priority in addition to other aspects such as social cohesion and consumer behavior risks. These are best captured in a recent analysis of social and political dimensions of greenhouse gas removals by Waller et al. (2020) embedded in a responsible development frame. These complement techno-economic and social-political acceptability facets as part of an overarching governance of climate futures. Micro-climate related chemical ecological profiles of such removals could enhance the complexity of assessments especially due to leakages inherent in biological systems. They could, however, be most resilient with characteristic nature-based adaptive abilities.

One of the earliest analyses of the factors influencing disclosures about greenhouse gases by companies around the world was by Lorenzo et al. (2009). They defined a large number of variables in the context of stakeholder and legitimacy theories with respect to litigation and abatement actions. Such aspects as the eagerness to stand out from competitors and regain legitimacy were seen as drivers of compliance. They also dealt with the dilemma of delays in action as a reflection of low commitment to comply and the tendency to omit information as a proxy indicator of lesser ethics. These raise questions about the dynamics of ethical values as the bedrock for leadership that recognizes the imperatives of reporting that can be verified.

5 Additional Insights Strengthening the Preventive Management Framework

While significant information is available in the public domain about climate change, especially through the IPCC and the Union of Concerned Scientists for instance, it is equally important to take note of the contrarians too. This interface is complicated further when political will could tend to constrain the precautionary approach. The call for mitigation and adaptation is emphatic with significant implications for business and consequences of inaction (Rebecca et al. 2018; Gillingham 2019; Benjamin and Brian 2020).

Averchenkova et al. (2015) deliberated on some important elements of a framework including drivers, responses, and outcomes that determine engagement with corporations/industry. They highlighted the need for a deeper understanding of conflicts that may dominate the interface between these elements and regulations across countries. Assistance to communities that suffer impacts is an important thrust for such multinational corporations. Several important deliberations on these elements and the need for evidence-based support to sustain momentum of change are highlighted in this reference.

The Ministry of Economy, Trade, and Industry, Japan encouraged its businesses to support adaptation in developing countries. This signifies a deep understanding of the premise stated above and the inspiration to deliver location/system specific resilience enhancement measures, and hence a large portfolio of twenty thrusts areas. The emissions gap report (UN Environment 2017) elaborated on the trajectories of preparedness with milestones and, importantly, calls for 'faster and deeper" reductions towards "greater de-carbonization".

The India's Intended Nationally Determined Contribution (INDC) presents a clear perspective of its development imperative and preparedness. This is further substantiated in its Draft National Energy Policy (2017), the report of the Committee presented by the Ministry of Power on optimal energy mix (2018) that is clean, affordable, reliable and sustainable, and an analysis of its options in the IEA-OECD publication about the country's energy outlook. A detailed analysis of coal-based thermal power plants to secure resource efficiency gains, tackle GHGs and locally significant pollution (Bhushan et al. 2015) invites a special attention in this context. The cited draft policy highlights climate change concerns and the need for greater policy clarity about energy pathways to internalize externalities and simultaneously serve India's developmental aspirations on four mutually reinforcing elements of public policy. They are energy access at affordable prices, improved security and independence, and increased sustainability, and economic growth. The INDCs of other countries also focus on their special circumstances of development.

As indicated above, the capacities to adopt and implement environmentally efficient practices are equally important. The financial institutions worldwide create the impetus to become cleaner on these fronts. The cases in point include Reprisk (2018) about Powering Past Coal Alliance of 20 countries and those related to the funding for pollution-intensive operations which is being curtailed. Compliance is given highest priority and is linked with reputational impacts. Financial stability in the light of climate risks is the main thrust, with a call for development of analytical tools and indicators on mainstreaming climate resilience. The UNEP Positive Impact and the Portfolio de-carbonization coalition are evolving metrices on climate progress of banks.

Deeper analyses of factors that determine the lack of willingness to report accurately about emissions are needed to determine the dynamics of compliance. Yona et al (2020) on the other hand create the scope to believe that omitting information need not always to be due to a propensity to conceal. This is because guidelines to define and document emissions, inventories, reporting methodologies and abatement correlates are not robust enough across tools despite the recent refinement processes. This could probably be a reflection of dilemmas that persist even at the level of nations about self-interests and collective action. Gulluscio et al. (2020) analyzed patterns of accounting and accountability about the SDGs and concluded that the methods followed are at best informal because principles of accounting are not robust enough. Strategic and operational planning interventions are therefore few and far between.

References

Asian Development Bank (ADB) (2020) Climate change coming soon to a court near you – Climate litigation in Asia and the Pacific and beyond. ADB Philippines

ADEC Innovations ESG Solutions (2017) 4 Steps to improving your CDP performance. CDP's rankings measure transparency of climate change impacts, as well as the efforts taken to mitigate and reduce those negative impacts, White Paper

American Geophysical Union (2020) COVID-19 lockdowns significantly impacting global air quality. ScienceDaily, www.sciencedaily.com/releases/2020/05/200511124444.htm, Accessed 23 Jan 2021

Artuso A (2020) Ethical obligations for climate change mitigation: basic principles, reply to objections, and implications. Int J Clim Change Impacts Responses 12(2):1–22

Averchenkova A, et al (2015) Frameworks to interpret MNC action, ESRC, Grantham Institute, Univ of Leeds and LSE. http://www.lse.ac.uk/GranthamInstitute/wp-content/uploads/2015/03/Working-Paper-183-Averchenkova-et-al.pdf, Accessed 23 Jan 2021

Benjamin MS, Brian CO (2020) Assessing the costs of historical inaction on climate change. Scientific reports. Nature Res 10:9173

Bhushan C, Priyavrat B , Sanjeev K, Angeline S, Sai S, Soundaram R, Abhishek R (2015) Heat on Power: Green Rating of Coal-based power plants. Centre for Science and Environment, New Delhi

Buckley T (2019) Over 100 Global Financial Institutions are Exiting Coal, With More to Come Every Two Weeks a Bank, Insurer or Lender Announces New Restrictions on Coal. IEEFA, Australia

Buckley T, Sanzillo T, Shah K (2019) Inaction is BlackRock's Biggest Risk During the Energy Transition Still Lagging in Sustainable Investing Leadership. IEEFA, Australia.

CERES (2020) Sweeping New Report Provides U.S. Financial Regulators with Key Action Steps to Protect Financial Markets from the Ongoing Climate Crisis. https://www.ceres.org/news-center/press-releases/sweeping-new-report-provides-us-financial-regulators-key-action-steps, Accessed 23 Jan 2021

Coeckelbergh M (2020) AI for climate: freedom, justice, and other ethical and political challenges. AI Ethics. https://doi.org/10.1007/s43681-020-00007-2
Dahlmann F, Branicki L, Brammer S (2019) Managing carbon aspirations: the influence of corporate climate change targets on environmental performance. J Bus Ethics 158:1–24
Demine V (2020) MSCI Managing climate change risks with high quality low carbon portfolios. https://www.morganstanley.com/ideas/climate-change-low-carbon-stock-investing, Accessed 23 Jan 2021
Dietz S, Byrne R, Gardiner D, Gostlow G, Jahn V, Nachmany M, Noels J, Sullivan R (2020) TPI State of Transition Report.
Fischhoff B, Scheufelec DA (2019) The science of science communication III. PNAS 116(16):7632–7633
Fischoff B, Davis AL (2015) Communicating scientific uncertainty. PNAS 111(4):13664–13671
Gitanjali GN (2017) Environmental justice in India. The National Green Tribunal. Earthscan/Routledge, London and New York.
Gillingham K (2019) Carbon calculus. In: Finance and Development, pp 6–11
Government of India (2017) Draft National Energy Policy NITI Aayog, available at: http://niti.gov.in/writereaddata/files/new_initiatives/NEP-ID_27.06.2017.pdf, Accessed 23 Jan 2021
Government of India (2018) Report of Government of India Ministry of Power, Central Electricity Authority Committee on Optimal Energy Mix in Power Generation on Medium and Long Term Basis. https://www.gita.org.in/Attachments/Reports/IndiaEnergyOutlook_WEO2015.pdf, Accessed 23 Jan 2021
Government of India (undated) India's Intended Nationally Determined Contribution: Working Towards Climate Justice. https://www4.unfccc.int/sites/ndcstaging/PublishedDocuments/India%20First/INDIA%20INDC%20TO%20UNFCCC.PDF, Accessed 23 Jan 2021
Gulluscio C, Puntillo P, Luciani V, Huisingh D (2020) Climate change accounting and reporting: a systematic literature review. Sustainability 12:5455. https://doi.org/10.3390/su12135455
Hepburn C, O'Callaghan B, Stern N, Stiglitz J, Zenghelis D (2020) Will COVID-19 fiscal recovery packages accelerate or retard progress on climate change? Oxf Rev Econ Policy 36(S1):S359–S381
van den Hove S, Menestrel M, Bettignies HC (2002) The oil industry and climate change: strategies and ethical dilemmas. Climate Policy 2:3–18
International Energy Agency (2020a) Global Energy Review 2020 – The impacts of the Covid-19 crisis on global energy demand and CO2 emissions
International Energy Agency (2020b) Renewable Energy Market Update – Outlook for 2020 and 2021
IPIEC, API Energy and the IOGP (2020) Sustainability reporting guidance for the oil and gas industry: Advancing environmental and social performance across oil and gas, 4th edition
Kelly A (2020) 300+ Business Leaders Tell Congress: We Can Build Back Better. https://sustainablebrands.com/read/collaboration-cocreation/300-business-leaders-tell-congress-we-can-build-back-better, Accessed 23 Jan 2021
KPMG (2020) The time has come, The KPMG Survey of Sustainability Reporting
Kunreuther H, Gupta S, Bosetti V, Cooke R, Dutt V, Ha-Duong M, Held H, Llanes-Regueiro J, Patt A, Shittu E, Weber E (2014) Integrated risk and uncertainty assessment of climate change response policies. In Edenhofer O, Pichs-Madruga R, Sokona Y, Farahani E, Kadner S, Seyboth K, Adler A, Baum I, Brunner S, Eickemeier P, Kriemann B, Savolainen J, Schlömer S, von Stechow C, Zwickel T, Minx JC (eds.) Climate change 2014: mitigation of climate change. Contribution of working group III to the Fifth Assessment Report of the Intergovernmental Panel on Climate Change. Cambridge University Press, Cambridge
Lorenzo JM, Rodríguez-Domínguez L, Álvarez IG, García Sánchez IM (2009) Factors influencing the disclosure of greenhouse gas emissions in companies world-wide. Manag Decis 47(7):11331157

MSCI (2020) Navigating the Transition: Managing Climate Risks and Opportunities, Morgan Stanley's Task Force on Climate-related Financial Disclosures Report. https://www.msci.com/documents/10199/06d20a9e-bd6b-4ea8-a7a2-31cfb0444273#:~:text=The%20MSCI%20India%20ESG%20Leaders,cap%20companies%20in%20Indian%20markets
MSCI (2021) MSCI India ESG Leaders Index (USD). Accessed 23 Jan 2021
Nadakavukaren KS, Cottier T (2012) Responsibility to Protect (R2P) and the Emerging Principle of Common Concern. Working Paper No 2012/29, Swiss National Centre of Competence in Research
NASA (undated) Climate Change: How Do We Know? https://climate.nasa.gov/evidence/, Accessed 23 Jan 2021
Oreskes N (2015) The fact of uncertainty, the uncertainty of facts and the cultural resonance of doubt. Phil Trans R Soc A 373:20140455
Patterson A, McLean C (2019) The precautionary principle at work: the case of neonicotinoids and the health of bees. Sci Public Policy 46(3):441–449
Pouyat RV (1999) Science and environmental policy – making them compatible. Policy Forum Biosci 49(4):281–286
Rebecca MH, Sophius AR, Polina D, Amram M (2018) Climate change in 2018: implications for business. Harvard Bus School 9(317):032
Reprisk (2018) https://www.reprisk.com, Accessed 23 Jan 2021
S&P Global Ratings (2019) ESG Industry Report Card: Oil and Gas. https://www.spglobal.com/_assets/documents/ratings/research/itt-2020-oil-and-gas.pdf
Science Based Targets (2020) Over 150 Global Corporations Urge World Leaders for Net-Zero Recovery From COVID-19. https://sciencebasedtargets.org/2020/05/18/uniting-business-and-governments-to-recover-better/, Accessed 23 Jan 2021
Setzer J, Byrnes R (2019) Global trends in climate change litigation: 2019 Snapshot. Policy report, LSE. Granthan Research Institute on Climate Change and the environment, Columbia Law School, Sabin Centre for Climate Change Law, University of Leeds, Centre for Climate Change Economics and Policy
Shaddick G, Thomas ML, Mudu P, Rugger G, Gumy S (2020) Half the world's population are exposed to increasing air pollution. Climate Atmos Sci **23**. https://doi.org/10.1038/s41612-020-0124-2
Sharma J, Ravindranath NH (2019) Applying IPCC 2014 framework for hazard-specific vulnerability assessment under climate change. Environ Res Commun 1:051004
Talbot D, Boiral O (2015) GHG reporting and impression management: an assessment of sustainability reports from the energy sector. J Bus Ethics 1–17
UN (2017). Climate Change Annual Report 2017. https://unfccc.int/resource/annualreport/, https://unfccc.int/news/ipcc-agrees-outlines-of-sixth-assessment-report, Accessed 23 Jan 2021
UN Environment (2017). The Emissions Gap Report 2017. A UN Environment Synthesis Report. https://wedocs.unep.org/bitstream/handle/20.500.11822/22070/EGR_2017.pdf, Accessed 23 Jan 2021
UNEPFI (undated), Portfolio Decarbonization Coalition. http://www.unepfi.org/publications/climate-change-publications/portfolio-decarbonization-coalition-3rd-annual-progress-report-launches-detailing-benefits-of-low-carbon-approach/, Accessed 23 Jan 2021
University of Texas at Austin, Texas Advanced Computing Center (2019) GRACE data contributes to understanding of climate change: Expanded investigation through 2021 to be highly dependent on TACC supercomputing systems. ScienceDaily. www.sciencedaily.com/releases/2019/05/190523130226.htm, Accessed 23 Jan 2021
Waller L, Rayner T, Chilvers J, Gough CA, Lorenzoni I, Jordan A, Vaughan M (2020) Contested framings of greenhouse gas removal and its feasibility: social and political dimensions. Wiley Interdisc Rev Climate Change 11(12)
World Business Council for Sustainable Development recently (WBSCD) (2018) Climate-related financial disclosure by oil and gas companies: implementing the TCFD recommendations, WBSCD

Wills Towers Watson (2020) ESG: transforming energy's risk landscape Energy Market Review. http://www.willistowerswatson.com/assets/image/Energy-market-report-2020.pdf, Accessed 23 Jan 2021

Von Winterfeldt D (2013) Bridging the gap between science and decision making. PNAS 10(3):14055–14061

World Economic Forum (2020) The Global Risks Report. WEF

WRI (undated), CAIT Climate Data Explorer. https://www.wri.org/our-work/project/cait-climate-data-explorer, Accessed 23 Jan 2021

Yona L, Cashore B, Jackson RB, Ometto J, Bradford MA (2020) Refining national greenhouse gas inventories. Ambio 49:1581–1586

Yung L, Louder E, Gallagher LA, Jones K, Wyborn C (2019) How methods for navigating uncertainty connect science and policy at the water – energy – food nexus. Front Environ Sci 7: Article 37, 1–17

Postface

Social-ecological systems are increasingly threatened by many risks dynamics such as environmental and climatic changes and related disasters, resource conflicts, social inequity, women exclusion, human displacements beyond ecosystems' carrying capacity, land-use change, and stakeholders' awareness and knowledge gaps. There are many determinants that determine the harmony of socio-ecological systems, including political setup, people's mindset and needs, and developmental phase. At the same time, environmental and climatic changes have become a matter of grave concern as the issue had been largely ignored over the years but now demands immediate attention. The current COVID-19 pandemic is a clear example of disharmony between social and ecological systems.

Until now, there was less evidences of direct linkages of social issues with environmental disturbances. However, the debate is evolving and the nature-human nexus is being increasingly explored both theoretically and practically. This book is a step in that direction, not only adding to existing literature on social-ecological systems and emerging risks, but also showing pathways and options towards disaster risk reduction, climate adaptation, and resilience of such systems. This book presents case studies on crucial social-ecological risks and their implications, showing that there are inherent linkages between ecosystem changes and social dynamics, and the usefulness of engaging local communities in the response process by taking into account their specific vulnerabilities and expectations.

The book starts with presenting the risk dynamics and ends with probable solutions to restore socio-ecological resilience. There are four main takeaways from this book. *First*, to improve the resilience of the socio-ecological systems, governance must be designed based on the context-specific understanding of the functioning of these systems. Different regions face different challenges in light of their national circumstances and adaptive capacity, which is often ignored when designing policies to manage the situation, especially through top-down processes. It is also seen that there is either a lack of such governance mechanisms or the existing of weak and inappropriate settings in some developing countries. *Second*, gender inequity is directly linked to socio-ecological resilience as women not only

M. Behnassi et al. (eds.), *Social-Ecological Systems (SES)*,
https://doi.org/10.1007/978-3-030-76247-6

bear the brunt of ecological imbalance, including disasters, but also play a key role in managing ecosystems and its resources. It is, therefore, important to rethink and reorganize the underlying value structures and systems of hegemonic masculinity without which it is difficult to attain resilience, viability, and security of the socio-ecological systems. *Third*, the ecosystem-based approaches, and more recently the nature-based solutions, need to be further refined to assure multiple benefits pertaining to the socio-ecological systems including disaster risk reduction, climate change adaptation, and sustainable utilization of resources. For this to happen, it is imperative to address the key knowledge gaps in ecosystem-based adaptation planning, and monitoring and evaluation of ecosystem services. *Fourth*, risks to the socio-ecological systems need to be further explored, especially in politically disturbed, climatically vulnerable, and resource deficient regions of the world as they may become the source of unmanageable human displacements in future, further challenging human security and the stability of concerned countries. There is a need to bring focus of decision-making processes at all scales on the links between environmental stress, resource scarcity, and conflicts.

The strength of this volume lies in the fact that social-ecological resilience related issues are still taking shape in the international arena of UNFCCC and CBD, and such case studies provide insights into the real challenges and opportunities to tackle them at the right time. These new challenges in light of changing political realities are deemed to present us with the dilemmas on which way to follow for a sustainable and secure future for all. We hope that this volume will initiate new points for discussion and offer readers with some thought-provoking insights.

Mohamed Behnassi
Himangana Gupta
Mahjoub El Haiba
Gopichandran Ramachandran

Biographical Notes of Authors

Agbleze, Linus Linus Agbleze is a researcher with a bachelor education in Geography and Resource Development from the University of Ghana as well as a Post-graduate certificate in Climate Change Management from the University of Edinburgh, United Kingdom. Currently, he is a Master student at United Nations University and University of Bonn studying environmental risk and human security with research interests spanning but not limited to the areas of climate change adaptation and mitigation, vulnerability and disaster risk assessment and management. His works have largely focused on Africa.

Ashfaq, Mohammad Mohammad Ashfaq is Additional General Manager with Corporate Planning Department of NTPC Ltd. and also heads the World Energy Council India (WEC India) Secretariat. NTPC is a leading power producer of India. NTPC is ranked No. 2 Independent Power Producer (IPP) in Platts Top 250 Global Energy Company rankings. WEC India is the country member of World Energy Council promoting sustainable supply and use of energy. In his current role as In-charge WEC India Secretariat, through WEC India work programs, he plays a pivotal role in contributing to the energy transition of the country and showcasing the energy transition efforts of the country at global platforms. As AGM (CP), he is looking after international cooperation related to energy sector for NTPC Ltd. through ministry of power, Government of India. As a post-graduate in engineering from IIT, Delhi and PGPMAX from ISB, Hyderabad, Md. Ashfaq comes with a rich experience of 27 years in the energy sector. Previous assignments with NTPC gave him extensive experience in power plant operations and maintenance where he contributed significantly to the improvement of performance parameters. Highly talented and committed, he is the winner of many awards. He is passionate about emerging technology areas in the energy sector. His ability to engage with stakeholders and to create a strong network is worth mentioning.

Chougrani, El Houcine Dr. Chougrani is an Associate Professor of International Law and Relations at the Faculty of Law, Economics, and Social Sciences, Cadi Ayyad University of Marrakech. He obtained a Doctorate in 2003 with a thesis entitled *The European Investment Bank and the protection of the Environment in the Mediterranean Are*a. He is the Director of the Moroccan Observatory for Future

M. Behnassi et al. (eds.), *Social-Ecological Systems (SES)*,
https://doi.org/10.1007/978-3-030-76247-6

Generations (MOROFUGE) and Member of the International Research Center for Crisis Management. His main research is international environmental law, and his publications cover water, climate justice, and the rights of future generations. His most recent publications in Arabic are: *The Rights of the Future Generations in the Light of Current Developments in the Arab World* (Arab Center for Research and Policy Studies, 2018); and *Contradictions in International Law. A Preliminary Analysis* (Centre for Arab Unity Studies, 2019). Dr. Chougrani has also an extensive professional experience as Accountant General at the Ministry of Economy and Finance (2002–2014).

Deafalla, Taisser H. H. Deafalla is a crisis mapper and Coach for human development at many national and international NGOs. Currently, she is a Senior Researcher at the Institute of Photogrammetry and Remote Sensing, Faculty of Environmental Sciences, University of Dresden, Germany. She has a professional experience in the areas of the forestry, environmental change and rural development where she worked with many local and international organizations. Moreover, she has participated in many conferences and international workshops and she has many scientific publications related to environmental change.

Defe, Rameck Defe is a Master of Philosophy student at Midlands State University researching on climate change impacts and mitigation and adaptation strategies in rural communities. He has worked as a Research Assistant doing annual assessments of projects done by NGOs like CARE Zimbabwe, PLAN International, ICRISAT in rural communities under the Zimbabwe Resilience Building Fund. He has contributed a research paper published with the Midlands State University Dyke on Occupational Safety Health and Environment. He has also contributed in research papers on enhancing disaster risk reduction in communities mainly focusing on climate change mitigation and adaptation strategies.

El Amrani, Moulay Hicham El Amrani is a Ph.D student at the Faculty of Law, Economics, and Social Sciences of Agadir, Ibn Zohr University, Morocco. His main research areas are: environment, human security, cyberspace, and new technologies. El Amrani is engaged in completing a Doctoral thesis entitled *Health governance in the perspective of the coviability of social-ecological systems*. He participated in several international conferences such as: the CERES International Conference on *Social-Ecological Systems - From Risks and Insecurity to Viability and Resilience* (Marrakech, 2019); the GRESS International conference on *Cyberspace Regulation and Data Protection* (Marrakech, 2019); the GRAPP International symposium on *the Challenges of the Social Question and Comparative Public Policies* (Marrakech, 2019); the AI2SD International Conference on *Advanced Intelligent Systems for Sustainable Development* (Tanger, 2018). He is also an active member of numerous environmental protection NGOs.

El-Abbas, Mustafa M. Dr. El-Abbas is an assistance professor at the Department of Forest Management, Faculty of Forestry, University of Khartoum, Sudan where he obtained his B.Sc. (Hons) and M.Sc. degrees (2001, 2006). Moreover, he has got a diploma of forest ecology and forest resource management, university of Helsinki,

Finland in 2006. Dr. El-Abbas awards his PhD in remote sensing and natural resource management from TU Dresden. Currently, he has been working as a Researcher at the Institute of Photogrammetry and Remote Sensing, Faculty of Environmental Sciences, University of Dresden, Germany. He has professional experience in the areas of remote sensing and GIS in general, and object-based approaches in particular, with an emphasis on natural resources management. Dr. El-Abbas participated in several international workshops and scientific forums and has many scientific publications.

Elkhair, Osman Dr. Elkhair is an assistance professor, senior Architect at Sudan, Mozambique, Gambia, Tanzania, Sweden and the Chairman of Newtech Consulting Group. He awards his PhD in Human Settlements Architecture from KTH Stockholm, Sweden. He has professional experience of more than 42 years in the areas of Architecture. Dr. Elkhair participated in several international workshops and scientific forums and has many scientific publications. Furthermore, he obtains the Ordre des Palmes Académiques from France.

Csaplovics, Elmar Dr. Csaplovics is a Professor of remote sensing at the Institute of Photogrammetry and Remote Sensing, Department of Geosciences, TU Dresden. He holds a Doctorate and a Habilitation in remote sensing from TU Vienna (1982, 1992), was a Post-Doc Research Fellow at the National Institute of Agronomic Research (INRA), Montpellier and at the Department of Geology, Geophysics and Geoinformatics, Free University Berlin from 1988-1992, and was appointed professor of remote sensing at the TU Dresden in 1993. Dr. Csaplovics was a Visiting Professor at the University College London (UCL) in 2007 and focuses on research into remote sensing and applied geoinformation analysis for spatio-temporal monitoring and assessment of land use/land cover with emphasis on wetlands and semi-arid lands (desertification and degradation), as well as into landscape history and world conservation monitoring in transnational project cooperation.

Ghislain, Slama Jules Dr. Ghislain is a Titular Professor at the Federal University of Rio de Janeiro (UFRJ) and Researcher at the Mechanical Post-Graduate Program-COPPE/UFRJ. Dr. Ghislain has a Master in Acoustics and Doctorate in Acoustics and Vibration Dynamics, LMA Marseille, France. Dr. Ghislain is supervisor of Masters and Doctorate thesis in Acoustics and Signal processing at COPPE/UFRJ. Currently, Dr. Ghislain works on environmental acoustics and airport noise and provides consultancy in airport noise and airport urban integration.

Gill, Gitanjali Nain Dr. Gill is a Professor of environmental law and has been an academic since 1993, formerly employed at India's leading university, Faculty of Law, Delhi University, where she experienced a distinguished academic career finishing her tenure as a Reader in 2011. Prior to joining Northumbria University, Professor Gill was a Visiting Fellow at BRASS (Business Relationships, Accountability, Sustainability and Society) Cardiff University, working on environmental and sustainability issues. She taught at the School of Law, Queen Mary University, London; Cardiff Law School. In August 2011 Professor Gill joined the School of Law, Northumbria University. She is widely published with over 50 peer-reviewed articles and book chapters, two books, and has delivered more than 80 international and national presentations. Currently, Professor Gill is engaged in completing a research monograph titled 'Just Sustainabilities, SDGs and a Transformative Indian Judiciary' (Routledge UK; commissioned; submission December 2021). The inter-disciplinary monograph covers global, regional and local dynamics and stresses the role of the judiciary under Goal 16 SDG as drivers of change. It has global relevance as nations are steered towards achieving SDGs. She is a peer-reviewer for leading publishing houses and journals including Hart Publishers, Routledge UK, Sage UK, Edward Elgar, Springer and Mary Ann Liebert USA. She acted as an expert contributor and reviewer on India for the 2016 United Nations Environment Programme book titled' Environmental Courts and Tribunals: A Guide for Policy-Makers'. In 2018, she was invited to the United Nations OHCHR Accountability and Remedy Project II to enhance the effectiveness of state based non-judicial mechanisms (such as ECTs including NGT) in cases of business-related human rights abuses. As a Member of the Expert Consultation Group, her NGT expertise was utilised for commenting, advising, providing feedback on policy objectives and elements of good state practice consultation draft prepared by OHCHR that assisted in the policy-framing of Resolution 38/13 by the Human Rights Council. Dr. Gill is a recipient of two prestigious British Academy/Leverhulme Research Grants. The first British Academy Grant (2013-16) examined the casework and environmental jurisprudence of the National Green Tribunal of India [NGT]. The project focused on the National Green Tribunal as one element of a reformist approach to environmental government and explored its working and effectiveness. In 2017, her research findings and conclusions were published in her book *Environmental Justice in India: The National Green Tribunal* (Routledge, Earthscan). The book was also published in Chinese by Routledge China and Social Science Press China (2019). The current second BA grant (2020–23) is titled 'Land Acquisition, Rehabilitation and Resettlement of Vulnerable Poor Communities: Criticalities and Scrutiny of Law in Gujarat, India'. The empirical research investigates the reach, effectiveness and impact of Rehabilitation and Resettlement (R&R) provisions in the Right to Fair Compensation and Transparency in Land Acquisition, Rehabilitation and Resettlement Act 2013 (LARR Act). In 2019, she was granted a GCRF Networking and Collaboration Fund award to undertake original empirical research in India on scientific uncertainty and the precautionary principle. Dr. Gill has held Visiting Professorships at several universities including the Vermont Law School, USA; Department of Comparative Law, University of Insubria, Como, Italy; Law School, Shanghai University of Finance and Economics, China; Charles University Prague, and Symbiosis Law School, Noida, Delhi. Dr. Gill has received several awards and fellowships. In 2018 received the prestigious Global Initiative of Academic Networks (GIAN) in Higher Education Award by the Ministry of Human Resources and Development, Government of India. She was also awarded the 2006 Fellowship on International and Comparative Environmental Law, Budapest, Hungary by the United Nations Institute for Training and Research [UNITAR]. It led to her appointment as an Associate Fellow with the United Nations Institute for

Training and Research [UNITAR]. In addition, she was awarded the 2003 Fellowship on Law Teaching and Research in Cardiff Law School, Cardiff University, UK.

She is a member of leading international and national professional societies: IUCN World Commission on Environmental Law; Socio-Legal Studies Association; United Kingdom Environmental Law Association; Global Network for the Study of Human Rights and the Environment.

Gueiros Teixeira, Suzana Dr. Gueiros Teixeira is an Associate Professor L1 at the Federal University of Rio de Janeiro and Currently a Researcher Professor at the Technology Center of Federal University of Rio de Janeiro. This research work has been developed while on the Laboratory of Advanced Systems Management and Sustainability SAGE-COPPE/UFRJ. Professor Rogerio Valle Center of Sustainability. She has a Master and Doctorate with both Production Engineering and Mechanical Engineering Supervisors. She has the specialized Course of Higher Studies on Politics and Strategy from the Defense Secretary War School - ESG – Brazilian Defense Secretary; Environmental Management Specialization Course from the Brazilian Committee from the United Nations Environmental Program (UNEP). Her current work is focusing on Sustainable Management under a Multi-Criteria approach with the emphasis on Soundscape Security, Human Factors Engineering, Life Cycle Assessment (LCA), Product Category Rules (PCR), Environmental Product Declaration (EPD), Defense, Security and Strategic Planning.

Gupta, Devangana Gupta is a Master student of Philosophy in the Department of Gandhian and Peace Studies, Panjab University, Chandigarh, India. She completed her Master of Arts from the same Department in 2018. She did her internship with the Indira Gandhi National Centre for the Arts, New Delhi on subjects related to Gandhian philosophy and culture. She has interests in sustainable development, culture (Buddhism and Jainism) and Gandhian philosophy, peace, conflict resolution, and poverty alleviation.

Lakeman, Silvana Dr. Lakeman holds an MA in International Relations from Leiden University, a BA in Economics and Politics from University College Dublin, and a PhD in Political Science from the University of Bremen, Germany where she has worked as an early-stage researcher and Marie Curie fellow at the Bremen International Graduate School of Social Sciences. Her research evaluates the involvement and inter-agency cooperation of organizations in the areas of migration, environmental and disaster displacement, climate change and disaster risk reduction. She has worked in both a research and advisory capacity for a range of international and non-governmental organizations. She.

Lokesh, Chandra Dube Lokesh is a Programme Officer with the National Communications Cell at the Ministry of Environment, Forest and Climate Change, Government of India. He was instrumental in preparing India's first and Second Biennial Update Report to UNFCCC. Earlier, as a Consultant at Emergent Ventures India, he developed a Framework for CDM Program of Activities in renewable energy and authored a new CDM methodology approved by the UNFCCC. As a GHG Auditor with TÜV NORD group, he validated/verified several mitigation projects. Lokesh is also a PhD candidate at TERI School of Advanced Studies, New Delhi.

Pinguelli Rosa, Luiz Dr. Pinguelli Rosa is a Titular Professor at the Federal University of Rio de Janeiro (UFRJ), Current Institutional Vice-Diretor of COPPE-UFRJ. He is Undergraduate in Physics from UFRJ, Master in Nuclear Engineering COPPE/UFRJ, and Doctor in Physics at Pontific Catholic University of Rio de Janeiro. Dr. Pinguelli Rosa is four times Director of COPPE Foi Diretor da COPPE/UFRJ and Ex-President of Eletrobrás – Brazilian Federal Energy Enterprise. He is also an Emeritus and Titular Professor of the Energy Planning Program, Professor of History of Sciences and Epistemology Techniques from UFRJ, and Executive Secretary of Brazilian Forum of Climate Change. In addition, he was a Visitor Professor in the Standford University (SLAC), California and in other universities in Pennsylvania, Grenoble, Cracóvia, Poland, Centre International Center for Environment and Development in Paris, France, etc. He is Member of the Pugwash Council (1999–2001) founded by Albert Einstein and Bertrand Russel, Winner of Nobel Peace at 1995.

Manuku, Mukoni Dr. Manuku is a Lecturer and Researcher at the Midlands State University Gender Institute in Zimbabwe. Her research interests revolve around gender, climate change, environmental education, and education for sustainable development.

Matsa, Mark Makomborero Professor Matsa is an Associate Professor of Climate Change, Environmental Change and Community development. He has worked with many farmer communities in various parts of Zimbabwe and has visited and exchanged notes with some small-scale farmers in a number of countries, including Namibia and the State of New Hampshire, USA. Prof Matsa also serves as a routine consultant in resilience building initiatives undertaken by NGOs like CARE Zimbabwe, PLAN, ICRISAT, LID Agency, Lutheran Development Services among others under the Zimbabwe Resilience Building Fund. He has published several books, book chapters and many scientific papers in reputable journals across the Geographical spectrum on Millennium Development Goals and Sustainable Development goals (SDGs) with a bias towards indigenous minority farmer communities who are at the knife edge of climate-induced hazards, shocks and stresses. Prof. Matsa has attended, facilitated, and presented research papers at various conferences in different parts of the world. He is also a seasoned consultant in environmental issues, including environmental impact assessment as well as monitoring and evaluation of projects.

Milagres Pereira, Carlos Eduardo Milagres Pereira is a Public Security Expert and Former Rio de Janeiro Military Police Officer Commander. He is currently Expert in non-lethal weapons, crowd-control, public security, policies and strategies. He is post-graduated in public policies, criminal justice, and public security (Federal Fluminense University), advance studies on policies and strategies (War

School – ESG- Defense Secretary). Among the many expertise courses are: Course of Failure identification, Global Positioning System, Brazilian Center of Global Positioning System and Topography, Public Administration and Multipliers (Public Service School), Public Security (Superior Core on Governmental Studies from UERJ), Strategy and Planning (Nacional School of Public Administration), Integrated Communication Systems (Technological University of Brasilia). He has partnership with Israel and France where he has undertaken technical study visits.

Pollmann, Olaf Dr. Pollmann is a civil engineering and natural scientist. He holds a Doctorate (Dr.-Ing./PhD) in environmental-informatics from the Technical University of Darmstadt, Germany and a second Doctorate (Dr. rer. nat./PhD) in sustainable resource management from the North-West University, South Africa. He is a visiting scientist and extraordinary senior lecturer at the North-West University and CEO of the company SCENSO - Scientific Environmental Solutions in Germany. He is also deputy head of the section "African Climate Service Centers" in West (WASCAL) and Southern Africa (SASSCAL) at the Project Management Agency (DLR-PT) on behalf of the Federal Ministry of Education and Research (BMBF). Dr. Pollmann is regularly requested to contribute, review and evaluate research-proposals, manuscripts, and programs from national and international universities, research institutions and governments. As an environmental expert, he has written several research papers in reputed international and national journals on current state of climate, food, and environmental issues.

Ramachandran, Nira Dr. Ramachandran is currently an independent researcher and consultant based in Delhi, India. In her last posting as ICSSR Senior Fellow at the Institute of Economic Growth, New Delhi her research centred on accelerating improvement in nutrition outcomes in India. Her interest areas include environment, food security, nutrition, gender, child labour, and related issues. She has held teaching/research positions at a number of prestigious institutions in India including the Indian Institute of Management, Bangalore, Bangalore University, Delhi University, and the LBS National Academy of Administration, Mussoorie. She has provided consultancy to a number of international organizations, including the FAO, UN World Food Programme, UNICEF and the World Bank. She holds a Ph. D from the Jawaharlal Nehru University, Delhi and has also been trained abroad. She has lectured extensively and conducted training programmes for research students, members of the Indian Civil Service, and Government officials from developing countries in the areas of food security, environmental sustainability, primary education, research methodology, and project management techniques. She has to her credit several articles in peer-reviewed journals as well as a number of books, including: *Persisting Undernutrition in India: Causes, Consequences and Possible Solutions* (Springer, 2014); *Towards a Food Secure India: Issues and Policies* (Institute for Human Development, 2003); *Coming to Grips with Rural Child Work: A Food Security Approach* (WFP, 2002); *Carrying Capacity of Mussoorie:Towards Establishing Benchmarks for Substainability* (Concept Publishing Company Co, 2001); and *Monitoring Sustainability: Indices and Techniques of Analysis* (Concept Publishing Co, 2000).

Sharma, Manish Dr. Sharma is the chairperson of the Faculty of Gandhian and Peace Studies at Panjab University, Chandigarh, India. He is the Faculty In-charge of Gandhi Bhawan Project with Getty Foundation, USA, and Holder of UNESCO Network Chair on Global Peace and Non Violence. He is also the Co-Secretary General of *Asia-Pacific Peace Research Association (APPRA), 2017–21*. He specializes in non-violence, peace research, Gandhian thought, research methodology, Buddhist Studies, rural development and Panchayati Raj in India.

Winniefridah, Matsa Dr. Winniefridah is a lecturer at the Gender Institute at Midlands State University in Zimbabwe. She is a PhD holder in socio-education with the University of South Africa (UNISA). Her areas of research interest are: gender and migration, gender and education; gender and culture (GAD issues); gender, environment and development (GED issues); climate change and women; and migration, women, and children (WID issues). She has published a number of book chapters and papers on education, gender and education, children's rights, gender and migration, gender and environmental/climate change as well as gender and culture in peer-reviewed journals. She is a member of organisations such as Msasa and legal resources foundation projects, Itai Trust, Organisation of Social Science Research in East Africa (OSSREA), UNESCO Teacher education capacity change projects for ESD, and Southern African Institute for Peace and Development (SIPD). She has presented papers at various international conferences.

Author Index

M. Behnassi et al. (eds.), *Social-Ecological Systems (SES)*,
https://doi.org/10.1007/978-3-030-76247-6

The manufacturer's authorised representative in the EU is Springer Nature Customer Service Centre GmbH, Europaplatz 3, 69115 Heidelberg, Germany. If you have any concerns regarding our products, please contact ProductSafety@springernature.com

Printed and bound by CPI Group (UK) Ltd, Croydon, CR0 4YY
30/09/2026
02225695-0008